Lecture Notes in Computer Science 11545

Commenced Publication in 1973
Founding and Former Series Editors:
Gerhard Goos, Juris Hartmanis, and Jan van Leeuwen

More information about this series at http://www.springer.com/series/7407

Miroslav Ćirić · Manfred Droste ·
Jean-Éric Pin (Eds.)

Algebraic Informatics

8th International Conference, CAI 2019
Niš, Serbia, June 30 – July 4, 2019
Proceedings

 Springer

Editors
Miroslav Ćirić (iD)
University of Niš
Niš, Serbia

Manfred Droste (iD)
University of Leipzig
Leipzig, Germany

Jean-Éric Pin
Université Paris Denis Diderot and CNRS
Paris, France

ISSN 0302-9743   ISSN 1611-3349 (electronic)
Lecture Notes in Computer Science
ISBN 978-3-030-21362-6   ISBN 978-3-030-21363-3 (eBook)
https://doi.org/10.1007/978-3-030-21363-3

LNCS Sublibrary: SL1 – Theoretical Computer Science and General Issues

This Springer imprint is published by the registered company Springer Nature Switzerland AG
The registered company address is: Gewerbestrasse 11, 6330 Cham, Switzerland

Preface

These proceedings contain the papers presented at the 8th International Conference on Algebraic Informatics (CAI 2019) held from June 30 to July 4, 2019, in Niš, Serbia, and organized under the auspices of the University of Niš and its Faculty of Science.

CAI is the biennial conference serving the community interested in the intersection of theoretical computer science, algebra, and related areas. As with the previous seven CAIs, the goal of CAI 2019 was to enhance the understanding of syntactic and semantic problems by algebraic models, as well as to propagate the application of modern techniques from computer science in algebraic computation.

This volume contains the abstracts of three invited lectures and 20 contributed papers that were presented at the conference. The invited lectures were given by Paul Gastin, Bane Vasić, and Franz Winkler. In total, 20 contributed papers were carefully selected from 35 submissions. The peer review process was single blind and each submission was reviewed by at least three, and on average 3.1, Program Committee members and additional reviewers. The papers report original unpublished research and cover a broad range of topics from automata theory and logic, cryptography and coding theory, computer algebra, design theory, natural and quantum computation, and related areas.

We are grateful to a great number of colleagues for making CAI 2019 a successful event. We would like to thank the members of the Steering Committee, the colleagues in the Program Committee and the additional reviewers for careful evaluation of the submissions, and all the authors for submitting high-quality papers. We would also thank Jelena Ignjatović, chair of the Organizing Committee, Ivan Stanković, who created and maintained the conference website, and all other members of the Organizing Committee, for a successful organization of the conference.

The reviewing process was organized using the EasyChair conference system created by Andrei Voronkov. We would like to acknowledge that this system helped greatly to improve the efficiency of the committee work.

Special thanks are due to Alfred Hofmann and Anna Kramer from Springer LNCS, who helped us to publish the proceedings of CAI 2019 in the LNCS series.

The sponsors of CAI 2019 are also gratefully acknowledged.

April 2019

Miroslav Ćirić
Manfred Droste
Jean-Éric Pin

Organization

CAI 2019 was organized by the Faculty of Sciences and Mathematics, University of Niš, Serbia.

Steering Committee

Symeon Bozapalidis	Aristotle University of Thessaloniki, Greece
Olivier Carton	Université Paris-Diderot, Paris, France
Manfred Droste	University of Leipzig, Germany
Zoltan Esik (Deceased)	University of Szeged, Hungary
Werner Kuich	Technical University of Vienna, Austria
Dimitrios Poulakis	Aristotle University of Thessaloniki, Greece
Arto Salomaa	University of Turku, Finland

Program Committee Chairs

Miroslav Ćirić	University of Niš, Serbia
Manfred Droste	University of Leipzig, Germany
Jean-Éric Pin	Université Paris-Diderot, CNRS, Paris, France

Program Committee

Claude Carlet	Université Paris 8, France
Charles Colbourn	Arizona State University, Tempe, USA
Zoltán Fülöp	University of Szeged, Hungary
Dora Giammarresi	Università di Roma Tor Vergata, Italy
Mika Hirvensalo	University of Turku, Finland
Lila Kari	University of Waterloo, Canada
Nataša Jonoska	University of South Florida, Tampa, USA
Dino Mandrioli	Politecnico di Milano, Italy
Miodrag Mihaljević	Mathematical Institute of the SASA, Belgrade, Serbia
Benjamin Monmege	Aix-Marseille Université, France
Lucia Moura	University of Ottawa, Canada
Dimitrios Poulakis	Aristotle University of Thessaloniki, Greece
Svetlana Puzynina	Saint Petersburg State University, Russia
George Rahonis	Aristotle University of Thessaloniki, Greece
Robert Rolland	Aix-Marseille Université, France
Kai Salomaa	Queen's University, Kingston, Canada
Rafael Sendra	University of Alcalá, Alcalá de Henares, Madrid, Spain
Dimitris Simos	SBA Research, Vienna, Austria
Branimir Todorović	University of Niš, Serbia

Bianca Truthe   Justus Liebig University, Giessen, Germany
Heiko Vogler   Technical University of Dresden, Germany
Mikhail Volkov   Ural Federal University, Ekaterinburg, Russia

Additional Reviewers

Johanna Björklund	Ludwig Kampel	Matteo Pradella
Eunice Chan	Nikos Karampetakis	Matthieu Rambaud
Siniša Crvenković	János Karsai	Ioannis Refanidis
Igor Dolinka	Hwee Kim	Ivan Szabolcs
Sven Dziadek	Christos Konaxis	Éric Schost
Margherita Maria Ferrari	Maria Madonia	David Sevilla
Kilian Gebhardt	Pavlos Marantidis	Jean-Marc Talbot
Gustav Grabolle	Pierrick Meaux	Pavlos Tzermias
Kishan Gupta	Irini-Eleftheria Mens	Sam van Gool
Tero Harju	Johann Mitteramskogler	Tamás Vinkó
Luisa Herrmann	Timothy Ng	Michael Wagner
Iiro Honkala	Paulina Paraponiari	Johannes Waldmann
Velimir Ilić	Erik Paul	Alfred Wassermann
Bryan Jurish	Martin Pavlovski	Markus Whiteland

Organizing Committee

Jelena Ignjatović (Chair)	Jelena Milovanović
Milan Bašić	Aleksandar Stamenković
Velimir Ilić	Stefan Stanimirović
Zorana Jančić	Ivan Stanković
Dejan Mančev	Lazar Stojković
Jelena Matejić	Aleksandar Trokicić
Ivana Micić	

Sponsoring Institutions

University of Niš
University of Niš – Faculty of Science
Ministry of Education, Science and Technological Development, Republic of Serbia

Abstracts of Invited Talks

Neural Network Decoding of Quantum LDPC Codes

Bane Vasić, Xin Xiao, and Nithin Raveendran

Department of Electrical and Computer Engineering,
Department of Mathematics, University of Arizona, Tucson
vasic@ece.arizona.edu
http://www2.engr.arizona.edu/~vasic

Quantum error correction (QEC) codes [1] are vital in protecting fragile qubits from decoherence. QEC codes are indispensable for practical realizations of fault tolerant quantum computing. Designing good QEC codes, and more importantly low-complexity high-performance decoders for those codes that can be constructed using lossy and noisy devices, is arguably the most important theoretical challenge in quantum computing, key-distribution and communications.

Quantum low-density parity check (QLDPC) codes [4] based on the stabilizer formalism [3] has led to a myriad of QLDPC codes whose constructions and decoding algorithms rely on classical LDPC codes and the theory of syndrome measurement based decoding of quantum stabilizer codes. QLDPC codes are a promising candidate for both quantum computing and quantum optical communications as they admit potentially simple local decoding algorithms, and the history of success in classical LDPC codes in admitting low-complexity decoding and near-capacity performance.

Traditional iterative message-passing algorithms for decoding of LDPC codes are based on *belief propagation* (BP) [5], and operate on a *Tanner graph* [6] of the code's parity check matrix. The BP, as an algorithm to compute marginals of functions on a graphical model, has its roots in the broad class of Bayesian inference problems [2]. While inference using BP is exact only on loop-free graphs (trees), and provides close approximations to exact marginals on loopy graphs with large girth, due to the topology of Tanner graphs of finite-length LDPC codes and additional constraints imposed by quantum version, the application of traditional BP for QEC codes in general, and for QLDPC codes in particular has some fundamental limitations.

Despite the promise of QLDPC codes for quantum information processing, they have several important current limitations. In this talk we will discuss these limitations and present a method to design practical low-complexity high-performance codes and decoders. Our approach is based on using neural networks (NN). The neural network performs the syndrome matching algorithm over a depolarizing channel with noiseless error syndrome measurements. We train our NN to minimize the bit error rate, which is an accurate metric to measure the performance of iterative decoders. In addition it uses straight through estimator (STE) technique to tackle the zero-gradient problem of the

Supported by the NSF under grants ECCS-1500170 and SaTC-1813401.

objective function and outperforms conventional min-sum algorithm up to an order of magnitude of logical error rate.

Keywords: Quantum error correction · Quantum low-density parity check codes · Iterative decoding · Neural networks · Neural network decoding

References

1. Calderbank, A.R., Shor, P.W.: Good quantum error-correcting codes exist. Phys. Rev. A **54**, 1098–1105 (1996)
2. Frey, B.J.: Graphical Models for Machine Learning and Digital Communication. MIT Press, Cambridge (1998)
3. Gottesman, D.: Class of quantum error-correcting codes saturating the quantum hamming bound. Phys. Rev. A **54**(3), 1862–1868 (1996)
4. MacKay, D., Mitchison, G., McFadden, P.: Sparse-graph codes for quantum error correction. IEEE Trans. Inf. Theory **50**(10), 2315–2330 (2004)
5. Pearl, J.: Probabilistic Reasoning in Intelligent Systems: Networks of Plausible Inference. Morgan Kaufmann Publishers Inc., San Francisco (1988)
6. Tanner, R.M.: A recursive approach to low complexity codes. IEEE Trans. Inf. Theory **27**(5), 533–547 (1981)

Algebraic Differential Equations –
Parametrization and Symbolic Solution

Franz Winkler

RISC, Johannes Kepler University Linz
franz.winkler@risc.jku.at

An algebraic differential equation (ADE) is a polynomial relation between a function, some of its partial derivatives, and the variables in which the function is defined. Regarding all these quantities as unrelated variables, the polynomial relation leads to an algebraic relation defining a hypersurface on which the solution is to be found. A solution in a certain class of functions, such as rational or algebraic functions, determines a parametrization of the hypersurface in this class. So in the algebro-geometric method we first decide whether a given ADE can be parametrized with functions from a given class; and in the second step we try to transform a parametrization into one respecting also the differential conditions.

This approach is called the algebro-geometric method for solving ADEs. It is relatively well understood for rational and algebraic solutions of single algebraic ordinary differential equations (AODEs). First steps are taken in a generalization to other types of solutions such as power series solution. Partial differential equations and systems of equations are the topic of current research.

References

1. Eremenko, A.: Rational solutions of first-order differential equations. Annales Academiae Scientiarum Fennicae **23**(1), 181–190 (1990)
2. Feng, R., Gao, X.S.: Rational general solutions of algebraic ordinary differential equations. In: Gutierrez, J. (ed.) Proceedings of the 2004 International Symposium on Symbolic and Algebraic Computation (ISSAC 2004), pp. 155–162. ACM Press, New York (2004)
3. Feng, R., Gao, X.S.: A polynomial time algorithm for finding rational general solutions of first order autonomous ODEs. J. Symb. Comput. **41**(7), 739–762 (2006)
4. Fuchs, L.: Über Differentialgleichungen, deren Integrale feste Verzweigungspunkte besitzen. Sitzungsberichte der Königlich Preuß ischen Akademie der Wissenschaften zu Berlin **11**(3), 251–273 (1884)
5. Grasegger, G., Lastra, A., Sendra, J.R., Winkler, F.: Rational general solutions of systems of first-order algebraic partial differential equations. J. Comput. Appl. Math. **331**, 88–103 (2018)
6. Grasegger, G., Vo, N.T.: An algebraic-geometric method for computing Zolotarev polynomials. In: Burr, M. (ed.) Proceedings of the International Symposium on Symbolic and Algebraic Computation (ISSAC 2017), pp. 173–180. ACM Press, New York (2017)
7. Kamke, E.: Differentialgleichungen: Lösungsmethoden und Lösungen I. B.G. Teubner, Stuttgart (1983)

8. Ngô, L.X.C., Sendra, J.R., Winkler, F.: Classification of algebraic ODEs with respect to rational solvability. Contemp. Math. **572**, 193–210 (2012)
9. Ngô, L.X.C., Sendra, J.R., Winkler, F.: Birational transformations preserving rational solutions of algebraic ordinary differential equations. J. Comput. Appl. Math. **286**, 114–127 (2015)
10. Ngô, L.X.C., Winkler, F.: Rational general solutions of first order non-autonomous parametrizable ODEs. J. Symb. Comput. **45**(12), 1426–1441 (2010)
11. Ngô, L.X.C., Winkler, F.: Rational general solutions of planar rational systems of autonomous ODEs. J. Symb. Comput. **46**(10), 1173–1186 (2011)
12. Sendra, J.R., Winkler, F., Pérez-D az, S.: Rational Algebraic Curves – A Computer Algebra Approach. Springer-Verlag, Heidelberg (2008)
13. Vo, N.T., Grasegger, G., Winkler, F.: Deciding the existence of rational general solutions for first-order algebraic ODEs. J. Symb. Comput. **87**, 127–139 (2018)
14. Winkler, F.: Polynomial Algorithms in Computer Algebra. Springer-Verlag, Wien (1996)

Contents

Invited Paper

Modular Descriptions of Regular Functions

Paul Gastin[✉]

LSV, ENS Paris-Saclay and CNRS, Université Paris-Saclay, Cachan, France
paul.gastin@lsv.fr

Abstract. We discuss various formalisms to describe string-to-string transformations. Many are based on automata and can be seen as operational descriptions, allowing direct implementations when the input scanner is deterministic. Alternatively, one may use more human friendly descriptions based on some simple basic transformations (e.g., copy, duplicate, erase, reverse) and various combinators such as function composition or extensions of regular operations.

We investigate string-to-string functions (which are ubiquitous). A preprocessing that erases comments from a program, or a micro-computation that replaces a binary string with its increment, or a syntactic fix that reorders the arguments of a function to comply with a different syntax, are all examples of string-to-string transformations/functions. We will discuss and compare various ways of describing such functions.

Operationally, we need to parse the input string and to produce an output word. The simplest such mechanism is to use a deterministic finite-state automaton (1DFA) to parse the input from left to right and to produce the output along the way. These are called *sequential* transducers, or one-way input-deterministic transducers (1DFT), see e.g. [17, Chapter V] or [14]. Transitions are labelled with pairs $a \mid u$ where a is a letter read from the input string and u is the word, possibly empty, to be appended to the output string. Sequential transducers allow for instance to strip comments from a latex file, see Fig. 1. Transformations that can be realized by a sequential transducer are called sequential functions. A very important property of sequential functions is that they are closed under composition. Also, each sequential function f can be realized with a canonical minimal sequential transducer $\mathcal{A}_f$ which can be computed from any sequential transducer $\mathcal{B}$ realizing f. As a consequence, equivalence is decidable for sequential transducers.

With a sequential transducer, it is also possible to increment an integer written in binary if the string starts with the least significant bit (lsb), see Fig. 2 left. On the other hand, increment is not a sequential function when the lsb is on the right. There are two possibilities to overcome this problem.

The first solution is to give up determinism when reading the input string. One-way input-nondeterministic finite-state transducers (1NFT) do not necessarily define functions. It is decidable in PTIME whether a 1NFT defines a

© Springer Nature Switzerland AG 2019
M. Ćirić et al. (Eds.): CAI 2019, LNCS 11545, pp. 3–9, 2019.
https://doi.org/10.1007/978-3-030-21363-3_1

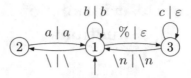

Fig. 1. A sequential transducer stripping comments from a latex file, where $a, b, c \in \Sigma$ are letters from the input alphabet with $b \notin \{\backslash, \%\}$ and $c \neq \backslash n$.

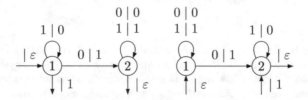

Fig. 2. Transducers incrementing a binary number.

function [18,16]. We are interested in *functional* 1NFT (f1NFT). This is in particular the case when the transducer is input-*unambiguous*. Actually, one-way, input-unambiguous, finite-state transducers (1UFT) have the same expressive power as f1NFT [19]. For instance, increment with lsb on the right is realized by the 1UFT on the right of Fig. 2. Transformations realized by f1NFT are called rational functions. They are easily closed under composition. The equivalence problem is undecidable for 1NFT [15] but decidable in PTIME for f1NFT [18,16]. It is also decidable in PTIME whether a f1NFT defines a sequential function, i.e., whether it can be realized by a 1DFT [7,19].

Interestingly, any rational function h can be written as $r \circ g \circ r \circ f$ where f, g are sequential functions and r is the *reverse* function mapping $w = a_1 a_2 \cdots a_n$ to $w^r = a_n \cdots a_2 a_1$ [12]. We provide a sketch of proof below.[1]

The other solution is to keep input-determinism but to allow the transducer to move its input head in both directions, i.e., left or right (two-way). So we consider two-way input-deterministic finite-state transducers (2DFT) [1]. To realize increment of binary numbers with the lsb on the right with a 2DFT, one has to

---

[1] Assume that h is realized with a 1UFT $\mathcal{B}$. Consider the unique accepting run $q_0 \xrightarrow{a_1|u_1} q_1 \cdots q_{n-1} \xrightarrow{a_n|u_n} q_n$ of $\mathcal{B}$ on some input word $w = a_1 \cdots a_n$. We have $h(w) = u_1 \cdots u_n$. Let $\mathcal{A}$ be the DFA obtained with the subset construction applied to the input NFA induced by $\mathcal{B}$. Consider the run $X_0 \xrightarrow{a_1} X_1 \cdots X_{n-1} \xrightarrow{a_n} X_n$ of $\mathcal{A}$ on w. We have $q_i \in X_i$ for all $0 \leq i \leq n$. The first sequential function f adorns the input word with the run of $\mathcal{A}$: $f(w) = (X_0, a_1) \cdots (X_{n-1}, a_n)$. The sequential transducer $\mathcal{C}$ realizing g is defined as follows. For each state q of $\mathcal{B}$ there is a transition $\delta = q \xrightarrow{(X,a)} p$ in $\mathcal{C}$ if there is a unique $p \in X$ such that $\delta' = p \xrightarrow{a} q$ is a transition in $\mathcal{B}$. Moreover, if δ' outputs u in $\mathcal{B}$ then δ outputs u^r in $\mathcal{C}$. Notice that $q_n \xrightarrow{(X_{n-1}, a_n)|u_n^r} q_{n-1} \cdots q_1 \xrightarrow{(X_0, a_1)|u_1^r} q_0$ is a run of $\mathcal{C}$ producing $u_n^r \cdots u_1^r = h(w)^r$. The result follows.

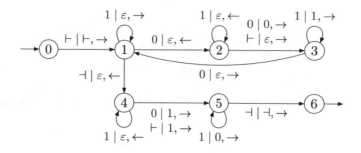

Fig. 3. Two-way transducer incrementing a binary number.

locate the last 0 digit, replace it with 1, keep unchanged the digits on its left and replace all 1's on its right with 0's. This is realized by the 2DFT of Fig. 3. We use $\vdash, \dashv \notin \Sigma$ for the end-markers so the input tape contains $\vdash w \dashv$ when given the input word $w \in \Sigma^*$.

Transformations realized by 2DFTs are called regular functions. They form a very robust class. Regular functions are closed under composition [8]. Actually, a 2DFT can be transformed into a *reversible* one of exponential size [10]. In a reversible transducer, computation steps can be deterministically reversed. As a consequence, the composition of two 2DFTs can be achieved with a single exponential blow-up. Also, contrary to the one-way case, input-nondeterminism does not add expressive power as long as we stay functional: given a f2NFT, one may construct an equivalent 2DFT [13]. Moreover, the equivalence problem for regular functions is still decidable [9].

In classical automata, whether or not a transition can be taken only depends on the input letter being scanned. This can be enhanced using regular look-ahead or look-behind. For instance, the f1NFT on the right of Fig. 2 can be made deterministic using regular look-ahead. In state 1, when reading digit 0, we move to state 2 if the suffix belongs to 1^* and we stay in state 1 otherwise, i.e., if the suffix belongs to $1^*0\{0,1\}^*$. Similarly, we choose to start in the initial state 2 (resp. 1) if the word belongs to 1^* (resp. $1^*0\{0,1\}^*$). More generally, any 1UFT can easily be made deterministic using regular look-ahead: if we have the choice between two transitions leading to states q_1 and q_2, choose q_1 (resp. q_2) if the suffix can be accepted from q_1 (resp. q_2). This query is indeed regular. Hence, regular look-ahead increases the expressive power of one-way deterministic transducers. But regular look-ahead and look-behind do not increase the expressive power of the robust class of regular functions realized by 2DFTs [13].

Regular functions are also those that can be defined with MSO transductions [13], but we will not discuss this here.

By using registers, we obtain another formalism defining string-to-string transformations. For instance incrementing a binary number with lsb on the right is realized by the one-way register transducer on Fig. 4. It uses two registers X, Y initialized with the empty string and 1 respectively and updated while reading the binary number. Register X keeps a copy of the binary number read so far, while Y contains its increment. The final output of the transducer

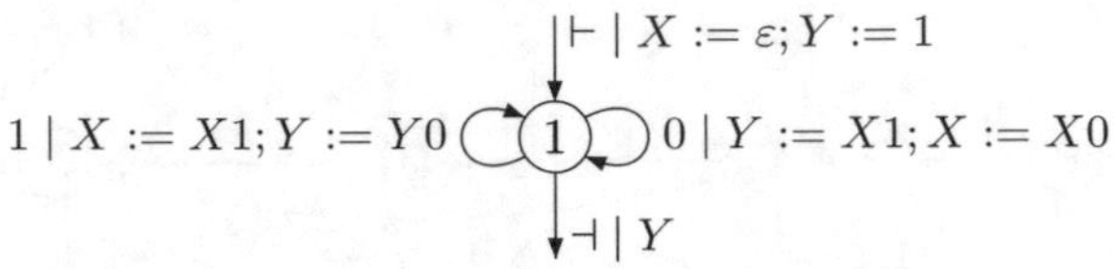

$$\vdash \mid X := \varepsilon; Y := 1$$

$$1 \mid X := X1; Y := Y0 \qquad \boxed{1} \qquad 0 \mid Y := X1; X := X0$$

$$\dashv \mid Y$$

Fig. 4. One-way register transducer incrementing a binary number.

$$\vdash \mid X := \varepsilon; Y := \varepsilon; Z := \varepsilon$$

$$1 \mid X := X1; Y := Y0; Z := Z \qquad \boxed{1} \longrightarrow \dashv \mid Z := 1Y$$

$$0 \mid Z := X; X := \varepsilon; Y := \varepsilon$$

$$1 \mid X := X1; Y := Y0; Z := Z \qquad \boxed{2} \qquad 0 \mid Z := Z0X; X := \varepsilon; Y := \varepsilon$$

$$\dashv \mid Z := Z1Y$$

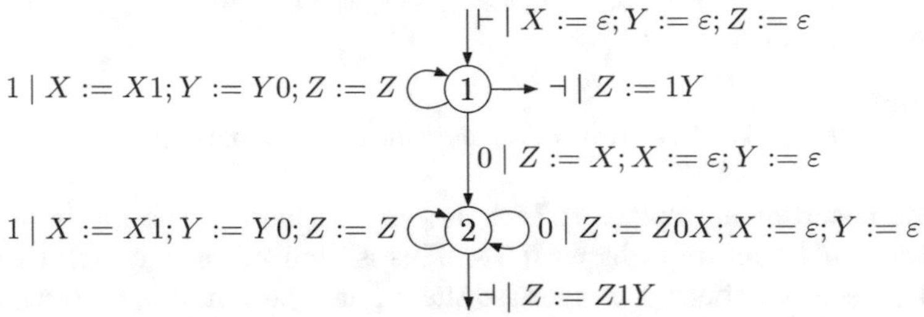

Fig. 5. Streaming string transducer incrementing a binary number.

is the string contained in register Y. This register automaton is a special case of "*simple programs*" defined in [8]. In these simple programs, a register may be reset to the empty string, copied to another register, or updated by appending a finite string. The input head is two-way and most importantly simple programs may be composed. Simple programs coincide in expressive power with 2DFTs [8], hence define once again the class of regular functions.

Notice that when reading digit 0, the transducer of Fig. 4 copies the string stored in X into Y without resetting X to ε. By restricting to one-way register automata with copyless updates (e.g., not of the form $Y := X1; X := X0$ where the string contained in X is duplicated) but allowing concatenation of registers in updates (e.g., $Z := Z0X; X := \varepsilon$), we obtain another kind of machines, called copyless streaming string transducers (SST), once again defining the same class of regular functions [3]. Continuing our example, incrementing a binary number with lsb on the right can be realized with the SST on Fig. 5. It uses three registers X, Y, Z initialized with the empty string and updated while reading the binary number. The final output of the transducer is the string contained in register Z.

The above machines provide a way of describing string-to-string transformations which is not modular. Describing regular functions in such devices is difficult, and it is even more difficult to understand what is the function realized by a 2DFT or an SST. We discuss now more compositional and modular descriptions of regular functions. Such a formalism, called regular list functions, was described in [6]. It is based on function composition together with some natural functions over lists such as reverse, append, co-append, map, etc. Here we choose to look in combinators derived from regular expressions.

The idea is to start from basic functions, e.g., $(1 \mid 0)$ means "read 1 and output 0", and to apply simple combinators generalizing regular expressions [4,2,11,5]. For instance, using the Kleene iteration, $(1 \mid 0)^*$ describes a function which replaces a sequence of 1's with a sequence of 0's of same length. Similarly, copy $:= ((0 \mid 0) + (1 \mid 1))^*$ describes a regular function which simply copies an input binary string to the output. Now, incrementing a binary number with lsb on the right is described with the expression increment0 $:=$ copy $\cdot (0 \mid 1) \cdot (1 \mid 0)^*$, assuming that the input string contains at least one 0 digit. If the input string belongs to 1^*, we may use the expression increment1 $:= (\varepsilon \mid 1) \cdot (1 \mid 0)^*$. Notice that such a regular transducer expression (RTE) defines simultaneously the *domain* of the regular function as a regular expression, e.g., dom(increment0) $= (0 + 1)^* 0 1^*$, and the output to be produced. The input regular expression explains how the input should be parsed. If the input regular expression is ambiguous, parsing the input word is not unique and the expression may be non functional. For instance, copy $\cdot (1 \mid 0)^*$ is ambiguous. The input word $w = 1011$ may be parsed as $10 \cdot 11$ or $101 \cdot 1$ or $1011 \cdot \varepsilon$ resulting in the outputs 1000 or 1010 or 1011 respectively. On the other end, increment $:=$ increment0 $+$ increment1 has an unambiguous input regular expression.

A 2DFT may easily duplicate the input word, defining the function $w \mapsto w\$w$, which cannot be computed with a sequential transducer or a f1NFT. In addition to the classical regular combinators ($+$ for disjoint union, $\cdot$ for unambiguous concatenation or Cauchy product, * for unambiguous Kleene iteration), we add the Hadamard product $(f \odot g)(w) = f(w) \cdot g(w)$ where the input word is read twice, first producing the output computed by f then the output computed by g. Hence the function duplicating its input can be simply written as duplicate $:=$ (copy $\cdot (\varepsilon \mid \$)) \odot$ copy. This can be iterated duplicating each $\#$-separated words in a string with $f := ($duplicate $\cdot (\# \mid \#))^*$. We have $f(u_1 \# u_2 \# \cdots u_n \#) = u_1 \$ u_1 \# u_2 \$ u_2 \# \cdots u_n \$ u_n \#$ when $u_1, \ldots, u_n$ are binary strings.

The Hadamard product also allows to exchange two strings $u \# v \mapsto v \# u$ where $u, v \in \{0, 1\}^*$. Let erase $:= ((0 \mid \varepsilon) + (1 \mid \varepsilon))^*$ and

$$\text{exchange} := \Big(\text{erase} \cdot (\# \mid \varepsilon) \cdot \text{copy} \cdot (\varepsilon \mid \#)\Big) \odot \Big(\text{copy} \cdot (\# \mid \varepsilon) \cdot \text{erase}\Big).$$

Again this can be iterated on the output of the function f defined above with the map

$$g := \text{erase} \cdot (\$ \mid \varepsilon) \cdot (\text{exchange} \cdot (\$ \mid \$))^* \cdot \text{erase} \cdot (\# \mid \varepsilon).$$

We have $g \circ f(u_1 \# u_2 \# \cdots u_n \#) = u_2 \# u_1 \$ u_3 \# u_2 \$ \cdots u_n \# u_{n-1} \$$. It turns out that the regular function $g \circ f$ cannot be described using the regular combinators $+, \cdot, *, \odot$. This is the reason for introducing a 2-chained Kleene iteration [4]: $[K, h]^{2+}$ first *unambiguously* parse an input word as $w = u_1 u_2 \cdots u_n$ with $u_1, \ldots, u_n \in K$ and then apply h to all consecutive pairs of factors, resulting in the output $h(u_1 u_2) h(u_2 u_3) \cdots h(u_{n-1} u_n)$. For instance, with the functions defined above, we can easily check that $g \circ f = [K, h]^{2+}$ with $K = \{0, 1\}^* \#$ and $h := $ exchange $\cdot (\# \mid \$)$.

Another crucial feature of 2DFTs is the ability to reverse the input $w \mapsto w^r$, e.g., $(1101000)^r = 0001011$. In regular transducer expressions, we introduce a *reversed* Kleene star $r\text{-}*$ which parse the input word from left to right but produce the output in reversed order. For instance, $f^{r\text{-}*}(w) = f(u_n) \cdots f(u_2) f(u_1)$

if the input word is parsed as $w = u_1u_2 \cdots u_n$. Hence, reversing a binary string is described with the RTE reverse $:= ((0 \mid 0) + (1 \mid 1))^{r\text{-}*}$. There is also a reversed version of the two-chained Kleene iteration. With the above notation, we get $[K, h]^{r\text{-}2+}(w) = h(u_{n-1}u_n) \cdots h(u_2u_3)h(u_1u_2)$.

Once again, we obtain an equivalent formalism for describing regular functions: the regular transducer expressions using $+$, $\cdot$, $\odot$, $*$, $r\text{-}*$, $2+$ and $r\text{-}2+$ as combinators [4,2,11,5]. Alternatively, as illustrated on an example above, we may remove the two-chained iterations if we allow function compositions, i.e., using the combinators $+$, $\cdot$, $\odot$, $\circ$, $*$ and $r\text{-}*$. Further, we may remove the Hadamard product if we provide duplicate as a basic function. Indeed, we can easily check that $f \odot g = (f \cdot (\$ \mid \varepsilon) \cdot g) \circ$ duplicate. Also, the reversed Kleene iteration may be removed if we use reverse as a basic function. We will see that $f^{r\text{-}*} = (f \circ \text{reverse})^* \circ \text{reverse}$. Indeed, assume that an input word is parsed as $w = u_1u_2 \cdots u_n$ when applying $f^{r\text{-}*}$ resulting in $f(u_n) \cdots f(u_2)f(u_1)$. Then, reverse$(w)$ is parsed as $u_n^r \cdots u_2^r u_1^r$ when applying $(f \circ \text{reverse})^*$. The result follows since $(f \circ \text{reverse})(u^r) = f(u)$.

To conclude, we have an expressively complete set of combinators $+$, $\cdot$, $*$ and $\circ$ when we allow duplicate and reverse as basic functions. We believe that this is a very convenient, compositional and modular, formalism for defining regular functions.

References

1. Aho, A.V., Ullman, J.D.: A characterization of two-way deterministic classes of languages. J. Comput. Syst. Sci. 4(6), 523–538 (1970)
2. Alur, R., D'Antoni, L., Raghothaman, M.: DReX: a declarative language for efficiently evaluating regular string transformations. In: Rajamani, S.K., Walker, D. (eds.) Proceedings of the 42nd Annual ACM SIGPLAN-SIGACT Symposium on Principles of Programming Languages - POPL 2015, pp. 125–137. ACM Press (2015)
3. Alur, R., Deshmukh, J.V.: Nondeterministic streaming string transducers. In: Aceto, L., Henzinger, M., Sgall, J. (eds.) ICALP 2011. LNCS, vol. 6756, pp. 1–20. Springer, Heidelberg (2011). https://doi.org/10.1007/978-3-642-22012-8_1
4. Alur, R., Freilich, A., Raghothaman, M.: Regular combinators for string transformations. In: Joint Meeting of the Twenty-Third EACSL Annual Conference on Computer Science Logic (CSL) and the Twenty-Ninth Annual ACM/IEEE Symposium on Logic in Computer Science (LICS), CSL-LICS 2014, Vienna, Austria, July 14–18, 2014, pp. 9:1–9:10 (2014)
5. Baudru, N., Reynier, P.-A.: From two-way transducers to regular function expressions. In: Hoshi, M., Seki, S. (eds.) DLT 2018. LNCS, vol. 11088, pp. 96–108. Springer, Cham (2018). https://doi.org/10.1007/978-3-319-98654-8_8
6. Bojańczyk, M., Daviaud, L., Krishna, S.N.: Regular and first-order list functions. In: Dawar, A., Grädel, E. (eds.) Proceedings of the 33rd Annual ACM/IEEE Symposium on Logic in Computer Science - LICS 2018, pp. 125–134. ACM Press (2018)
7. Choffrut, Ch.: Une caractérisation des fonctions séquentielles et des fonctions sous-séquentielles en tant que relations rationnelles. Theoret. Comput. Sci. 5(3), 325–337 (1977)

8. Chytil, M.P., Jákl, V.: Serial composition of 2-way finite-state transducers and simple programs on strings. In: Salomaa, A., Steinby, M. (eds.) ICALP 1977. LNCS, vol. 52, pp. 135–147. Springer, Heidelberg (1977). https://doi.org/10.1007/3-540-08342-1_11

9. Culik, K., Karhumäki, J.: The equivalence of finite valued transducers (on HDT0L languages) is decidable. Theoret. Comput. Sci. **47**, 71–84 (1986)

10. Dartois, L., Fournier, P., Jecker, I., Lhote, N.: On reversible transducers. In: Chatzigiannakis, I., Indyk, P., Kuhn, F., Muscholl, A. (eds.) 44th International Colloquium on Automata, Languages, and Programming (ICALP 2017), volume 80 of Leibniz International Proceedings in Informatics (LIPIcs), pp. 113:1–113:12, Dagstuhl, Germany, Schloss Dagstuhl-Leibniz-Zentrum fuer Informatik (2017)

11. Dave, V., Gastin, P., Krishna, S.N.: Regular transducer expressions for regular transformations. In: Hofmann, M., Dawar, A., Grädel, E. (eds.) Proceedings of the 33rd Annual ACM/IEEE Symposium on Logic in Computer Science (LICS 2018), pp. 315–324. ACM Press, Oxford (2018)

12. Elgot, C.C., Mezei, J.E.: On relations defined by generalized finite automata. IBM J. Res. Dev. **9**(1), 47–68 (1965)

13. Engelfriet, J., Hoogeboom, H.J.: MSO definable string transductions and two-way finite-state transducers. ACM Trans. Comput. Log. **2**(2), 216–254 (2001)

14. Filiot, E., Reynier, P.-A.: Transducers, logic and algebra for functions of finite words. SIGLOG News **3**(3), 4–19 (2016)

15. Griffiths, T.V.: The unsolvability of the equivalence problem for lambda-free non-deterministic generalized machines. J. ACM **15**(3), 409–413 (1968)

16. Gurari, E.M., Ibarra, O.H.: A note on finite-valued and finitely ambiguous transducers. Math. Syst. Theory **16**(1), 61–66 (1983)

17. Sakarovitch, J.: Elements of Automata Theory. Cambridge University Press, New York (2009)

18. Schützenberger, M.P.: Sur les relations rationnelles. In: Brakhage, H. (ed.) GI-Fachtagung 1975. LNCS, vol. 33, pp. 209–213. Springer, Heidelberg (1975). https://doi.org/10.1007/3-540-07407-4_22

19. Weber, A., Klemm, R.: Economy of description for single-valued transducers. Inf. Comput. **118**(2), 327–340 (1995)

Contributed Papers

Enhancing an Attack to DSA Schemes

Marios Adamoudis[1], Konstantinos A. Draziotis[2]([⊠]), and Dimitrios Poulakis[1]

[1] Department of Mathematics, Aristotle University of Thessaloniki,
Thessaloniki, Greece
`marios.p7@hotmail.com, poulakis@math.auth.gr`
[2] Department of Informatics, Aristotle University of Thessaloniki,
Thessaloniki, Greece
`drazioti@csd.auth.gr`

Abstract. In this paper, we improve the theoretical background of the attacks on the DSA schemes of a previous paper, and we present some new more practical attacks.

Keywords: Public key cryptography · Digital Signature Algorithm · Elliptic Curve Digital Signature Algorithm · Closest Vector Problem · LLL algorithm · BKZ algorithm · Babai's Nearest Plane Algorithm

MSC 2010: 94A60 · 11T71 · 11Y16

1 Introduction

In August 1991, the U.S. government's National Institute of Standards and Technology (NIST) proposed the Digital Signature Algorithm (DSA) for digital signatures [13,15]. This algorithm has become a standard [6] and was called Digital Signature Standard (DSS). In 1998, an elliptic curve analogue called Elliptic Curve Digital Signature Algorithm (ECDSA) was proposed and standardized, see [10]. In the first subsection we recall the outlines of DSA and ECDSA.

1.1 The DSA and ECDSA Schemes

First, let us summarize DSA. The signer chooses a prime p of size between 1024 and 3072 bits with increments of 1024, as recommended in FIPS 186-3 [6, p. 15]. Also, he chooses a prime q of size 160, 224 or 256 bits, with $q|p-1$ and a generator g of the unique order q subgroup G of the multiplicative group $\mathbb{F}_p^*$ of the prime finite field $\mathbb{F}_p$. Furthermore, he selects a randomly $a \in \{1, \ldots, q-1\}$ and computes $R = g^a \bmod p$. The public key of the signer is (p, q, g, R) and his private key a. He also publishes a hash function $h : \{0,1\}^* \to \{0, \ldots, q-1\}$. To sign a message $m \in \{0,1\}^*$, he selects randomly $k \in \{1, \ldots, q-1\}$ which is the ephemeral key, and computes $r = (g^k \bmod p) \bmod q$ and $s = k^{-1}(h(m) +$

$ar) \bmod q$. The signature of m is the pair (r, s). The signature is valid if and only if we have:

$$r = ((g^{s^{-1}h(m)\bmod q} R^{s^{-1}r \bmod q}) \bmod p) \bmod q.$$

For the ECDSA the signer selects an elliptic curve E over $\mathbb{F}_p$, a point $P \in E(\mathbb{F}_p)$ with order a prime q of size at least 160 bits. According to FIPS 186-3, the prime p must be in the set $\{160, 224, 256, 512\}$. Further, he selects randomly $a \in \{1, \ldots, q-1\}$ and computes $Q = aP$. The public key of the signer is (E, p, q, P, Q) and his private key a. He also publishes a hash function $h : \{0, 1\}^* \to \{0, \ldots, q-1\}$. To sign a message m, he selects randomly $k \in \{1, \ldots, q-1\}$ which is the ephemeral key and computes $kP = (x, y)$ (where x and y are regarded as integers between 0 and $p - 1$). Next, he computes $r = x \bmod q$ and $s = k^{-1}(h(m) + ar) \bmod q$. The signature of m is (r, s). For its verification one computes

$$u_1 = s^{-1}h(m) \bmod q, \quad u_2 = s^{-1}r \bmod q, \quad u_1P + u_2Q = (x_0, y_0).$$

He accepts the signature if and only if $r = x_0 \bmod q$.

The security of the two systems is relied on the assumption that the only way to forge the signature is to recover either the secret key a, or the ephemeral key k (in this case is very easy to compute a). Thus, the parameters of these systems were chosen in such a way that the computation of discrete logarithms is computationally infeasible.

1.2 Our Contribution

Except the attacks in discrete logarithm problem, we have attacks based on the equality $s = k^{-1}(h(m) + ar) \bmod q$ which use lattice reduction techniques [1–5, 11, 12, 16–19]. In this paper, we also use this equality and following the ideas of [19], we improve the efficiency of attacks on the DSA schemes described in it.

The attack described in [19] is based on a system of linear congruences of a particular form which has at most a unique solution below a certain bound, which can be computed efficiently. Thus, in case where the length of a vector, having as coordinates the secret and the ephemeral keys of some signed message is quite small, the secret key can be computed. More precisely in this work, we also consider the system of linear congruences of [19] and we improve the upper bound under which it has at most one solution. This extension provides an improvement of the attack [19], which also remains deterministic. Thus, when some signed messages are available, we can construct a such system whose solution has among its coordinates the secret key, and so it is possible to find it in practical time.

Furthermore, an heuristic improvement based on our attack is given. We update experimental results on (EC)DSA based on known bits of the ephemeral keys. In fact we prove that if we know 1 bit of a suitable multiple of the ephemeral keys for 206 signatures, we can find the secret key with success rate 62%. The previous best result was of Liu and Nguyen [12], where they provided a probabilistic attack based on enumeration techniques, where managed to find the secret key if they know 2 bits of 100 ephemeral keys. The attack provided in [12] first reduces the problem of finding the secret key, to the hidden number problem (HNP) and then reduces HNP to a variant of CVP (called Bounded Decoded Distance problem : BDD).

The Structure of the Paper. The paper is organized as follows. In Sect. 2, we recall some basic results about lattices, and the Babai's Nearest Plane Algorithm. In Sect. 3, we prove some result which we need for the presentation of our attacks. Our attacks are presented in Sects. 4. Some experimental results are given in Sect. 5. Finally, Sect. 6 concludes the paper.

2   Background on Lattices

In this section, we collect several well-known facts about lattices which form the background to our algorithms.

Let $\mathbf{b}_1, \mathbf{b}_2, \ldots, \mathbf{b}_n$ linearly independent vectors of $\mathbb{R}^m$. The set

$$\mathcal{L} = \left\{ \sum_{j=1}^{n} \alpha_j \mathbf{b}_j : \alpha_j \in \mathbb{Z}, 1 \leq j \leq n \right\}$$

is called a *lattice* and the set $\mathcal{B} = \{\mathbf{b}_1, \ldots, \mathbf{b}_n\}$ a basis of $\mathcal{L}$. All the bases of $\mathcal{L}$ have the same number of elements n which is called *dimension* or *rank* of $\mathcal{L}$. If $n = m$, then the lattice $\mathcal{L}$ is said to have *full rank*. We denote by M the $n \times m$-matrix having as rows the vectors $\mathbf{b}_1, \ldots, \mathbf{b}_n$. If $\mathcal{L}$ has full rank, then the *volume* of the lattice $\mathcal{L}$ is defined to be the positive number $|\det M|$ which is independent from the basis $\mathcal{B}$. It is denoted by $vol(\mathcal{L})$ or $\det \mathcal{L}$ (see also [7]). If $\mathbf{v} \in \mathbb{R}^m$, then $\|\mathbf{v}\|$ denotes, as usually, the Euclidean norm of $\mathbf{v}$. We denote by $LLL(M)$, the application of well-known LLL-algorithm on the rows of M. Finally, we denote by $\lambda_1(\mathcal{L})$ the smaller of the lengths of vectors of $\mathcal{L}$.

We define the approximate Closest Vector Problem $(CVP_{\gamma_n}(L))$ as follows: Given a lattice $\mathcal{L} \subset \mathbb{Z}^m$ of rank n and a vector $\mathbf{t} \in \mathbb{R}^m$, find a vector $\mathbf{u} \in \mathcal{L}$ such that, for every $\mathbf{u}' \in \mathcal{L}$ we have:

$$\|\mathbf{u} - \mathbf{t}\| \leq \gamma_n \|\mathbf{u}' - \mathbf{t}\| \quad \text{(for some } \gamma_n \geq 1\text{)}.$$

We say that we have a CVP oracle, if we have an efficient probabilistic algorithm that solves CVP_{γ_n} for $\gamma_n = 1$. To solve CVP_{γ_n}, we usually use Babai's algorithm [7, Chapter 18] (which has polynomial running time). In fact, combining this algorithm with LLL algorithm, we solve $CVP_\gamma(\mathcal{L})$ for some lattice $\mathcal{L} \subset \mathbb{Z}^m$, for $\gamma_n = 2^{n/2}$ and $n = rank(\mathcal{L})$ in polynomial time.

Babai's Nearest plane Algorithm:

```
INPUT: A n × m-matrix M with rows the vectors of a basis B    =
{bᵢ}₁≤ᵢ≤ₙ ⊂ ℤᵐ of the lattice L and a vector t ∈ ℝᵐ
OUTPUT: x ∈ L such that ||x − t|| ≤ 2ⁿ/²dist(L, t).
```

01. $M \leftarrow LLL(M)$ $(\delta = 3/4)$ # we can also use $BKZ_\beta(M)$
02. $M^* = \{(\mathbf{b}_j^*)_j\} \leftarrow GSO(M)$ # GSO : Gram-Schimdt Orthogonalization
03. $\mathbf{b} \leftarrow \mathbf{t}$
04. For $j = n$ to 1
05. $c_j \leftarrow \left\lceil \dfrac{\mathbf{b} \cdot \mathbf{b}_j^*}{||\mathbf{b}_j^*||^2} \right\rfloor$ #$\lceil x \rfloor = \lfloor x + 0.5 \rfloor$
06. $\mathbf{b} \leftarrow \mathbf{b} - c_j \mathbf{b}_j$
07. Return $\mathbf{t} - \mathbf{b}$

Note that there is a variant of CVP, called BDD, where we search for vectors $\mathbf{u}$ such that $\|\mathbf{u} - \mathbf{t}\| \leq \lambda_1(L)/2$. Further, there are enumeration algorithms that compute all the lattice vectors within distance R from the target vector, see [8,9]. These algorithms are not of polynomial time with respect to the rank of the lattice.

3   Auxiliary Results

In this section we prove some results that we need for the description of our attack.

Proposition 1. *Let n, q and A_j be positive integers satisfying*

$$\frac{q^{\frac{j}{n+1}+f_q(n)}}{2} < A_j < \frac{q^{\frac{j}{n+1}+f_q(n)}}{1.5} \quad (j = 1, \ldots, n), \tag{1}$$

where $f_q(n)$ is a positive real number such that

$$f_q(n) + \frac{n}{n+1} < 1 \tag{2}$$

and

$$\frac{q^{1+2f_q(n)}}{1.5} < q - \frac{1}{2} q^{\frac{n}{n+1}+f_q(n)} \tag{3}$$

Let $\mathcal{L}$ be the lattice generated by the vectors

$$\mathbf{b}_0 = (-1, A_1, \ldots, A_n), \mathbf{b}_1 = (0, q, 0, \ldots, 0), \ldots, \mathbf{b}_n = (0, \ldots, 0, q).$$

Then, for all non-zero $\mathbf{v} \in \mathcal{L}$, we have:

$$\|\mathbf{v}\| > \frac{1}{2} q^{\frac{n}{n+1}+f_q(n)}.$$

Proof. Suppose that there is a vector $\mathbf{v} \in L \setminus \{\mathbf{0}\}$ such that

$$\|\mathbf{v}\| \leq \frac{1}{2} q^{\frac{n}{n+1}+f_q(n)}. \tag{4}$$

Then, the inequality (2) yields:

$$\|\mathbf{v}\| < \frac{q}{2} < q. \tag{5}$$

Since $\mathbf{v} \in L$, there are integers $x_0, x_1, \ldots, x_n$ such that

$$\mathbf{v} = x_0 \mathbf{b}_0 + \cdots + x_n \mathbf{b}_n = (-x_0, x_0 A_1 + x_1 q, \ldots, x_0 A_n + x_n q).$$

Thus, we deduce:

$$|x_0|, |x_0 A_j + x_j q| \leq \frac{1}{2} q^{\frac{n}{n+1} + f_q(n)}.$$

If $x_0 = 0$, then we get the vector $\mathbf{v} = (0, x_1 q, \ldots, x_n q)$ with length $> q$. On the other hand, (5) implies $\|\mathbf{v}\| < q$. Thus, we have a contradiction, and so we deduce that $x_0 \neq 0$.

Since $1 \leq |x_0| \leq \frac{1}{2} q^{n/(n+1) + f_q(n)}$, we distinguish the following two cases:

(i) There is $k \in \{1, 2, \ldots, n-1\}$ such that

$$q^{\frac{k-1}{n+1}} < |x_0| < q^{\frac{k}{n+1}}.$$

By (1), we obtain:

$$\frac{1}{2} q^{\frac{n+1-k}{n+1} + f_q(n)} \leq A_{n+1-k} \leq \frac{1}{1.5} q^{\frac{n+1-k}{n+1} + f_q(n)}.$$

Multiplying the two previous inequalities, we get:

$$\frac{1}{2} q^{\frac{n}{n+1} + f_q(n)} < |x_0| A_{n+1-k} < \frac{1}{1.5} q^{1 + f_q(n)}.$$

By (3), we have:

$$\frac{q^{1+f_q(n)}}{1.5} < \frac{q^{1+2f_q(n)}}{1.5} < q - \frac{1}{2} q^{\frac{n}{n+1} + f_q(n)}.$$

It follows:

$$\frac{1}{2} q^{\frac{n}{n+1} + f_q(n)} < |x_0| A_{n+1-k} < q - \frac{1}{2} q^{\frac{n}{n+1} + f_q(n)} \tag{6}$$

If $x_0 A_{n+1-k} + q x_{n+1-k} = 0$, then $|x_0| A_{n+1-k} = q |x_{n+1-k}| > q$, which contradicts the above inequality. Thus, we have $x_0 A_{n+1-k} + q x_{n+1-k} \neq 0$.

Since $\|\mathbf{v}\| \geq |x_0 A_{n+1-k} + q x_{n+1-k}|$, we get:

$$\|\mathbf{v}\| \geq \|x_0| A_{n+1-k} - q |x_{n+1-k}\| \geq q |x_{n+1-k}| - |x_0| A_{n+1-k}. \tag{7}$$

Assume that $x_{n+1-k} \neq 0$. It follows: $\|\mathbf{v}\| \geq q - |x_0| A_{n+1-k}$. Then, using the right part of inequality (6), we get:

$$\|\mathbf{v}\| > \frac{1}{2} q^{\frac{n}{n+1} + f_q(n)},$$

which contradicts (4). Hence, we have $x_{n+1-k} = 0$. By (7), we have: $\|\mathbf{v}\| \geq |x_0| A_{n+1-k}$. Using the left part of inequality (6) we obtain:

$$\|\mathbf{v}\| > \frac{1}{2} q^{\frac{n}{n+1} + f_q(n)},$$

which is a contradiction.

(ii) We have:

$$q^{\frac{n-1}{n+1}} < |x_0| < q^{\frac{n}{n+1}+f_q(n)}.$$

Further, (1) gives:

$$\frac{1}{2}\, q^{\frac{1}{n+1}+f_q(n)} \le A_1 \le \frac{1}{1.5}\, q^{\frac{1}{n+1}+f_q(n)}.$$

Multiplying the two inequalities we obtain:

$$\frac{1}{2}\, q^{\frac{n}{n+1}+f_q(n)} < |x_0|A_1 < \frac{1}{1.5}\, q^{1+2f_q(n)}.$$

By (3), we have:

$$\frac{1}{1.5}\, q^{1+2f_q(n)} < q - \frac{1}{2}\, q^{\frac{n}{n+1}+f_q(n)}.$$

Combining the two above inequalities, we deduce:

$$\frac{1}{2}\, q^{\frac{n}{n+1}+f_q(n)} < |x_0|A_1 < q - \frac{1}{2}\, q^{\frac{n}{n+1}+f_q(n)},$$

which is relation (6) with $k = n$. Proceeding, as previously, we obtain a contradiction. Thus, the result follows.

Remark 1. Proposition 1 is an improvement of [19, Lemma 1], since the obtained lower bound is better than the lower bound $\|v\| > \frac{q^{n/(n+1)}}{8}$, obtained in [19, Lemma 1]. Furthermore, the number of signed messages in [19, Lemma 1], for q 160-bits, are less than $\ln \ln q \approx 4$. In the previous proposition the number of messages can be much larger. Note that a largest lower bound will allow us, as we shall see, to attack larger DSA keys (for fixed n and q).

Using the terminology of Proposition 1, we prove the following.

Proposition 2. *Let q and A_i, B_i $(i = 1, \ldots, n)$ be positive integers with A_i as in Proposition 1. Then, the system of congruences*

$$y_i + A_i x + B_i \equiv 0 \,(\mathrm{mod}\, q) \quad (i = 1, \ldots, n)$$

has at most one solution $\mathbf{v} = (x, y_1, \ldots, y_n)$ such that

$$\|\mathbf{v}\| < \frac{1}{4}\, q^{\frac{n}{n+1}+f_q(n)}.$$

If such $\mathbf{v}$ exists, then $\mathbf{v} = \mathbf{w} - \mathbf{b}$, where $\mathbf{b} = (0, B_1, \ldots, B_n)$, and $\mathbf{w}$ is a vector obtained by using a CVP oracle for the lattice $\mathcal{L}$ of Proposition 1 and $\mathbf{b}$.

Proof. Let $\mathbf{v} = (x, y_1, \ldots, y_n)$ be a solution of the system with

$$\|\mathbf{v}\| < \frac{1}{4}\, q^{\frac{n}{n+1}+f_q(n)}.$$

We denote by $\mathcal{L}$ the lattice spanned by the rows of the $(n+1) \times (n+1)$ matrix

$$
\begin{bmatrix}
-1 & A_1 & A_2 & \dots & A_n \\
0 & q & 0 & \dots & 0 \\
0 & 0 & q & \dots & 0 \\
\vdots & \vdots & \vdots & \ddots & \vdots \\
0 & 0 & 0 & \dots & q
\end{bmatrix}
\tag{8}
$$

and set $\mathbf{b} = (0, B_1, \dots, B_n)$. Since $y_i + A_i x + B_i \equiv 0 \,(\mathrm{mod}\, q)$ there is $z_i \in \mathbb{Z}$, such that $y_i + B_i = -A_i x + z_i q$. Set $\mathbf{u} = \mathbf{v} + \mathbf{b} = (x, y_1 + B_1, \dots, y_n + B_n)$. Then $\mathbf{u} = (x, -A_1 x + z_1 q, \dots, -A_n x + z_n q)$ belongs to $\mathcal{L}$ and we have

$$
\|\mathbf{u} - \mathbf{b}\| = \|\mathbf{v}\| < \frac{1}{4} q^{\frac{n}{n+1} + f_q(n)}.
$$

On the other hand, using the CVP-oracle, we compute $\mathbf{w} \in \mathcal{L}$ such that

$$
\|\mathbf{w} - \mathbf{b}\| \le \|\mathbf{u} - \mathbf{b}\| < \frac{1}{4} q^{\frac{n}{n+1} + f_q(n)}.
$$

Thus, we get: $\|\mathbf{w} - \mathbf{u}\| \le \|\mathbf{w} - \mathbf{b}\| + \|\mathbf{b} - \mathbf{u}\| < \frac{1}{2} q^{\frac{n}{n+1} + f_q(n)}$. Since $\mathbf{w} - \mathbf{u} \in \mathcal{L}$, Proposition 1 implies $\mathbf{w} = \mathbf{u}$. So, the CVP oracle outputs the vector $\mathbf{u}$ and so we can compute $\mathbf{v} = \mathbf{u} - \mathbf{b}$.

To get an idea how to choose the sequence $f_q(n)$ we use the following well known result.

Lemma 1. *(Hermite). For every full rank lattice $\mathcal{L} \subset \mathbb{R}^n$ we have:*

$$
\lambda_1(\mathcal{L}) \le \sqrt{n} \, (\det \mathcal{L})^{1/n}.
$$

In our case we get $\lambda_1(L) \le \sqrt{n+1} \, q^{\frac{n}{n+1}}$. The lower bound of Proposition 1 is of the form $\frac{1}{2} q^x$. So, we get:

$$
\frac{1}{2} q^x \le \lambda_1(L) \le \sqrt{n+1} \, q^{\frac{n}{n+1}}.
$$

Solving with respect to x, we obtain:

$$
x \le \frac{\ln 2 + \frac{1}{2} \ln (n+1)}{\ln q} + \frac{n}{n+1} = \frac{\ln (4(n+1))}{2 \ln q} + \frac{n}{n+1}.
$$

Thus, we get an upper bound for x. For instance we may select

$$
f_q(n) = \frac{\ln 2 + \ln (n+1)}{2n \ln q}.
$$

In general, we can choose

$$
f_q(n) = g_{q,b,c,d}(n) = \frac{c \ln (n+1)}{b n^d \ln q},
\tag{9}
$$

where b, c, d are chosen such that $0 < g_{q,b,c,d}(n) < \frac{\ln (4(n+1))}{2 \ln q}$.

4   The Attack

In this section we describe our attack. Let m_i $(i = 1, \ldots, n)$ be messages signed with (EC)DSA system and (r_i, s_i) their signatures. Then, there are $k_i \in \{1, \ldots, q - 1\}$ such that $r_i = (g^{k_i} \bmod p) \bmod q$ (resp. $r_i = x_i \bmod q$ and $k_i P = (x_i, y_i)$) and $s_i = k_i^{-1}(h(m_i) + ar_i) \bmod q$. It follows that

$$k_i + C_i a + D_i \equiv 0 \pmod{q} \quad (i = 1, \ldots, n),$$

where $C_i = -r_i s_i^{-1} \bmod q$ and $D_i = -s_i^{-1} h(m_i) \bmod q$. Multiplying both sides by $C_i^{-1} \bmod q$, we get: $C_i^{-1} k_i + a + C_i^{-1} D_i \equiv 0 \pmod{q}$.

Now, we pick integers A_i satisfying

$$\frac{q^{\frac{i}{n+1} + f_q(n)}}{2} < A_i < \frac{q^{\frac{i}{n+1} + f_q(n)}}{1.5}$$

and we multiply by A_i both sides of the above congruence. So, we get:

$$A_i C_i^{-1} k_i + A_i a + A_i C_i^{-1} D_i \equiv 0 \pmod{q}.$$

Set $B_i = A_i C_i^{-1} D_i \bmod q$ $(i = 1, \ldots, n)$. Then, the vector

$$\mathbf{s} = (a, A_1 C_1^{-1} k_1 \bmod q, \ldots, A_n C_n^{-1} k_n \bmod q)$$

satisfies the system

$$y_i + A_i x + B_i \equiv 0 \pmod{q} \quad (i = 1, \ldots, n). \tag{10}$$

We set $\mathbf{b} = (0, B_1, \ldots, B_n)$ and $M_{n,q} = \frac{1}{4} q^{\frac{n}{n+1} + f_q(n)}$, for some sequence $f_q(n)$ satisfying the hypotheses of Proposition 1. The vectors

$$\mathbf{b_0} = (-1, A_1, \ldots, A_n), \quad \mathbf{b_1} = (0, q, 0, \ldots, 0), \ldots, \mathbf{b_n} = (0, \ldots, 0, q),$$

form a basis of $\mathbb{R}^{n+1}$. We denote by $\mathcal{L}$ the lattice spanned by $\mathbf{b_0}, \mathbf{b_1}, \ldots, \mathbf{b_n}$. If $\|\mathbf{s}\| < M_{n,q}$, then Proposition 2 implies that $\mathbf{s} = \mathbf{w} - \mathbf{b}$, where $\mathbf{w}$ is a vector obtained by using a CVP oracle for $\mathcal{L}$ and $\mathbf{b}$. Thus, we can compute the secret key a which is the first coordinate of $\mathbf{s}$.

An Improvement of the Attack. The previous attack, which is deterministic, needs quite short solution vector (and so small secret key) in order to succeed. Now, we remark that if $\mathbf{u} \in \mathcal{L}$, then $\mathbf{u} - \mathbf{b}$ is a solution of the system (10). Indeed, since $\mathbf{u} \in \mathcal{L}$, there are integers $l_0, \ldots, l_n$ such that $\mathbf{u} = l_0 \mathbf{b_0} + \cdots + l_n \mathbf{b_n}$, and so we get:

$$\mathbf{u} - \mathbf{b} = (-l_0, l_0 A_1 + l_1 q - B_1, \ldots, l_0 A_n + l_n q - B_n) = (x, y_1, \ldots, y_n).$$

Thus, we obtain that $y_i + A_i x + B_i = l_i q \equiv 0 \pmod{q}$. So, the vectors of the form $\mathbf{x} - \mathbf{b}$, where $\mathbf{x}$ is in the set

$$\mathcal{A}_{\gamma_n}(\mathbf{b}) = \{\mathbf{u} \in L : ||\mathbf{u} - \mathbf{b}|| < \gamma_n M_{n,q}\},$$

are solutions of the system (10), for some positive number γ_n.

The parameters we usually use in a (EC)DSA system, provides a solution $\mathbf{s} = (x, y_1, \dots, y_n)$, where all the entries are 160-bits integers and smaller than q. In this case, we compute experimentally that the value of $\gamma_n = \frac{||\mathbf{s}||}{M_{n,q}}$ is on average approximately equal to 38. For larger values of γ_n, Babai's algorithm (or some other approximation algorithm) may succeed in finding in polynomial time such a vector. Indeed, we checked (see Sect. 5) that Babai's algorithm managed to find a solution for $\gamma_n \in [44, 52]$. We get such values of γ_n in the case where we have smaller keys. We provide the details of these experiments in Sect. 5.

This attack is non deterministic, since $\gamma_n > 1$ and so Proposition 2 does not hold. If $\gamma_n = 1$, then Proposition 2 holds, and the attack is deterministic. Below we present our attack.

Babai's Attack

Input : A public key (p, q, g, R) of a DSA scheme or a public key (E, p, q, P, Q) of a ECDSA scheme. Further, m signed messages are given.
Output : The secret key or Fail.
1. Choose $f_q(n)$.
2. Pick the maximum integer $n > 0$ such that there is an integer in the intervals

$$I_i = \left(\frac{q^{i/(n+1)+f_q(n)}}{2}, \frac{q^{i/(n+1)+f_q(n)}}{1.5} \right), \quad 1 \le i \le n.$$

3. If $n \le m$, then go to the next step, else return fail.
4. Choose randomly A_i from I_i.
5. Set $\mathbf{b} = (0, B_1, \dots, B_n)$ and construct the system

$$y_i + A_i x + B_i \equiv 0 \pmod{q} \quad (i = 1, \dots, n).$$

6. Construct the lattice $\mathcal{L}$, generated by the rows of the DSA matrix (8).
7. Compute $B = LLL(A)$.
8. Apply Babai's Nearest Plane Algorithm in the rows of matrix B with target vector $\mathbf{b}$. Let $\mathbf{s}$ be the output.
9. If the first coordinate s_1 of $\mathbf{s}$ satisfy $g^{s_1} = R$, (respectively $Q = s_1 P$) in $\mathbb{F}_p^*$, return s_1, else return fail.

The attacker, say Eve, has to make the choice of n and f_q. The choice of n is not random. A minimal condition is to choose it, in such a way that the interval I_i contains at least one integer. Then, she can construct a system of the form $y_i + A_i x + B_i \equiv 0 \pmod{q}$, $1 \le i \le n$, and $A_i \in I_i$. So, in practice the difficult part for Eve, is to find n signed messages. So Eve must be an active attacker and uses the DSA system as a signing oracle.

Remark 2. (i) Proposition 2 may not be satisfied, but the attack may return the secret key. This is because the hypotheses of Proposition 2 are only necessary and not sufficient.

(ii) The attack may fail in the sense that will not compute the secret key. The probability of success depends on the choices of A_i.

(iii) In the 7th step we can use BKZ algorithm with a suitable blocksize instead of LLL. This is something common in case we want a more reduced basis.

5 Experimental Results

For the following experiments we used the computer algebra system Sagemath [20][1]. We generated 100 DSA-systems of the form (10), $y_i + A_i x + B_i \equiv 0$ (mod q), $1 \leq i \leq n$, having a solution $\mathbf{s} = (x, y_1, \ldots, y_n)$ where x is the secret key. Once q is fixed to a 160 bit prime, we choose a natural number $n = n_0$ such that the intervals I_i contain at least one integer and a sequence $f_q(n)$, such that the value $f_q(n_0)$ satisfies the inequalities (2) and (3) of Proposition 1. Then, we choose the secret key x and y_i (we call y_i's *derivative ephemeral keys*), such that Proposition 2 is satisfied. In fact we pick the solution vector $\mathbf{s}$ such that

$$\frac{q^{n/(n+1)}}{16} < \|\mathbf{s}\| < \frac{q^{n/(n+1)+f_q(n)}}{4}.$$

The attack in [19], it may fail for the previous experiments, since the norm of solution we are looking for, is larger than $\frac{q^{n/(n+1)}}{16}$ and $n \gg \ln(\ln(q))$. But our attack will succeed since Proposition 2 is satisfied. The attack is deterministic. A choice that satisfies the previous is q : a 160-bit prime, $n_0 = 14$, $f_q(n) = \frac{\ln(n+1)}{n \ln q}$. We generated 100 DSA systems with the previous parameters and secret key 147 bits and derivative ephemeral keys 145 bits. As a CVP oracle, we used Babai algorithm. In all the instances we found the secret key, as our attack suggests. We tested our improvement (Babai's attack), which is no longer deterministic. We summarize the results in Table 1.

Remark 3. In the Example studied in [19] some typographic errors occurred in the values of the quantities A_1, $h(m_2)$, $h(m_3)$, s_2 and s_3. The correct values are $A_1 = 32D_1$ (and so, we have $l_1 = a^{-1}k_1 32 \mod q < 2^{96}$), and

$$h(m_2) = 43284768763225798962704594566716554599305078933 9,$$
$$h(m_3) = 102247883422181353858596598828981363231626289233,$$
$$s_2 = 1286644068312084224467989193436769265471767284571,$$
$$s_3 = 1357235540051781293143720232752751840677247754090.$$

---

[1] The code can be found in https://github.com/drazioti/python_scripts/tree/master/paper_dsa.

Table 1. We set $n = 206$ and $f_q(n) = \frac{170 \ln{(n+1)}}{n \ln q}$. We generated 100 random DSA systems for each row. The pair (α, β) at the first column, means that we pick the secret key uniformly from $[2^{\alpha-1}, 2^\alpha - 1]$ and the derivative ephemeral keys from $[2^{\beta-1}, 2^\beta - 1]$ (and fixed 160 bit prime q). For preprocessing we used BKZ with blocksize 70. The second column is the average value of γ_n. The last column contains the percentage that Babai's attack succeeds in finding the solution i.e. the secret key.

bits:(Skey, Der.Ep.keys)	γ_n	suc.rate
$(158, 157)$	43.81	17%
$(158, 155)$	51.34	100%
$(157, 157)$	44.15	23.3%
$(157, 156)$	48.68	100%

A Further Heuristic Improvement. We can further improve the previous results. The idea is to use another target vector instead of $\mathbf{b} = (0, B_1, \ldots, B_n)$. We consider the following vector

$$\mathbf{b} = (\varepsilon, \varepsilon + B_1, \ldots, \varepsilon + B_n),$$

where $\varepsilon = 2^{159} - 2^{157}$. That is the new target vector is equal with the previous plus the vector $(\varepsilon, \ldots, \varepsilon)$. The reason for picking such a target vector, is because the entries of the solution vector are balanced, i.e. of the same bit lengths. We test the previous idea, by choosing the solution vectors

$$\mathbf{s} \in \{2^{159}, \ldots, 2^{160} - 1\} \times \{2^{158}, \ldots, 2^{159} - 1\}^n.$$

That is the secret key has $160-$bits and the derivative ephemeral keys $159-$bits. We considered 100 DSA such systems with $n = 206$ and $f_q(n)$ as in Table 1, preprocessing BKZ-85, and our algorithm found the secret keys in 62 instances. The time execution per example was on average 2 min in an I3 Intel CPU. So, having one least (or most) significant bit for 206 derivative ephemeral keys, we can find the secret key. This result improves, in some sense, the result [12, Sect. 4.4], where with 100 signatures and knowing 2 least significant bits of the ephemeral keys, they computed the secret key with success rate 23% and in 4185 s on average per instance.

6   Conclusion

Following the ideas of [19], we have presented new attacks on DSA schemes. First, we have improved the bound for the lengths of the secret and ephemeral keys and we replaced the use of Micciancio-Voulgaris algorithm [14], by the polynomial complexity Babai's algorithm. This allowed us to find larger keys than in [19]. Our attack remained deterministic. Furthermore, we presented an heuristic extension of our attack for finding even larger secret keys. Finally, we improved the state of the art attack in (EC)DSA presented in [12]. Several experiments were described.

References

1. Bellare, M., Goldwasser, S., Micciancio, D.: "Pseudo-random" number generation within cryptographic algorithms: the DDS case. In: Kaliski, B.S. (ed.) CRYPTO 1997. LNCS, vol. 1294, pp. 277–291. Springer, Heidelberg (1997). https://doi.org/10.1007/BFb0052242
2. Blake, I.F., Garefalakis, T.: On the security of the digital signature algorithm. Des. Codes Cryptogr. **26**(1–3), 87–96 (2002)
3. Draziotis, K.A., Poulakis, D.: Lattice attacks on DSA schemes based on Lagrange's algorithm. In: Muntean, T., Poulakis, D., Rolland, R. (eds.) CAI 2013. LNCS, vol. 8080, pp. 119–131. Springer, Heidelberg (2013). https://doi.org/10.1007/978-3-642-40663-8_13
4. Draziotis, K.A.: (EC)DSA lattice attacks based on Coppersmith's method. Inform. Proc. Lett. **116**(8), 541–545 (2016)
5. Faugère, J.-L., Goyet, C., Renault, G.: Attacking (EC)DSA given only an implicit hint. In: Knudsen, L.R., Wu, H. (eds.) SAC 2012. LNCS, vol. 7707, pp. 252–274. Springer, Heidelberg (2013). https://doi.org/10.1007/978-3-642-35999-6_17
6. FIPS PUB 186-3, Federal Information Processing Standards Publication, Digital Signature Standard (DSS)
7. Galbraith, S.: Mathematics of Public Key Cryptography. Cambridge University Press, Cambridge (2012)
8. Hanrot, G., Pujol, X., Stehlé, D.: Algorithms for the shortest and closest lattice vector problems. In: Chee, Y.M., et al. (eds.) IWCC 2011. LNCS, vol. 6639, pp. 159–190. Springer, Heidelberg (2011). https://doi.org/10.1007/978-3-642-20901-7_10
9. Hanrot, G., Stehlé, D.: Improved analysis of Kannan's shortest lattice vector algorithm. In: Menezes, A. (ed.) CRYPTO 2007. LNCS, vol. 4622, pp. 170–186. Springer, Heidelberg (2007). https://doi.org/10.1007/978-3-540-74143-5_10
10. Johnson, D., Menezes, A.J., Vanstone, S.A.: The elliptic curve digital signature algorithm (ECDSA). Int. J. Inf. Secur. **1**, 36–63 (2001)
11. Howgrave-Graham, N.A., Smart, N.P.: Lattice attacks on digital signature schemes. Des. Codes Cryptogr. **23**, 283–290 (2001)
12. Liu, M., Nguyen, P.Q.: Solving BDD by enumeration: an update. In: Dawson, E. (ed.) CT-RSA 2013. LNCS, vol. 7779, pp. 293–309. Springer, Heidelberg (2013). https://doi.org/10.1007/978-3-642-36095-4_19
13. Menezes, A.J., van Oorschot, P.C., Vanstone, S.A.: Handbook of Applied Cryptography. CRC Press, Boca Raton (1997)
14. Micciancio, D., Voulgaris, P.: A deterministic single exponential time algorithm for most lattice problems based on Voronoi cell computations. In: Proceedings of the 42nd ACM Symposium on Theory of Computing - STOC 2010, pp. 351–358. ACM (2010)
15. National Institute of Standards and Technology (NIST). FIPS Publication 186: Digital Signature Standard, May 1994
16. Nguyen, P.Q., Shparlinski, I.E.: The insecurity of the digital signature algorithm with partially known nonces. J. Cryptology **15**, 151–176 (2002)
17. Nguyen, P.Q., Shparlinski, I.E.: The insecurity of the elliptic curve digital signature algorithm with partially known nonces. Des. Codes Cryptogr. **30**, 201–217 (2003)

18. Poulakis, D.: Some lattice attacks on DSA and ECDSA. Appl. Algebra Eng. Commun. Comput. **22**, 347–358 (2011)
19. Poulakis, D.: New lattice attacks on DSA schemes. J. Math. Cryptol. **10**(2), 135–144 (2016)
20. Sage Mathematics Software, The Sage Development Team (version 8.1). http://www.sagemath.org

Constraint Satisfaction Through GBP-Guided Deliberate Bit Flipping

Mohsen Bahrami and Bane Vasić[(✉)]

Electrical and Computer Engineering, University of Arizona,
Tucson, AZ 85721, USA
bahrami@email.arizona.edu, vasic@ece.arizona.edu

Abstract. In this paper, we consider the problem of transmitting binary messages over data-dependent two-dimensional channels. We propose a *deliberate bit flipping* coding scheme that removes channel harmful configurations prior to transmission. In this method, user messages are encoded with an error correction code, and therefore the number of bit flips should be kept small not to overburden the decoder. We formulate the problem of minimizing the number of bit flips as a binary constraint satisfaction problem, and devise a *generalized belief propagation* guided method to find approximate solutions. Applied to a data-dependent binary channel with the set of 2-D isolated bit configurations as its harmful configurations, we evaluated the performance of our proposed method in terms of uncorrectable bit-error rate.

Keywords: Probabilistic inference · Graphical models ·
Generalized belief propagation

1 Introduction

Many of probabilistic inference problems can be reformulated as the computation of marginal probabilities of a joint probability distribution over the set of solutions of a constraint satisfaction problem (CSP) [1,2]. A CSP consists of a number of variables and a number of constraints, where each constraint specifies admissible values of a subset of variables. A solution to a CSP is an assignment of variables satisfying all the constraints. Message passing algorithms have been successfully used for solving hard CSPs [3]. Traditional low-complexity approximate algorithms for solving these problems are based on belief propagation (BP) [4,5] which operate on factor graphs. BP, as an algorithm to compute marginals over a factor graph, has its roots in the broad class of Bayesian inference problems [6]. It is well known that the BP algorithm gives exact inference only on cycle-free graphs (trees). It has been also observed that in some applications BP surprisingly can provide close approximations to exact marginals on loopy graphs. However, an understanding of the behavior of BP in the latter case is far from complete. Moreover, it is known that BP does not perform well on

M. Ćirić et al. (Eds.): CAI 2019, LNCS 11545, pp. 26–37, 2019.
https://doi.org/10.1007/978-3-030-21363-3_3

graphs which contain a large number of short cycles. A new class of message-passing algorithm called generalized belief propagation (GBP) is introduced in [7] to solve the problem of computing marginal probability distributions on factor graphs with short cycles. The algorithm relies on the extension of cluster variation method [8,9], which is called the region graph method. The GBP algorithm provides approximate marginals by minimizing the Gibbs free energy using region graph method. In GBP, messages are sent among clusters of variables nodes instead of the node-to-node message passing fashion in BP and SP. More recently GBP has been shown empirically to have good performance, in either accuracy or convergence properties, for certain applications [10,11].

In this paper, we consider the problem of transmitting a binary message over a data-dependent communication channel and recovering it back at the receiver side. This problem is one of the most fundamental problems in communication theory, and can be considered as an instance of a CSP. Shannon in his seminal work [12] introduced two coding schemes for reliable transmission of information over a noisy channel, namely error correction coding and constrained coding. The first method protects user messages against random errors, which are independent of input data, by introducing redundancy in the messages prior to transmission. On the other hand, a constrained coding method assumes that channel solely introduces errors in response to specific patterns in input messages, and removing these problematic patterns makes the channel noiseless. Recent advances in emerging data storage technologies like magnetic recording systems [13,14], optical recording devices [15] and flash memory drives [16] necessitate to study two-dimensional coding (2-D) techniques for reliable storage of information. In these systems, user information bits are arranged into 2-D arrays for storing over the recording channel, and occurrences of specific patterns in input arrays are the significant cause of errors during read-back process. These systems require the use of some form of error-correction coding in addition to constrained coding of the input data or symbol sequences. It is therefore natural to investigate the interplay between these two forms of coding and the possibilities for efficiently combining their functions into a single coding operation. For this purpose, we introduce a generic 2-D channel with a set of harmful configurations to model patterning effects on an information bit from its neighboring bits in a 2-D channel input array. In this model, information bits contained in the harmful configurations are more vulnerable to errors than the other bits. Different 2-D constrained coding methods have been proposed to remedy the patterning effects in data-dependent 2-D channels, e.g., [17–22]. The goal of most of these methods is to achieve tighter bounds on the Shannon noiseless channel capacity of constraint. However, these schemes are non-linear in nature, and their encoder/decoder has a memory. Therefore, combinations of these methods with an error-correction coding scheme are challenging, and even a small number of bit errors can result multiple errors and severely degrade the performance of an error correction decoder. As an alternative coding scheme to address the non-linear effects of conventional 2-D constrained coding schemes, we present a *deliberate bit flipping* (DBF) coding scheme for data-dependent

2-D channels, where passing through channel specific patterns in inputs are the main cause of errors. The user message is first encoded by an error correction code, and is arranged into a 2-D array as an input to the channel. The idea is to completely eliminate a constrained encoder and, instead, to remove the harmful configurations by deliberately flipping the selected bits prior to transmission. The DBF method relies on the error correction capability of the error correction code (ECC) being used so that it should be able to correct both deliberate errors and channel errors. Therefore, it is crucial to keep the number of flipped bits small in order not to overburden the error correction decoder.

The problem of minimizing the number of deliberate bit flips for removing a set of configurations from a 2-D array is an instance of a CSP, where variables are arranged into a 2-D array, and constraints are defined locally over a set of neighboring variables. Assignments to variables are chosen from encoded messages of information bits (the codewords of ECC being used), and a constraint is violated if the realization of the neighboring variables involved in the constraint belongs to the given set of configurations. An initial realization of variables may violate some of constraints, and the goal is to change values of minimum number of variables to make all the constraints satisfied. This is equivalent to removing the forbidden configurations entirely from the 2-D array by flipping minimum number of bits. Using a factor graph representation, we devise a constrained combinatorial formulation for minimizing the number of bit flips in the DBF scheme for removing a given set of configurations. We find an approximate solution by reformulating the minimization problem as a 2-D maximum *a posteriori* (MAP) problem using a probabilistic graphical model. In this framework, patterns which do not contain harmful configurations are assumed to be uniformly distributed, and each pattern containing a harmful configuration has zero probability. The GBP algorithm, as a MAP inference method, is used to find the approximate solution for the 2-D MAP problem. Applied to a data-dependent 2-D channel with 2-D *isolated bit patterns* as the set of harmful patterns for the channel, we have shown the performance of DBF method in terms of uncorrectable bit-error rate.

The organization of the paper is as follows. Section 2 introduces the data-dependent 2-D channel model. The DBF coding scheme is presented in Sect. 3. Section 4 explains the probabilistic formulation devised for minimizing the number of bit flips in DBF coding scheme. Numerical results are given in Sect. 5.

2   Channel Model

In this section, we present a data-dependent 2-D communication channel which transmits binary rectangular patterns and produces as an output a binary pattern. Passing through the channel, information bits belong to a predefined set of configurations are more prone to errors than the other bits. The channel is characterized by this set of binary configurations, which is called the set of harmful configurations and is denoted by $\mathcal{F}$.

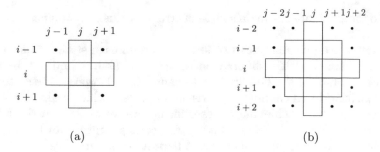

Fig. 1. Fig. shows (a) $Q^+(i,j)$ and (b) $\mathcal{P}_{i,j}$ over the lattice $\mathbb{Z}^2$ for the case of cross-shaped polyomino.

The set of channel input patterns and the set of channel output patterns are denoted by $\mathcal{X}$ and $\mathcal{Y}$. An input pattern $\mathbf{x} = [x_{i,j}]$ is chosen uniformly and randomly from $\mathcal{X}$, and is transmitted through the channel. A pattern $\mathbf{y} = [y_{i,j}] \in \mathcal{Y}$ is observed through the channel. The input pattern $\mathbf{x}$ can be considered as a square binary tiling of a rectangle, where each information bit $x_{i,j}$ on the 2-D input pattern represents a colored tile (0 (1) refers to a white (black) tile). The channel is data-dependent, and for each tile $x_{i,j}$, error is characterized by a Bernoulli random variable which depends on the realization of polyominoes having intersection with this tile. A polyomino of order k is constructed by joining k square tiles. Here we consider cross-shaped polyominoes of order 5 which are defined over the 2-D lattice $\mathbb{Z}^2$ as the following

$$Q^+(i,j) = \{(i,j-1),(i-1,j),(i,j),(i,j+1),(i+1,j)\}. \tag{1}$$

The set of cross-shaped polyominoes that have intersection with tile $x_{i,j}$ over an $m \times n$ rectangle is identified by

$$\mathcal{P}_{i,j} = \bigcup_{(i',j')\in Q^+(i,j)} Q^+(i',j'). \tag{2}$$

Figure 1 shows $Q^+(i,j)$ and $\mathcal{P}_{i,j}$ on a 2-D lattice $\mathbb{Z}^2$.

The received tile $y_{i,j}$ is characterized by

$$y_{i,j} = x_{i,j} \oplus z_{i,j}, \tag{3}$$

where $z_{i,j}$ is a Bernoulli random variable which depends on the realization of $\mathcal{P}_{i,j}$, $\mathbf{x}_{\mathcal{P}_{i,j}}$, and is defined by

$$z_{i,j} \sim \begin{cases} \mathrm{Bern}(\alpha_b), & \mathbf{x}_{\mathcal{P}_{i,j}} \in \mathcal{F}, \\ \mathrm{Bern}(\alpha_g), & \mathbf{x}_{\mathcal{P}_{i,j}} \notin \mathcal{F}. \end{cases} \tag{4}$$

Passing through the channel, colors of input tiles belong to $\mathcal{F}$ invert with probability α_b, while colors of other tiles invert with probability α_g. Since patterns belong to the set $\mathcal{F}$ are the main source of errors for this communication channel, we have $\alpha_b \gg \alpha_g$.

The introduced channel has two states where in each state acts as a binary symmetric channel with a different cross-over probability, and can be considered as an instance of the Gilbert-Elliot channel [23]. However, the state transitions in the introduced channel depends on input data which makes the problem of designing capacity achieving codes difficult. As we explain in the following section, we introduce a deliberate bit flipping coding strategy for this communication channel to overcome the effects of harmful configurations.

3   Deliberate Bit Flipping Coding Method

In this section, we characterize the deliberate bit flipping coding strategy for removing harmful configurations from 2-D channel input patterns before transmission through a data-dependent 2-D channel.

A user binary message $\mathbf{m}$ of length K is given. The message $\mathbf{m}$ is first encoded by an error correction code with rate $R = \frac{K}{N}$, and we have the codeword $\mathbf{c}$ of length N. The codeword is arranged into a 2-D array $\mathbf{x} = [x_{i,j}]$ of size $m \times n$, where $x_{i,j} = c_{(i-1)m+j}$ and $N = m \times n$. For each tile $x_{i,j}$, a 2-D constraint is defined over polyominoes having intersection with this tile. The 2-D constraint $\mathbb{S}$ forbids some of the configurations of $\mathcal{P}_{i,j}$, where the set of these configurations are denoted by $\mathcal{F}$. These configurations are essentially harmful configurations for the channel, and they must be removed before transmission. We use a deliberate error insertion approach to remove the harmful configurations from the input pattern $\mathbf{x}$ before transmission through the channel. Whenever there is a configuration from the list $\mathcal{F}$ in the input pattern $\mathbf{x}$, the color of selected tiles in $\mathbf{x}$ are inverted to remove the forbidden configurations. In the following, we present an example to highlight the basic ideas behind the DBF method for removing a set of predefined configurations from a 7×7 random binary pattern. In this example, the set of 2-D isolated bit patterns are required to be removed form the given random pattern.

Example 1. A 7×7 random binary pattern $\mathbf{x}$ as shown in Fig. 2 is given. The goal is to use the DBF scheme to remove the 2-D isolated bit configurations. We assume zero entries (white tiles) outside of $\mathbf{x}$, i.e., $x_{i,j} = 0$, while $i < 1$, $j < 1$, $i > 7$, or $j > 7$. There are two isolated bit patterns in $\mathbf{x}$, which are $\mathbf{x}_{Q+(3,6)}$ and $\mathbf{x}_{Q+(7,7)}$. Passing through the channel, the tiles whose belong to these two patterns are more prone to errors than the other tiles. These tiles are $(2,6)$, $(3,5)$, $(3,6)$, $(3,7)$, $(4,6)$, $(6,7)$, $(7,6)$ and $(7,7)$. For instance, for the tile $(2,6)$,

$$\mathcal{P}_{2,6} = \bigcup_{(i',j') \in Q+(2,6)} Q^+(i',j'). \tag{5}$$

Since $Q^+(3,6) \subset \mathcal{P}_{2,6}$ and $\mathbf{x}_{Q+(3,6)}$ is a 2-D isolated bit pattern, we have $\mathbf{x}_{\mathcal{P}_{2,6}}$ contains a 2-D isolated bit pattern. Similarly, we can verify this for the rest of tiles in $\mathbf{x}$. 2-D isolated bit configurations can be removed form $\mathbf{x}$ by inverting the colors of tiles $(3,6)$ and $(7,7)$.

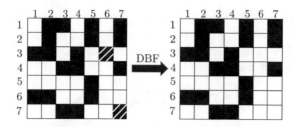

Fig. 2. In order to remove the 2-D isolated bit patterns from the given 7×7 binary pattern, the colors of tiles $(3, 6)$ and $(7, 7)$ are inverted.

In the DBF method, the main role is to select tiles whose colors need to be inverted for removing the harmful configurations. We define a *tile-selection* function to determine these tiles.

Definition 1 (Tile-Selection Function). The tile-selection function $\theta : \mathcal{X} \rightarrow \{0, 1\}^{m \times n}$ selects tiles whose colors need to be inverted for removing the harmful configurations from the pattern $\mathbf{x}$.

Using θ, $\mathbf{e}^{\mathrm{DBF}}$ is defined to identify the locations of tiles whose colors are inverted,

$$\mathbf{c}^{\mathrm{DBF}} = \theta(\mathbf{x}) = [c_{i,j}^{\mathrm{DBF}}], \tag{6}$$

where $e_{i,j}^{\mathrm{DBF}} = 1$ if the color of (i, j)-th tile is inverted, otherwise, $e_{i,j}^{\mathrm{DBF}} = 0$. Therefore, $\mathbf{x} \oplus \mathbf{e}^{\mathrm{DBF}}$ does not contain any harmful configurations from the list $\mathcal{F}$. Furthermore, the number of tiles whose colors are inverted is $w_H(\mathbf{e}_{\mathrm{DBF}})$. Now, instead of $\mathbf{x}$, we send $\mathbf{x} \oplus \mathbf{e}^{\mathrm{DBF}}$ over the channel, and the received pattern is $\mathbf{y} = \mathbf{x} \oplus \mathbf{e}^{\mathrm{DBF}} \oplus \mathbf{e}^{\mathrm{CH}}$, where $\mathbf{e}^{\mathrm{CH}}$ indicates the locations of tiles whose colors are inverted due to channel errors. A decoder $\psi : \{0, 1\}^{m \times n} \rightarrow \mathcal{X}$ maps a received pattern $\mathbf{y}$ to a pattern $\hat{\mathbf{x}}$ in the input set $\mathcal{X}$. In the following, we define the average probability of error and the capacity of the method.

Definition 2 (Average Probability of Error). $\lambda_{\mathbf{m}} = p(\hat{\mathbf{m}} \neq \mathbf{m}|\mathbf{m})$ is the probability that the decoded message $\hat{\mathbf{m}}$ is different from the actual message $\mathbf{m}$. The average probability of error is defined by

$$p_e^{(N)} = p(\hat{\mathbf{m}} \neq \mathbf{m}) = \sum_{\mathbf{m} \in \mathcal{M}} \lambda_{\mathbf{m}} p(\mathbf{m}) = \frac{1}{2^{\lceil NR \rceil}} \sum_{\mathbf{m}} \lambda_{\mathbf{m}}. \tag{7}$$

Definition 3 (Achievable Rate and Capacity). A rate R is said to be achievable if for some N and $\epsilon_N > 0$, $p_e^{(N)} \leq \epsilon_N$. The capacity is defined as the supremum over all achievable rates.

In this communication system with DBF method, there are two types of error. The first type of error is the deliberate errors which are introduced before transmission through the channel, and the second type is the random channel errors. If we assume that the main cause of errors are the presence of harmful patterns in input patterns, removing the harmful configurations makes the channel almost noiseless. Therefore, for the Hamming distance between the input and received patterns, we have

$$d_H(\mathbf{x}, \mathbf{y}) = w_H(\mathbf{x} \oplus \mathbf{y}) \simeq w_H(\mathbf{e}^{\mathrm{DBF}}). \tag{8}$$

Without loss of generality, if we use a bounded-distance decoder, it should be ideally decode all the messages that

$$d_H\left(\mathbf{x}(\mathbf{m}), \mathbf{y}\right) \simeq w_H(\mathbf{e}^{\mathrm{DBF}}) \leq \lfloor \frac{d_{\min} - 1}{2} \rfloor, \tag{9}$$

where $d_{\min}$ is the minimum distance of the code. Therefore, the main obstacle for using the DBF method for removing harmful configurations is to keep the number of deliberate errors small enough not to overburden the decoder. For a given binary user message and a set of forbidden configurations, we are interested in finding $\hat{\mathbf{x}}$, that minimizes $w_H(\hat{\mathbf{x}} \oplus \mathbf{x})$ and $\hat{\mathbf{x}} \in \mathbb{S}$. This minimization problem can be considered as a constrained combinatorial optimization problem. Finding a binary pattern which satisfies a certain local constraints (which do not contain a predefined set of 2-D configurations), and has the minimum Hamming distance with the input binary pattern $\mathbf{x}$ via an exhaustive search can be computationally prohibitive for large patterns. This problem can be regarded as an instance of the Levenshtine distance problem [24], which is known to be a hard combinatorial problem. In the following section, we present a probabilistic graphical model, and reformulate the problem as a maximum *a-posetriori* (MAP) problem to find an approximation solution for the problem.

4   A Probabilistic Formulation for DBF Method

In this section, we present a probabilistic formulation for the problem of minimizing the number of bit flips in the DBF scheme. In this framework, the set of input patterns which do not contain any harmful configurations has uniform distribution, while the patterns containing harmful configurations have zero probability. For a given random input pattern, the problem originally is to find the pattern which does not contain any harmful configurations, and has the minimum Hamming distance with the given input pattern. We translate this problem into the problem of finding the most likely pattern (that does not contain any harmful configurations) to the given pattern using a binomial expression.

An input pattern $\mathbf{x}$ is given. For each tile $x_{i,j}$ over $\mathbf{x}$, existence of harmful configurations is determined based on the configuration of $\mathcal{P}_{i,j}$, $\mathbf{x}_{\mathcal{P}_{i,j}}$. Therefore, the problem of finding $\hat{\mathbf{x}} \in \mathbb{S}$ which has the minimum $w_H(\hat{\mathbf{x}} \oplus \mathbf{x})$ can be break down locally over each $\mathcal{P}_{i,j}$. We define a *local distortion function* D over $\mathcal{P}_{i,j}$'s to determine the Hamming distance between $\hat{\mathbf{x}}_{\mathcal{P}_{i,j}}$ and $\mathbf{x}_{\mathcal{P}_{i,j}}$. The function $D : \{0,1\}^{|\mathcal{P}_{i,j}|} \times \{0,1\}^{|\mathcal{P}_{i,j}|} \to \mathbb{N}$ is defined over the tiles indexed by $\mathcal{P}_{i,j}$ as follows

$$D\left(\hat{\mathbf{x}}_{\mathcal{P}_{i,j}}, \mathbf{x}_{\mathcal{P}_{i,j}}\right) = \begin{cases} w_H\left(\hat{\mathbf{x}}_{\mathcal{P}_{i,j}} \oplus \mathbf{x}_{\mathcal{P}_{i,j}}\right), & \hat{\mathbf{x}}_{\mathcal{P}_{i,j}} \notin \mathcal{F}, \\ \infty, & \hat{\mathbf{x}}_{\mathcal{P}_{i,j}} \in \mathcal{F}, \end{cases} \tag{10}$$

where the patterns belonging to $\mathcal{F}$ are specified by ∞. One may use the outputs of this function over the tiles in $\mathbf{x}$ to find $\mathbf{x}^\star \in \mathbb{S}$ which has the minimum Hamming distance with $\mathbf{x}$. This process can be intractable for large patterns as it needs to compute the output of D for each tile, which has $2^{|\mathcal{P}_{i,j}|}$ different configurations, and take exponentially large memory just to store. In the following, we present a probabilistic formulation to find approximate solution for this problem using GBP algorithm.

We use a binomial probability expression to reformulate the distortion indicator function defined in Eq. (10), and present a probabilistic formulation for the problem of minimizing the Hamming distance. We assume that the color of each tile contained in a harmful configuration is inverted with the probability $0 < \lambda \leq 1$. For each tile $x_{i,j}$, we define a function $D_p : \{0,1\}^{\mathcal{P}_{i,j}} \times \{0,1\}^{\mathcal{P}_{i,j}} \to \mathbb{R}^0, 1$ over the tiles tiles indexed by $\mathcal{P}_{i,j}$,

$$D_p(\mathbf{x}_{\mathcal{P}_{i,j}}, \hat{\mathbf{x}}_{\mathcal{P}_{i,j}}) = \begin{cases} \lambda^{w_H(\mathbf{e}_{\mathcal{P}_{i,j}})}(1-\lambda)^{|\mathcal{P}_{i,j}|-w_H(\mathbf{e}_{\mathcal{P}_{i,j}})}, & \hat{\mathbf{x}}_{\mathcal{P}_{i,j}} \notin \mathcal{F}, \\ 0, & \hat{\mathbf{x}}_{\mathcal{P}_{i,j}} \in \mathcal{F}, \end{cases} \tag{11}$$

where $\mathbf{e}_{\mathcal{P}_{i,j}} = \hat{\mathbf{x}}_{\mathcal{P}_{i,j}} \oplus \mathbf{x}_{\mathcal{P}_{i,j}}$, and $|\mathcal{P}_{i,j}|$ indicates the number of tiles in $\mathcal{P}_{i,j}$. This function is called as the *local probabilistic distortion* function. For each tile $(i,j) \in \mathcal{A}_{m,n}$, the distortion now is defined as the probability of having a distorted pattern $\mathbf{x}_{\mathcal{P}_{i,j}}$ which has the Hamming distance $w_H(\hat{\mathbf{x}}_{\mathcal{P}_{i,j}} \oplus \mathbf{x}_{\mathcal{P}_{i,j}})$ with $\hat{\mathbf{x}}_{\mathcal{P}_{i,j}} \notin \mathcal{F}$. When $\hat{\mathbf{x}}_{\mathcal{P}_{i,j}} \in \mathcal{F}$, this probability is set to be zero, as the first constraint is to find $\hat{\mathbf{x}} \in \mathbb{S}$.

For a given input pattern $\mathbf{x}$ and a set of harmful configurations $\mathcal{F}$, the goal is now to find $\hat{\mathbf{x}} \in \mathbb{S}$ that maximizes $p(\hat{\mathbf{x}}|\mathbf{x})$, which is equivalent to finding $\hat{\mathbf{x}}$ that minimizes $w_H(\hat{\mathbf{x}} \oplus \mathbf{x})$. In another word, we are interested in finding

$$\hat{\mathbf{x}} = \arg\max_{\hat{\mathbf{x}} \in \mathbb{S}} \{p(\hat{\mathbf{x}}|\mathbf{x})\}. \tag{12}$$

The *a-posteriori* probability $p(\hat{\mathbf{x}}|\mathbf{x})$ for a fixed λ is factored into

$$p(\hat{\mathbf{x}}|\mathbf{x}) = \frac{p(\mathbf{x}|\hat{\mathbf{x}})\,p(\hat{\mathbf{x}})}{p(\mathbf{x})} \stackrel{(a)}{\propto} p(\mathbf{x}|\hat{\mathbf{x}}) \stackrel{(b)}{=} \prod_{(i,j)} p\left(\mathbf{x}_{\mathcal{P}_{i,j}}|\hat{\mathbf{x}}_{\mathcal{P}_{i,j}}\right),$$

$$\stackrel{(c)}{=} \prod_{(i,j)} D_p(\mathbf{x}_{\mathcal{P}_{i,j}}, \hat{\mathbf{x}}_{\mathcal{P}_{i,j}}), \tag{13}$$

where (a) comes from this assumption that the set of patterns which do not contain harmful configurations has uniform distribution, (b) is established as harmful configurations can be determined locally over $\mathcal{P}_{i,j}$'s, and (c) is obtained according to the local probabilistic distortion function, Eq. (11). Therefore, we have

$$p\left(\hat{\mathbf{x}}|\mathbf{x}\right) = \frac{1}{Z(\mathbf{x})} \prod_{(i,j)\in\mathcal{A}_{m,n}} D_p(\mathbf{x}_{\mathcal{P}_{i,j}}, \hat{\mathbf{x}}_{\mathcal{P}_{i,j}}), \qquad (14)$$

where $Z(\mathbf{x})$ is the partition function and defined by

$$Z(\mathbf{x}) = \sum_{\mathbf{x}} \prod_{(i,j)\in\mathcal{A}_{m,n}} D_p(\mathbf{x}_{\mathcal{P}_{i,j}}, \hat{\mathbf{x}}_{\mathcal{P}_{i,j}}). \qquad (15)$$

Providing either exact or approximate solutions for the marginal probabilities in general is a NP-hard problem [3], as we need to take sum over exponential number of variables. In [7,25], it is shown that region-based approximation (RBA) method provides an approximate solution for the partition function by minimizing the region-based free energy (as an approximation to the variational free energy). Therefore, GBP as a method for finding approximate solution for region-based free energy can be used to solve the problem of minimizing the number of bit flips in the DBF scheme.

5 Numerical Results

In this section, we present the numerical results, and explain how the DBF method relies on the error correction capability of the code being used. We first provide an example of a short BCH code with incorporating DBF method.

5.1 Example of BCH-[5,7,15] Code

Consider the user messages of length 5, $\mathbf{m}_1 = (0,1,0,0,0)$, $\mathbf{m}_2 = (1,0,0,0,0)$, $\mathbf{m}_3 = (0,1,1,1,1)$ and $\mathbf{m}_4 = (0,1,1,0,1)$. The messages are encoded by the BCH-$[5,7,15]$ code, and the codewords are

$$\mathbf{c}_1 = (0,1,0,0,0,1,1,1,0,1,0,1,1,0)\,, \ \mathbf{c}_2 = (1,0,0,0,0,1,0,1,0,0,1,1,0,1,1)\,,$$
$$\mathbf{c}_3 = (0,1,1,1,1,0,1,0,1,1,0,0,1,0,0)\,, \ \mathbf{c}_4 = (0,1,1,0,1,1,1,0,0,0,0,1,0,1,0)\,.$$

The codewords are arranged into 3×5 arrays as four different binary patterns. These patterns are shown in Fig. 3. We want to remove forbidden configurations by the 2-D n.i.b. constraint entirely from the patterns with flipping minimum number of bits. We only focus on these four patterns out of 32 possible binary patterns with BCH-$[15,5,7]$ code as they present all different possible bit flipping scenarios for removing 2-D isolated bit patterns.

In Fig. 3(a), the pattern does not contain any of the 2-D isolated bit configurations, therefore there is no need to invert the tile colors, and $w_H(\mathbf{e}_{(a)}) = 0$. The pattern in Fig. 3(b) contains single 2-D isolated bit pattern, which is $\mathbf{x}_{Q^+(2,3)}$. This 2-D isolated bit pattern can be removed by inverting the color of any one of the tiles in $Q^+(2,3)$, and therefore the minimum $w_H(\mathbf{e}_{(b)}) = 1$. For the pattern in Fig. 3(c), there are two overlapping 2-D isolated bit patterns, which are $\mathbf{x}_{Q^+(2,3)}$ and $\mathbf{x}_{Q^+(3,3)}$. These two isolated bit patterns can be removed simultaneously by inverting either the color of tile $(2,3)$ or $(3,3)$, and therefore for this

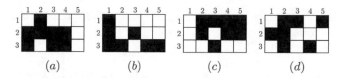

Fig. 3. The input patterns for this example. Outside of these patterns are filled with white tiles (zero entries).

case also the minimum $w_H(\mathbf{e}) = 1$. In Fig. 3(d), the pattern contains two non-overlapping 2-D isolated bit patterns, which are $\mathbf{x}_{Q+(1,5)}$ and $\mathbf{x}_{Q+(3,4)}$. At least colors of two tiles over this input pattern should be inverted, and for this case the minimum $w_H(\mathbf{e}_{(d)}) = 2$. For the systematic BCH-5, 7, 15 code (where the codewords are arranged into 3×5 arrays and the first row is equipped with the user bits), in average it needs to flip 0.6563 bits/pattern to remove the forbidden configurations by the 2-D n.i.b. constraint.

5.2 Uncorrectable Bit-Error Rate

In this section, we present the statistics of the number of bit flips required for removing 2-D isolated bit configurations from a random 2-D pattern of size 32×32, and also compute the uncorrectable bit-error rate (UBER) using these statistics.

The statistics of the number of bit flips are obtained by applying GBP-based DBF method for removing 2-D isolate bit patterns from a set of 8000 random 2-D patterns of size 32×32. Similar to the other examples, we assume white tiles or zero entries outside of patterns. Figure 4(a) shows the occurrence probability of the number of flipped bits.

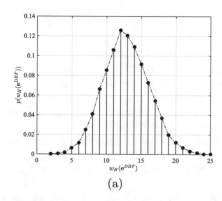
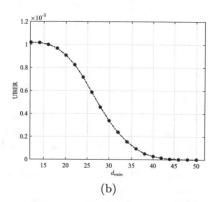

Fig. 4. (a) Fig. presents an occurrence probability of the number of bit flips for removing 2-D isolated bit configurations from a random 2-D pattern of size 32×32. (b) The UBER for the DBF scheme with BCH codes of length 1024 is given.

Using these occurrence probabilities, we can compute the UBER for an ECC being used, as follows

$$\text{UBER} = \left[\sum_{w_H(\mathbf{e}^{\text{DBF}}) > \lfloor \frac{d_{\min}-1}{2} \rfloor} p\left(w_H(\mathbf{e}^{\text{DBF}})\right) \right] / NR, \qquad (16)$$

where $d_{\min}$ is the minimum distance of code, $N = m \times n$ is the size of the pattern (length of the code), and R is the rate of the ECC. In fact, we compute the UBER under the assumptions that the channel only introduces errors in response to presences of 2-D isolated bit configurations, and removing these configurations make the channel noiseless. In our introduced channel, this is the case when $\alpha_g = 0$ and $\alpha_b \neq 0$. As an example, we use BCH codes of length 1024 with different code rates, and draw the UBER for these codes in Fig. 4(b).

6   Conclusions

We have presented a deliberate bit flipping coding scheme for data-dependent 2-D channels. For this method, we have shown that the main obstacle is the number of deliberate errors which are introduced for removing harmful configurations before transmission through the channel. We have devised a combinatorial optimization formulation for minimizing the number of bit flips, and have explained how this problem can be related to a binary constraint satisfaction problem. Finally, through an example, we have presented uncorrectable bit-error rate results of incorporating DBF for removing 2-D isolated-bit configurations from 2-D patterns of certain size.

Acknowledgment. This work is funded by the NSF under grants ECCS-1500170 and SaTC-1813401. A comprehensive version of this paper has been submitted to IEEE Transactions on Communications [26].

References

1. Jordan, M.: Learning in Graphical Models. MIT Press, Cambridge (1999)
2. Kschischang, F., Frey, B., Loeliger, H.-A.: Factor graphs and the sum-product algorithm. IEEE Trans. Inf. Theory **47**(2), 498–519 (2001)
3. Wainwright, M.J., Jordan, M.I.: Graphical models, exponential families, and variational inference. Found. Trends Mach. Learn. **1**, 1–305 (2008)
4. Pearl, J.: Probablisitic Reasoning in Intelligent Systems. Kaufmann, San Francisco (1988)
5. Gallager, R.G.: Low density parity check codes. Ph.D. dissertation, Cambridge (1963)
6. Frey, B.J.: Graphical Models for Machine Learning and Digital Communication. MIT Press, Cambridge (1998)
7. Yedidia, J.S., Freeman, W.T., Weiss, Y.: Constructing free energy approximations and generalized belief propagation algorithms. IEEE Trans. Inform. Theory **51**, 2282–2312 (2005)

8. Kikuchi, R.: A theory of cooperative phenomena. Phys. Rev. Online Arch. (PROLA) **81**(6), 988 (1951)
9. Morita, T.: Foundations and applications of cluster variation method and path probability method. Publication Office, Progress of Theoretical Physics (1994)
10. Shamai, S., Ozarow, L.H., Wyner, A.D.: Information rates for a discrete-time gaussian channel with intersymbol interference and stationary inputs. IEEE Trans. Inf. Theory **37**(6), 1527–1539 (1991)
11. Sabato, G., Molkaraie, M.: Generalized belief propagation for the noiseless capacity and information rates of run-length limited constraints. IEEE Trans. Commun. **60**(3), 669–675 (2012)
12. Shannon, C.E.: A mathematical theory of communication. Bell Syst. Tech. J. **27**, 379–423 (1948)
13. Wood, R., Williams, M., Kavcic, A., Miles, J.: The feasibility of magnetic recording at 10 terabits per square inch on conventional media. IEEE Trans. Magn. **45**(2), 917–923 (2009)
14. Garani, S.S., Dolecek, L., Barry, J., Sala, F., Vasić, B.: Signal processing and coding techniques for 2-D magnetic recording: an overview. Proc. IEEE **106**(2), 286–318 (2018)
15. Immink, K.A.S.: A survey of codes for optical disk recording. IEEE J. Sel. Areas Commun. **19**(4), 756–764 (2001)
16. Dolecek, L., Cassuto, Y.: Channel coding for nonvolatile memory technologies: theoretical advances and practical considerations. Proc. IEEE **105**(9), 1705–1724 (2017)
17. Halevy, S., Chen, J., Roth, R.M., Siegel, P.H., Wolf, J.K.: Improved bit-stuffing bounds on two-dimensional constraints. IEEE Trans. Inf. Theory **50**(5), 824–838 (2004)
18. Roth, R.M., Siegel, P.H., Wolf, J.K.: Efficient coding schemes for the hard-square model. IEEE Trans. Inf. Theory **47**(3), 1166–1176 (2001)
19. Forchhammer, S., Laursen, T.V.: Entropy of bit-stuffing-induced measures for two-dimensional checkerboard constraints. IEEE Trans. Inf. Theory **53**(4), 1537–1546 (2007)
20. Tal, I., Roth, R.M.: Bounds on the rate of 2-D bit-stuffing encoders. IEEE Trans. Inf. Theory **56**(6), 2561–2567 (2010)
21. Sharov, A., Roth, R.M.: Two dimensional constrained coding based on tiling. IEEE Trans. Inf. Theory **56**(4), 1800–1807 (2010)
22. Krishnan, A.R., Vasić, B.: Lozenge tiling constrained codes. Facta Univ. Ser. Electron. Energ. **4**, 1–9 (2014)
23. Mushkin, M., Bar-David, I.: Capacity and coding for the Gilbert-Elliott channels. IEEE Trans. Inf. Theory **35**(6), 1277–1290 (1989)
24. Levenshtein, V.I.: Binary codes capable of correcting deletions, insertions, and reversals. Soviet Phys. Dokl. **10**(8), 707–710 (1966)
25. Pakzad, P., Anantharam, V.: Kikuchi approximation method for joint decoding of LDPC codes and partial-response channels. IEEE Trans. Commun. **54**(7), 1149–1153 (2006)
26. Bahrami, M., Vasić, B.: A deliberate bit flipping coding scheme for data-dependent two-dimensional channels. IEEE Trans. Commun. (2019, under review)

On the Diffusion Property of the Improved Generalized Feistel with Different Permutations for Each Round

Tsonka Baicheva[1,2] and Svetlana Topalova[1(✉)]

[1] Institute of Mathematics and Informatics, Bulgarian Academy of Sciences,
P.O.Box 323, 5000 Veliko Tarnovo, Bulgaria
{tsonka,svetlana}@math.bas.bg
[2] D. A. Tsenov Academy of Economics,
2 Emanuil Chakarov Street, 5250 Svishtov, Bulgaria

Abstract. Suzaki and Minematsu (LNCS, 2010) present a comprehensive study of the diffusion property of the improved generalized Feistel structure (GFS_π) which is a generalization of the classical Feistel cipher. They study the case when one and the same permutation is applied at each round and finally remark that the usage of different permutations at the different rounds might lead to better diffusion in return for a larger implementation cost, but that it is an open question whether multiple permutations can really improve the diffusion property.

We give a positive answer to this question. For cyphers with 10, 12, 14 and 16 subblocks we present examples of permutations (different at each round) leading to GFS_π with better diffusion than the one which can be obtained if the same permutation is applied at all rounds. The examples were found by a computer-aided search which is described in the present paper.

Keywords: Block cypher · Improved generalized Feistel structure · Diffusion

1 Introduction

1.1 Previous Results

The generalized Feistel structure (GFS) is a generalization of the classical Feistel network. It divides a plaintext into k subblocks for some $k > 2$. In [8] the GFS of type 2 is suggested where the Feistel transformation is applied for every consecutive two subblocks and then a cyclic shift of the subblocks is applied.

The research of the first author was partially supported by the Bulgarian National Science Fund under Contract 12/8, 15.12.2017 and of the second author by the National Scientific Program "Information and Communication Technologies for a Single Digital Market in Science, Education and Security (ICTinSES)", financed by the Ministry of Education and Science.

© Springer Nature Switzerland AG 2019
M. Ćirić et al. (Eds.): CAI 2019, LNCS 11545, pp. 38–49, 2019.
https://doi.org/10.1007/978-3-030-21363-3_4

GFS of type 2 can be easily implemented and examples of such ciphers are RC6 [4], HIGHT [3] and CLEFIA [5]. Unfortunately this type of a Feistel structure has low diffusion for large k and needs a large number of rounds.

Suzaki and Minematsu [6] suggested the improved generalized Feistel structure (GFS_π) as a modification of the generalized Feistel structure of type 2. This modification allows improvement of the diffusion property. Further investigations and new ideas on the improved generalized Feistel structures are presented in [1,2,7,9]. We briefly describe here how a block cipher based on GFS_π works.

The block length is $k.n$ and the block has k parts of the same length (n-bits). We shall denote by $P_1, P_2, \ldots P_k$ the subblocks of the input plain text block. The block is encrypted (decrypted) in R rounds (rounds $1, 2, \ldots R$). We denote the parts of the input block to round r by $X_1^r, X_2^r, \ldots X_k^r$, and those of the output block by $Y_1^r, Y_2^r, \ldots Y_k^r$, where $X_i^1 = P_i$ and $X_i^r = Y_i^{r-1}$ for $i = 1, 2, \ldots k$ and $r = 2, 3, \ldots R$.

The subblocks are grouped two by two for the round transformation, namely X_i^r and X_{i+1}^r (for each odd i) are in one pair. The value of X_i^r remains unchanged, while the value of a nonlinear function $f(X_i^r)$ is added bitwise to the value of X_{i+1}^r. Then a permutation π_r is applied to the subblocks.

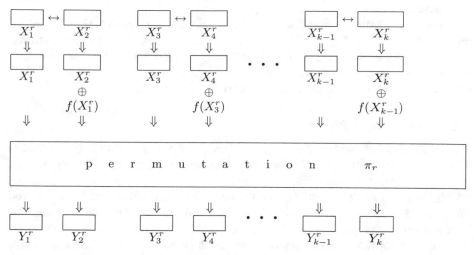

The inverse permutation π_r^{-1} is used for the decryption. If the permutation applied at each round is the cyclic shift the generalized Feistel structure of type 2 (GFS_2) is obtained.

Note that no permutation is applied at the last round or, if it is more convenient, we can think that the permutation at this round is the identity permutation $(1)(2)\ldots(k)$.

Full diffusion is achieved if each subblock depends on all input subblocks. Denote by r_e and r_d the smallest rounds after which full diffusion is achieved by encryption and decryption respectively. We shall call the bigger of them *full diffusion round* and will further denote it by R_d, namely

$$R_d = max(r_e, r_d).$$

For GFS_2 with k subblocks $R_d = k$. The diffusion round of GFS_π depends on the permutations used at the different rounds.

In [6] Suzaki and Minematsu construct for all even $k \leq 16$ all permutations π which lead to the smallest R_d if one and the same permutation π is applied at each round. They notice that the permutations with smallest R_d are *even-odd* ones, namely map odd to even numbers and vice versa. They next prove that for even-odd permutations

$$R_d \geq R_D,$$

where $a_{R_D-1} < \dfrac{k}{2} \leq a_{R_D}$ and a_{R_D} is the R_D-th element of the Fibonacci sequence (starting with $0, 1, 1, 2, 3, 5, 8, 13, 21, 34, 55, 89, 144, 233, \ldots$). For $k \leq 16$ the values of R_D are

k	2	4	6	8	10	12	14	16
R_D	2	4	5	6	6	7	7	7

Note that this lower bound is derived by assuming that the permutations are even-odd, but not by assuming that the permutation is one and the same at all rounds. That is why the same bound can be applied in the multiple permutation case which we consider.

Suzaki and Minematsu show by exhaustive computer search that $R_d = R_D$ for $k \leq 8$ and $R_d = R_D + 1$ for $10 \leq k \leq 16$ if one and the same permutation π is applied at each round. In conclusion they point out that the consideration of a different permutation at each round might lead to structures with better diffusion. This was our motivation to start the present work.

1.2 The Present Paper

We consider GFS_π with different permutations at the different rounds. That is why for all even values of $k \leq 16$ we construct permutation sequences $\pi_1, \pi_2, \ldots, \pi_{R_D-1}$ and examine their diffusion property until we find permutation sequences which lead to a GFS_π with R_D rounds and such that full diffusion is achieved after the last round. The investigation is computer-aided. All possible permutation sequences are, however, $(k!)^{R_D-1}$ and this is too much for our computer. Thus only a representative part of them must be tested. The challenge in this problem is to determine a small enough representative part. In Sect. 2 we explain the basic ideas of our construction method. Section 3 contains a summary of the results and a comparison to the previously known ones.

2 Construction Method

2.1 Notations and Dependence Matrices

Consider an R-round GFS_π with k subblocks. Denote by π_r the permutation at round $r < R$. To simplify the notations, when it is clear from the context which round r is concerned, we shall use X_i instead of X_i^r and Y_i instead of Y_i^r. Let

$$\Pi_r = \pi_r \pi_{r-1} \ldots \pi_1.$$

The notation means that Π_r is a permutation obtained by superposition of these r permutations, where π_1 is applied first, then $\pi_2, \ldots \pi_r$. We use the notation $Y_i \leftarrow P_j$, which means that $\Pi_r(j) = i$, namely as a result of the permutations at the first r rounds the j-th subblock moved to the i-th place.

We construct even-odd permutations π_r for each round $r < R_D$. As a measure of the diffusion after the r-th round we consider the dependence matrix $M = (m_{ij})_{k \times k}$, where m_{ij} is 1 if the value of the i-th subblock depends on the value of the j-th subblock of the plaintext, and 0 if not. In the following examples we denote by M_r the dependence matrix after round r, and by F_r the dependence matrix after the Feistel transformation at round r (before applying π_r). For each odd i the value of the subblock X_i remains unchanged by the Feistel transformation, while the value of a nonlinear function $f(X_i)$ is added bitwise to the value of X_{i+1}. That is why after the Feistel transformation the subblock X_{i+1} starts depending on all subblocks on which X_i depends. Therefore the odd rows of F_r are the same as those of M_{r-1}, while for even i the elements of the i-th row of F_r are obtained as disjunction ('or') of the corresponding elements of rows $i - 1$ and i of M_{r-1}. We denote this by

$$F_r = M_{r-1}^{\oplus}.$$

The matrix M_r is then obtained by applying π_r on the rows of F_r. Namely:

$$M_r = \pi_r F_r = \pi_r M_{r-1}^{\oplus}.$$

Note that at the full diffusion round R_D each of the matrices F_r and M_r is the all-one $k \times k$ matrix.

We also use a related to F_r and M_r matrix, which we call *base*. The *round r-th base* is a matrix $B_r = (b_{ij}^r)_{k \times k}$ which can be obtained by applying the permutation Π_r^{-1} to the rows of M_r, i.e.

$$B_r = \Pi_r^{-1} M_r.$$

Note that the rows of B_r are the same as the rows of F_r or M_r, but sorted by the P_i (while those of F_r and M_r are sorted by the X_i and Y_i respectively). Namely, the i-th row of the base shows the dependencies accumulated by subblock P_i while being encrypted during the previous rounds. The base *does not know* which place P_i moved to, but it holds all the necessary information for a study of the diffusion property. As we will point out below, several dependence matrices (and the corresponding to them permutation sequences) can have one and the same base. Since we want to reduce the number of the tested possible permutation sequences, the base is of major importance both for designing and for explaining our algorithm.

Example 1. Consider a four round GFS_π with $k = 4$ and the permutation sequence presented in Table 1. Before the first round the dependence matrix is the same as the base (every subblock depends on itself only):

$\mathbf{M_0}$	P_1	P_2	P_3	P_4
$X_1 = P_1$	1	0	0	0
$X_2 = P_2$	0	1	0	0
$X_3 = P_3$	0	0	1	0
$X_4 = P_4$	0	0	0	1

At the first round $\pi_1 = (1, 2)(3, 4)$. The dependencies at round 1 are presented below, where F_1 is obtained from M_0 by changing 0-s to 1-s in rows 2 and 4 if there are 1-s in the corresponding columns of rows 1 and 3 respectively, M_1 is obtained from F_1 by applying π_1 on its rows ($\pi_1(1) = 2$, $\pi_1(2) = 1$, $\pi_1(3) = 4$, $\pi_1(4) = 3$, i.e. row 1 becomes 2-nd, row 2 first, row 3 fourth and row 4 third), and B_1 is obtained by sorting the rows of F_1 or M_1 by the P_i.

$\mathbf{F_1}$	P_1	P_2	P_3	P_4
$X_1 \leftarrow P_1$	1	0	0	0
$X_2 \leftarrow P_2$	1	1	0	0
$X_3 \leftarrow P_3$	0	0	1	0
$X_4 \leftarrow P_4$	0	0	1	1

$\mathbf{M_1}$	P_1	P_2	P_3	P_4
$Y_1(X_2) \leftarrow P_2$	1	1	0	0
$Y_2(X_1) \leftarrow P_1$	1	0	0	0
$Y_3(X_4) \leftarrow P_4$	0	0	1	1
$Y_4(X_3) \leftarrow P_3$	0	0	1	0

$\mathbf{B_1}$	P_1	P_2	P_3	P_4
P_1	1	0	0	0
P_2	1	1	0	0
P_3	0	0	1	0
P_4	0	0	1	1

At the second round $\pi_2 = (1, 4, 3, 2)$. The dependencies at round 2 follow, where F_2 is obtained from M_1 by changing 0-s to 1-s in rows 2 and 4 if there are 1-s in the corresponding columns of rows 1 and 3 respectively, M_2 is obtained from F_2 by applying π_2 on its rows ($\pi_1(1) = 4$, $\pi_1(4) = 3$, $\pi_1(3) = 2$, $\pi_1(2) = 1$, i.e. row 1 becomes 4-th, row 4 moves to the third place, row 3 to the second and row 2 to the first), and B_2 is obtained by sorting the rows of F_2 or M_2 by the P_i.

$\mathbf{F_2}$	P_1	P_2	P_3	P_4
$X_1 \leftarrow P_2$	1	1	0	0
$X_2 \leftarrow P_1$	1	1	0	0
$X_3 \leftarrow P_4$	0	0	1	1
$X_4 \leftarrow P_3$	0	0	1	1

$\mathbf{M_2}$	P_1	P_2	P_3	P_4
$Y_1(X_2) \leftarrow P_1$	1	1	0	0
$Y_2(X_3) \leftarrow P_4$	0	0	1	1
$Y_3(X_4) \leftarrow P_3$	0	0	1	1
$Y_4(X_1) \leftarrow P_2$	1	1	0	0

$\mathbf{B_2}$	P_1	P_2	P_3	P_4
P_1	1	1	0	0
P_2	1	1	0	0
P_3	0	0	1	1
P_4	0	0	1	1

At the third round $\pi_3 = (1, 2)(3, 4)$ and the dependencies are:

$\mathbf{F_3}$	P_1	P_2	P_3	P_4
$X_1 \leftarrow P_1$	1	1	0	0
$X_2 \leftarrow P_4$	1	1	1	1
$X_3 \leftarrow P_3$	0	0	1	1
$X_4 \leftarrow P_2$	1	1	1	1

$\mathbf{M_3}$	P_1	P_2	P_3	P_4
$Y_1(X_2) \leftarrow P_4$	1	1	1	1
$Y_2(X_1) \leftarrow P_1$	1	1	0	0
$Y_3(X_4) \leftarrow P_2$	1	1	1	1
$Y_4(X_3) \leftarrow P_3$	0	0	1	1

$\mathbf{B_3}$	P_1	P_2	P_3	P_4
P_1	1	1	0	0
P_2	1	1	1	1
P_3	0	0	1	1
P_4	1	1	1	1

At the forth round $\pi_4 = (1)(2)(3)(4)$. This is the full diffusion round. Every output subblock is affected by all plaintext subblocks.

$\mathbf{F_4}$	P_1	P_2	P_3	P_4
$X_1 \leftarrow P_4$	1	1	1	1
$X_2 \leftarrow P_1$	1	1	1	1
$X_3 \leftarrow P_2$	1	1	1	1
$X_4 \leftarrow P_3$	1	1	1	1

$\mathbf{M_4}$	P_1	P_2	P_3	P_4
$Y_1(X_1) \leftarrow P_4$	1	1	1	1
$Y_2(X_2) \leftarrow P_1$	1	1	1	1
$Y_3(X_3) \leftarrow P_2$	1	1	1	1
$Y_4(X_4) \leftarrow P_3$	1	1	1	1

$\mathbf{B_4}$	P_1	P_2	P_3	P_4
P_1	1	1	1	1
P_2	1	1	1	1
P_3	1	1	1	1
P_4	1	1	1	1

2.2   The Backtrack Search

The construction of the permutation sequences is based on backtrack search with *suitability* tests on the partial solutions.

Suppose we have constructed permutations for the first $r - 1$ rounds and now we are choosing solutions for π_r. For that purpose our algorithm generates all permutations for which the two restrictions described below hold. Then the suitability test is applied to each of them and if it is positive, the current partial solution is extended further with permutations for the remaining rounds. If the suitability test is negative, the next possibility for the r-th permutation is considered. If in this case no more untested possibilities for π_r exist, we return back and test the next possibility for π_{r-1}.

If the dependence matrix of the current solution at round R_D has only nonzero elements, we check whether for decryption the full diffusion round is R_D too. Namely, we compute the full diffusion round of a GFS_π with a permutation sequence $\pi_{R_D-1}^{-1}, \pi_{R_D-2}^{-1}, \ldots \pi_1^{-1}$. If it is R_D, the algorithm stops because a solution attaining the lower bound for even-odd permutations has been obtained.

2.3   Restrictions

Allowing different permutations at the different rounds makes the search space very large. It is therefore infeasible to test all possibilities and the probability to come upon a good one (with respect to diffusion) in reasonable time is quite small. To solve the problem we impose some restrictions which narrow the search space.

Restriction 1. *We construct only even-odd permutations.*

This means that a nonlinear part (possibly with new dependencies) is added to each subblock exactly once in any two successive rounds. The even-odd property leads to the best diffusion in the one-and-the-same permutation case considered in [6]. We expect it to yield solutions with the best diffusion in the multiple permutation case too.

Restriction 2. *We construct only round permutations for which:*
at odd rounds: $\pi_r(2t - 1) = 2t$, $t = 1, 2, \ldots \frac{k}{2}$
at even rounds: $\pi_r(2t) = 2t - 1$, $t = 1, 2, \ldots \frac{k}{2}$

In this way we actually determine the movement of the odd plaintext subblocks during the whole encryption process. Such a restriction does not cut off diffusion solutions because the diffusion property depends only on the way the subblocks are grouped in pairs for the Feistel transformation, not on the exact place of each pair. That is why we can fix the position of one element of each pair. Since we construct even-odd permutations, without loss of generality we choose to fix the movement of the odd plaintext subblocks (i.e. the i-th one will be i-th at odd rounds, and $i+1$-st at even rounds). By this requirement we actually cut off many permutations for which the dependence matrices are different, but the base is the same as in the cases considered by our algorithm. Any of the solutions which are cut off this way, can be obtained by permutations of the pairs in the solutions we find. This is illustrated in the following example.

Example 2. The GFS_π from Example 1 has a permutation sequence $(1,2)(3,4)$, $(1,4,3,2)$, $(1,2)(3,4)$. Now we will compare it to a GFS_π with a permutation sequence $(1,2)(3,4)$, $(1,2,3,4)$, $(1,2)(3,4)$.

The dependencies at round 1 are the same as in Example 1. At the next rounds:

$$\pi_2 = (1,2,3,4)$$

$\mathbf{F_2}$	P_1	P_2	P_3	P_4	$\mathbf{M_2}$	P_1	P_2	P_3	P_4	$\mathbf{B_2}$	P_1	P_2	P_3	P_4
$X_1 \leftarrow P_2$	1	1	0	0	$Y_1(X_4) \leftarrow P_3$	0	0	1	1	P_1	1	1	0	0
$X_2 \leftarrow P_1$	1	1	0	0	$Y_2(X_1) \leftarrow P_2$	1	1	0	0	P_2	1	1	0	0
$X_3 \leftarrow P_4$	0	0	1	1	$Y_3(X_2) \leftarrow P_1$	1	1	0	0	P_3	0	0	1	1
$X_4 \leftarrow P_3$	0	0	1	1	$Y_4(X_3) \leftarrow P_4$	0	0	1	1	P_4	0	0	1	1

$$\pi_3 = (1,2)(3,4)$$

$\mathbf{F_3}$	P_1	P_2	P_3	P_4	$\mathbf{M_3}$	P_1	P_2	P_3	P_4	$\mathbf{B_3}$	P_1	P_2	P_3	P_4
$X_1 \leftarrow P_3$	0	0	1	1	$Y_1(X_2) \leftarrow P_2$	1	1	1	1	P_1	1	1	0	0
$X_2 \leftarrow P_2$	1	1	1	1	$Y_2(X_1) \leftarrow P_3$	0	0	1	1	P_2	1	1	1	1
$X_3 \leftarrow P_1$	1	1	0	0	$Y_3(X_4) \leftarrow P_4$	1	1	1	1	P_3	0	0	1	1
$X_4 \leftarrow P_4$	1	1	1	1	$Y_4(X_3) \leftarrow P_1$	1	1	0	0	P_4	1	1	1	1

$$\pi_4 = (1)(2)(3)(4)$$

$\mathbf{F_4}$	P_1	P_2	P_3	P_4	$\mathbf{M_4}$	P_1	P_2	P_3	P_4	$\mathbf{B_4}$	P_1	P_2	P_3	P_4
$X_1 \leftarrow P_2$	1	1	1	1	$Y_1(X_1) \leftarrow P_3$	1	1	1	1	P_1	1	1	1	1
$X_2 \leftarrow P_3$	1	1	1	1	$Y_2(X_2) \leftarrow P_2$	1	1	1	1	P_2	1	1	1	1
$X_3 \leftarrow P_4$	1	1	1	1	$Y_3(X_3) \leftarrow P_1$	1	1	1	1	P_3	1	1	1	1
$X_4 \leftarrow P_1$	1	1	1	1	$Y_4(X_4) \leftarrow P_4$	1	1	1	1	P_4	1	1	1	1

Denote by p_i the encrypted P_i, no matter where it moved to and consider the pairs (denoted $\leftrightarrow$) for the Feistel transformation at each round. The pairs can be seen from the successive pairs of rows of F_r. We explicitly show them here.

For Example 1:

round	$X_1 \leftrightarrow X_2$	$X_3 \leftrightarrow X_4$
1	$\mathbf{p_1} \leftrightarrow p_2$	$\mathbf{p_3} \leftrightarrow p_4$
2	$p_2 \leftrightarrow \mathbf{p_1}$	$p_4 \leftrightarrow \mathbf{p_3}$
3	$\mathbf{p_1} \leftrightarrow p_4$	$\mathbf{p_3} \leftrightarrow p_2$
4	$p_4 \leftrightarrow \mathbf{p_1}$	$p_2 \leftrightarrow \mathbf{p_3}$

For this permutation sequence:

round	$X_1 \leftrightarrow X_2$	$X_3 \leftrightarrow X_4$
1	$p_1 \leftrightarrow p_2$	$p_3 \leftrightarrow p_4$
2	$p_2 \leftrightarrow p_1$	$p_4 \leftrightarrow p_3$
3	$p_3 \leftrightarrow p_2$	$p_1 \leftrightarrow p_4$
4	$p_2 \leftrightarrow p_3$	$p_4 \leftrightarrow p_1$

We can see that at each round the same elements are in one pair for both sequences. Therefore as far as diffusion is concerned, the two second round permutations $(1,2,3,4)$ and $(1,4,3,2)$ have the same effect. This can also be seen from the bases which are the same in both examples. The requirements of Restriction 2 hold for Example 1 (see the bold elements above which show the required movement of the odd plaintext subblocks), but do not hold for the second permutation sequence. That is why for $\pi_1 = (1,2)(3,4)$ our algorithm will not construct $\pi_2 = (1,2,3,4)$.

The number of all tested permutations at each round is reduced by Restriction 1 from $k!$ to $(\frac{k}{2}!)^2$ (for $k = 16$ this means from 20922789888000 to 1625702400), and further to $\frac{k}{2}!$ by Restriction 2. (for $k = 16$ this is 40320). Thus by applying the upper restrictions we search for best diffusion solutions among $(\frac{k}{2}!)^{R_D-1}$ permutation sequences (for $k = 16$ this is $40320^6 \approx 4.10^{27}$). We further reduce this number by applying a suitability test.

2.4   Suitability Test

Denote by o_r and e_r the number of 1-s in all odd and, respectively, all even rows of F_r. Their values can be at most $k^2/2$ which is the number of all elements in the odd (even) rows. The Feistel transformation at each round can add new dependences only to the even subblocks. Since we construct even-odd permutations, the odd (even) subblocks at the beginning of round r are mapped by π_r to even (odd) subblocks, and so are the rows of F_r - the odd (even) ones are mapped to even (odd) rows of M_r (and F_{r+1} respectively). The best possible scenario for the diffusion is when all dependences of the odd subblocks are new for the corresponding even subblocks, i.e. the Feistel transformation adds the greatest possible number of new dependences to the even subblocks (this means that the greatest possible number of 1-s is added to the even rows of M_r to obtain F_r). More precisely for $r \geq 2$:

$$o_r = e_{r-1} \tag{1}$$

$$e_r \leq o_{r-1} + e_{r-1} \tag{2}$$

$$e_r \leq \frac{k^2}{2} \tag{3}$$

The best scenario implies equality in either (2) or (3). The suitability test at round r counts the exact value of o_{r+1} and e_{r+1}. Then using (1), (2) and (3) for the remaining rounds, it calculates upper bounds on the number of the 1-s in the odd (even) rows of F_{R_D}. If $o_{R_D} < k^2/2$ or $e_{R_D} < k^2/2$, there will be zero elements in F_{R_D} and it cannot be the full diffusion round. Therefore the current permutation sequence cannot be extended to a sequence with full diffusion round R_D. In this case we skip the current solution for π_r and construct the next possibility for it.

Example 4. Consider a GFS_π with $k = 8$. Suppose we have constructed $\pi_1 = (1,2)(3,4)(5,6)(7,8)$. Then the dependencies after round 1 are

$\mathbf{F_1}$	P_1 P_2 P_3 P_4 P_5 P_6 P_7 P_8	$\mathbf{M_1}$	P_1 P_2 P_3 P_4 P_5 P_6 P_7 P_8
$X_1 \leftarrow P_1$	1 0 0 0 0 0 0 0	$Y_1(X_2) \leftarrow P_2$	1 1 0 0 0 0 0 0
$X_2 \leftarrow P_2$	1 1 0 0 0 0 0 0	$Y_2(X_1) \leftarrow P_1$	1 0 0 0 0 0 0 0
$X_3 \leftarrow P_3$	0 0 1 0 0 0 0 0	$Y_3(X_4) \leftarrow P_4$	0 0 1 1 0 0 0 0
$X_4 \leftarrow P_4$	0 0 1 1 0 0 0 0	$Y_4(X_3) \leftarrow P_3$	0 0 1 0 0 0 0 0
$X_5 \leftarrow P_5$	0 0 0 0 1 0 0 0	$Y_5(X_6) \leftarrow P_6$	0 0 0 0 1 1 0 0
$X_6 \leftarrow P_6$	0 0 0 0 1 1 0 0	$Y_6(X_5) \leftarrow P_5$	0 0 0 0 1 0 0 0
$X_7 \leftarrow P_7$	0 0 0 0 0 0 1 0	$Y_7(X_8) \leftarrow P_8$	0 0 0 0 0 0 1 1
$X_8 \leftarrow P_8$	0 0 0 0 0 0 1 1	$Y_8(X_7) \leftarrow P_7$	0 0 0 0 0 0 1 0

We are now constructing all possibilities for π_2. To each of them we apply the suitability test. Suppose we constructed $\pi_2 = (1,2)(3,4)(5,6)(7,8)$. Then:

$\mathbf{F_2}$	P_1 P_2 P_3 P_4 P_5 P_6 P_7 P_8	$\mathbf{M_2}$	P_1 P_2 P_3 P_4 P_5 P_6 P_7 P_8
$X_1 \leftarrow P_2$	1 1 0 0 0 0 0 0	$Y_1(X_2) \leftarrow P_1$	1 1 0 0 0 0 0 0
$X_2 \leftarrow P_1$	1 1 0 0 0 0 0 0	$Y_2(X_1) \leftarrow P_2$	1 1 0 0 0 0 0 0
$X_3 \leftarrow P_4$	0 0 1 1 0 0 0 0	$Y_3(X_4) \leftarrow P_3$	0 0 1 1 0 0 0 0
$X_4 \leftarrow P_3$	0 0 1 1 0 0 0 0	$Y_4(X_3) \leftarrow P_4$	0 0 1 1 0 0 0 0
$X_5 \leftarrow P_6$	0 0 0 0 1 1 0 0	$Y_5(X_6) \leftarrow P_5$	0 0 0 0 1 1 0 0
$X_6 \leftarrow P_5$	0 0 0 0 1 1 0 0	$Y_6(X_5) \leftarrow P_6$	0 0 0 0 1 1 0 0
$X_7 \leftarrow P_8$	0 0 0 0 0 0 1 1	$Y_7(X_8) \leftarrow P_7$	0 0 0 0 0 0 1 1
$X_8 \leftarrow P_7$	0 0 0 0 0 0 1 1	$Y_8(X_7) \leftarrow P_8$	0 0 0 0 0 0 1 1

$\mathbf{F_3}$	P_1 P_2 P_3 P_4 P_5 P_6 P_7 P_8
$X_1 \leftarrow P_1$	1 1 0 0 0 0 0 0
$X_2 \leftarrow P_2$	1 1 0 0 0 0 0 0
$X_3 \leftarrow P_3$	0 0 1 1 0 0 0 0
$X_4 \leftarrow P_4$	0 0 1 1 0 0 0 0
$X_5 \leftarrow P_5$	0 0 0 0 1 1 0 0
$X_6 \leftarrow P_6$	0 0 0 0 1 1 0 0
$X_7 \leftarrow P_7$	0 0 0 0 0 0 1 1
$X_8 \leftarrow P_8$	0 0 0 0 0 0 1 1

From F_3 we count that at round 3: $o_3 = 8, e_3 = 8$. Then using (1), (2) and (3) we calculate:

at round 4: $o_4 = 8, e_4 \leq 16$,
at round 5: $o_5 \leq 16, e_5 \leq 24$,
at round 6: $o_6 \leq 24, e_6 \leq 32$.

You can see that $o_6 < 32$, i.e. there will be zero elements in F_6. Since we want to construct a permutation sequence with full diffusion round 6, we reject the solution $(1,2)(3,4)(5,6)(7,8)$ for π_2. Note that if a suitability test is not applied, this solution will be considered and extended to $(4!)^3 = 13824$ solutions for $\pi_1, \pi_2, \pi_3, \pi_4, \pi_5$, which start by $\pi_1 = (1,2)(3,4)(5,6)(7,8)$ and $\pi_2 = (1,2)(3,4)(5,6)(7,8)$.

Suppose $\pi_2 = (1, 4, 3, 2)(5, 8, 7, 6)$. Then

$\mathbf{F_2}$	P_1	P_2	P_3	P_4	P_5	P_6	P_7	P_8
$X_1 \leftarrow P_2$	1	1	0	0	0	0	0	0
$X_2 \leftarrow P_1$	1	1	0	0	0	0	0	0
$X_3 \leftarrow P_4$	0	0	1	1	0	0	0	0
$X_4 \leftarrow P_3$	0	0	1	1	0	0	0	0
$X_5 \leftarrow P_6$	0	0	0	0	1	1	0	0
$X_6 \leftarrow P_5$	0	0	0	0	1	1	0	0
$X_7 \leftarrow P_8$	0	0	0	0	0	0	1	1
$X_8 \leftarrow P_7$	0	0	0	0	0	0	1	1

$\mathbf{M_2}$	P_1	P_2	P_3	P_4	P_5	P_6	P_7	P_8
$Y_1(X_2) \leftarrow P_1$	1	1	0	0	0	0	0	0
$Y_2(X_3) \leftarrow P_4$	0	0	1	1	0	0	0	0
$Y_3(X_4) \leftarrow P_3$	0	0	1	1	0	0	0	0
$Y_4(X_1) \leftarrow P_2$	1	1	0	0	0	0	0	0
$Y_5(X_6) \leftarrow P_5$	0	0	0	0	1	1	0	0
$Y_6(X_7) \leftarrow P_8$	0	0	0	0	0	0	1	1
$Y_7(X_8) \leftarrow P_7$	0	0	0	0	0	0	1	1
$Y_8(X_5) \leftarrow P_6$	0	0	0	0	1	1	0	0

$\mathbf{F_3}$	P_1	P_2	P_3	P_4	P_5	P_6	P_7	P_8
$X_1 \leftarrow P_1$	1	1	0	0	0	0	0	0
$X_2 \leftarrow P_4$	1	1	1	1	0	0	0	0
$X_3 \leftarrow P_3$	0	0	1	1	0	0	0	0
$X_4 \leftarrow P_2$	1	1	1	1	0	0	0	0
$X_5 \leftarrow P_5$	0	0	0	0	1	1	0	0
$X_6 \leftarrow P_8$	0	0	0	0	1	1	1	1
$X_7 \leftarrow P_7$	0	0	0	0	0	0	1	1
$X_8 \leftarrow P_6$	0	0	0	0	1	1	1	1

The suitability test will be positive. In particular it will give:

round 3: $o_3 = 8, e_3 = 16$,
round 4: $o_4 = 16, e_4 \leq 24$,
round 5: $o_5 \leq 24, e_5 \leq 32$,
round 6: $o_6 \leq 32, e_6 \leq 32$.

So this solution will be extended.

The suitability test and the imposed restrictions allow us to cut off big branches of the search tree and make it possible to construct the permutations which are presented in the next section.

3   Summary of the Results

In Table 1 we present round permutations for GFS_π of $4, 6, 8, 10, 12, 14$ and 16 subblocks. They can be used to construct GFS_π with R_D rounds which have a full diffusion round equal to the lower bound R_D for even-odd permutations. If one and the same permutation is applied at each round this bound cannot be attained for $10, 12, 14$ and 16 subblocks. Files with the dependence matrices at the different rounds of the GFS_π from Table 1 can be downloaded from http://www.moi.math.bas.bg/~svetlana.

We have to point out here that if the presented permutation sequences are used as the first $R_D - 1$ permutations in a GFS_π with more than R_D rounds, then the diffusion round for encryption will be R_D, but the diffusion round for decryption may not be R_D.

Table 1. Permutation sequences for k subblocks with a full diffusion round equal to the lower bound for even-odd permutations

k=4
$\pi_1 = (1,2)(3,4)$
$\pi_2 = (1,4,3,2)$
$\pi_3 = (1,2)(3,4)$

k=6
$\pi_1 = (1,2)(3,4)(5,6)$
$\pi_2 = (1,6,5,4,3,2)$
$\pi_3 = (1,2,5,6,3,4)$
$\pi_4 = (1,6,5,2)(3,4)$

k=8
$\pi_1 = (1,2)(3,4)(5,6)(7,8)$
$\pi_2 = (1,4,3,2)(5,8,7,6)$
$\pi_3 = (1,2)(3,4)(5,6)(7,8)$
$\pi_4 = (1,8,7,2)(3,6,5,4)$
$\pi_5 = (1,2,5,6,7,8)(3,4)$

k=10
$\pi_1 = (1,2)(3,4)(5,6)(7,8)(9,10)$
$\pi_2 = (1,10,9,8,7,6,5,4,3,2)$
$\pi_3 = (1,2,5,6,9,10,3,4,7,8)$
$\pi_4 = (1,6,5,10,9,4,3,8,7,2)$
$\pi_5 = (1,2,7,8,9,10,3,4)(5,6)$

k=12
$\pi_1 = (1,2)(3,4)(5,6)(7,8)(9,10)(11,12)$
$\pi_2 = (1,4,3,2)(5,8,7,12,11,10,9,6)$
$\pi_3 = (1,2)(3,4)(5,6,11,12)(7,8)(9,10)$
$\pi_4 = (1,10,9,4,3,8,7,2)(5,12,11,6)$
$\pi_5 = (1,2,9,10,5,6)(3,4,7,8,11,12)$
$\pi_6 = (1,2)(3,12,11,4)(5,6)(7,10,9,8)$

k=14
$\pi_1 = (1,2)(3,4)(5,6)(7,8)(9,10)(11,12)(13,14)$
$\pi_2 = (1,4,3,2)(5,14,13,12,11,10,9,8,7,6)$
$\pi_3 = (1,2,7,8,11,12)(3,4,5,6,13,14,9,10)$
$\pi_4 = (1,8,7,12,11,6,5,10,9,4,3,14,13,2)$
$\pi_5 = (1,2,3,4,7,8,9,10,11,12,13,14,5,6)$
$\pi_6 = (1,14,13,4,3,2)(5,12,11,6)(7,10,9,8)$

k=16
$\pi_1 = (1,2)(3,4)(5,6)(7,8)(9,10)(11,12)(13,14)(15,16)$
$\pi_2 = (1,4,3,2)(5,8,7,6)(9,12,11,10)(13,16,15,14)$
$\pi_3 = (1,2,11,12)(3,4,7,8)(5,6,15,16)(9,10,13,14)$
$\pi_4 = (1,6,5,2)(3,10,9,4)(7,14,13,8)(11,16,15,12)$
$\pi_5 = (1,2,3,4)(5,6,7,8)(9,10,11,12)(13,14,15,16)$
$\pi_6 = (1,16,15,2)(3,14,13,8,7,10,9,6,5,12,11,4)$

The present results show that the allowance to use different round permutations (without requiring that any two of them are different) can lead to a better diffusion property of the improved generalized Feistel structures.

We cannot say much for GFS_π of more than 16 subblocks. The method which we describe here, is weak to cover the next cases. A different computational approach seems to be needed for bigger k, because our construction algorithm, even well-tuned, remains exponential.

We obtain the results by setting two restrictions on the properties of the constructed permutations. Restriction 2 does not lead to any loss of generality, but Restriction 1 does. It implies the usage of even-odd permutations only. Although this is a well working strategy in the cases we cover, and the even-odd case seems to be intuitively the best choice, we may consider the question of the existence of sequences of permutations which are not even-odd, but might attain Suzaki and Minematsu's even-odd diffusion bound, or might even be better than it. We cannot give a precise answer to this question. We suppose that the diffusion of any sequence cannot be better than the even-odd diffusion bound, but that this bound can also be attained by sequences in which not all permutations are even-odd.

References

1. Berger, T., Francq, J., Minier, M., Thomas, G.: Extended generalized Feistel networks using matrix representation to propose a new lightweight block cipher: Lilliput. IEEE Trans. Comput. **65**(7), 2074–2089 (2016)
2. Berger, T.P., Minier, M., Thomas, G.: Extended generalized feistel networks using matrix representation. In: Lange, T., Lauter, K., Lisoněk, P. (eds.) SAC 2013. LNCS, vol. 8282, pp. 289–305. Springer, Heidelberg (2014). https://doi.org/10.1007/978-3-662-43414-7_15
3. Hong, D., et al.: HIGHT: a new block cipher suitable for low-resource device. In: Goubin, L., Matsui, M. (eds.) CHES 2006. LNCS, vol. 4249, pp. 46–59. Springer, Heidelberg (2006). https://doi.org/10.1007/11894063_4
4. Rivest, R.L., Robshaw, M.J.B., Sidney, R., Yin, Y.L.: The RC6 block cipher. http://people.csail.mit.edu/rivest/pubs/RRSY98.pdf. Accessed 24 Jan 2019
5. Shirai, T., Shibutani, K., Akishita, T., Moriai, S., Iwata, T.: The 128-bit block-cipher CLEFIA (extended abstract). In: Biryukov, A. (ed.) FSE 2007. LNCS, vol. 4593, pp. 181–195. Springer, Heidelberg (2007). https://doi.org/10.1007/978-3-540-74619-5_12
6. Suzaki, T., Minematsu, K.: Improving the generalized Feistel. In: Hong, S., Iwata, T. (eds.) FSE 2010. LNCS, vol. 6147, pp. 19–39. Springer, Heidelberg (2010). https://doi.org/10.1007/978-3-642-13858-4_2
7. Zhang, L., Wu, W.: Analysis of permutation choices for enhanced generalised Feistel structure with SP-type round function. IET Inf. Secur. (2016). https://doi.org/10.1049/iet-ifs.2015.0433
8. Zheng, Y., Matsumoto, T., Imai, H.: On the construction of block ciphers provably secure and not relying on any unproved hypotheses. In: Brassard, G. (ed.) CRYPTO 1989. LNCS, vol. 435, pp. 461–480. Springer, New York (1990). https://doi.org/10.1007/0-387-34805-0_42
9. Wang, Y., Wu, W.: New criterion for diffusion property and applications to improved GFS and EGFN. Des. Codes Crypt. **81**(3), 393–412 (2016)

Fast Computing the Algebraic Degree
of Boolean Functions

Valentin Bakoev$^{(\boxtimes)}$ⓘ

"St. Cyril and St. Methodius" University of Veliko Tarnovo, Veliko Tarnovo, Bulgaria
`v.bakoev@ts.uni-vt.bg`

Abstract. Here we consider an approach for fast computing the algebraic degree of Boolean functions. It combines fast computing the ANF (known as ANF transform) and thereafter the algebraic degree by using the weight-lexicographic order (WLO) of the vectors of the n-dimensional Boolean cube. Byte-wise and bitwise versions of a search based on the WLO and their implementations are discussed. They are compared with the usual exhaustive search applied in computing the algebraic degree. For Boolean functions of n variables, the bitwise implementation of the search by WLO has total time complexity $O(n.2^n)$. When such a function is given by its truth table vector and its algebraic degree is computed by the bitwise versions of the algorithms discussed, the total time complexity is $\Theta((9n-2).2^{n-7}) = \Theta(n.2^n)$. All algorithms discussed have time complexities of the same type, but with big differences in the constants hidden in the Θ-notation. The experimental results after numerous tests confirm the theoretical results—the running times of the bitwise implementation are dozens of times better than the running times of the byte-wise algorithms.

Keywords: Boolean function · Algebraic Normal Form ·
Algebraic degree · Weight-Lexicographic Order ·
WLO sequence generating · Byte-wise algorithm ·
WLO masks generating · Bitwise algorithm

1 Introduction

Boolean functions are of great importance in the modern cryptography, coding theory, digital circuit theory, etc. When they are used in the design of block ciphers, pseudo-random numbers generators (PRNG) in stream ciphers etc., they should satisfy certain cryptographic criteria [5–7]. One of the most important cryptographic parameters is the **_algebraic degree_** of a Boolean function or vectorial Boolean function, called also an S-box. This degree should be higher in order the corresponding Boolean function (or S-box, or PRNG) to be resistant to various types of cryptanalytic attacks. The process of generating such Boolean

This work was partially supported by the Research Fund of the University of Veliko Tarnovo (Bulgaria) under contract FSD-31-340-14/26.03.2019.

© Springer Nature Switzerland AG 2019
M. Ćirić et al. (Eds.): CAI 2019, LNCS 11545, pp. 50–63, 2019.
https://doi.org/10.1007/978-3-030-21363-3_5

functions needs this parameter, as well as the other important cryptographic parameters, to be computed as fast as possible. In this way, more Boolean functions can be generated and a better choice among them can be done.

Let f be a Boolean function of n variables given by its Truth Table vector denoted by $TT(f)$. There are **two main approaches** for computing the algebraic degree of f. The **first** one uses the Algebraic Normal Form (ANF) representation of f and selects the monomial of the highest degree in it. The **second** approach uses only the $TT(f)$, its weight, support, etc., without computing the ANF of f. In [5,6,8,10] it is proven that if $TT(f)$ has an odd weight, then the algebraic degree of f is maximal. This condition holds for the half of all Boolean functions and it can be verified very easily. The algorithms proposed in [8] work only with the $TT(f)$ and use this property. They are fast for just over half of all Boolean functions of n variables. However, when these algorithms are compared with an algorithm of the first type (i.e., based on ANF), the computational results set some questions about the efficiency of algorithms used for computing the ANF and thereafter the algebraic degree. This is one of the reasons that motivated us to do a more comprehensive study of the first approach—fast computing the algebraic degree of Boolean functions by their ANFs. We have already done three basic steps in this direction discussed in Sects. 3 and 4.2. Here we represent the next step which is a natural continuation of the previous ones. It includes a bitwise implementation of the ANF Transform (ANFT) followed by a bitwise computing the algebraic degree by using masks for one special sequence representing the weight-lexicographic order (WLO) of the vectors of Boolean cube.

The paper is structured as follows. The basic notions are given in Sect. 2. In Sect. 3 we outline some preliminary results about the enumeration and distribution of Boolean functions of n variables according to their algebraic degrees, as well as the WLO of the vectors of the Boolean cube and the corresponding sequences. At the beginning of Sect. 4, an algorithm for computing the algebraic degree of Boolean function by using the WLO sequence is discussed. Section 4.2 starts with a comment on the preliminary results about the bitwise ANF transform. Thereafter, a search by using masks for the WLO sequence is considered. Section 5 shows a scheme of computations and used algorithms. The time complexities of the algorithms under consideration are summarized and the experimental results after numerous tests are given. They are used for comparison of the byte-wise and bitwise implementations of the proposed algorithms. The **general conclusion** is: in computing the algebraic degree of a Boolean function it is worth to use the bitwise implementation of proposed algorithms instead of the byte-wise one—it is tens of times faster. In the last section, some ideas about the forthcoming steps of this study are outlined. Experiments in one of these directions have already begun and their first results are good.

2   Basic Notions

Here $\mathbb{N}$ denotes the set of natural numbers. We consider that $0 \in \mathbb{N}$ and $\mathbb{N}^+ = \mathbb{N}\backslash\{0\}$ is the set of positive natural numbers.

Usually, the *n-dimensional Boolean cube* is defined as $\{0,1\}^n =$ $\{(x_1, x_2, \ldots, x_n)| \ x_i \in \{0,1\}, \forall i = 1, 2, \ldots, n\}$, i.e., it is the set of all n-dimensional binary vectors. So $|\{0,1\}^n| = |\{0,1\}|^n = 2^n$. Further, we use the following alternative, inductive and constructive definition.

Definition 1. (1) The set $\{0,1\} = \{(0), (1)\}$ is called *one-dimensional Boolean cube* and its elements (0) and (1) are called *one-dimensional binary vectors*.
(2) Let $\{0,1\}^{n-1} = \{\alpha_0, \alpha_1, \ldots, \alpha_{2^{n-1}-1}\}$ be the $(n-1)$-*dimensional Boolean cube* and $\alpha_0, \alpha_1, \ldots, \alpha_{2^{n-1}-1}$ be its $(n-1)$-*dimensional binary vectors*.
(3) The *n-dimensional Boolean cube* $\{0,1\}^n$ is built by taking the vectors of $\{0,1\}^{n-1}$ twice: firstly, each vector of $\{0,1\}^{n-1}$ is prefixed by zero, and thereafter each vector of $\{0,1\}^{n-1}$ is prefixed by one:

$$\{0,1\}^n = \{(0, \alpha_0), (0, \alpha_1), \ldots, (0, \alpha_{2^{n-1}-1}),$$
$$(1, \alpha_0), (1, \alpha_1), \ldots, (1, \alpha_{2^{n-1}-1})\}.$$

For an arbitrary vector $\alpha = (a_1, a_2, \ldots, a_n) \in \{0,1\}^n$, the natural number $\#\alpha = \sum_{i=1}^{n} a_i.2^{n-i}$ is called a *serial number* of the vector α. So $\#\alpha$ is the natural number having n-digit binary representation $a_1 a_2 \ldots a_n$. A *(Hamming) weight* of α is the natural number $wt(\alpha)$, equal to the number of non-zero coordinates of α, i.e., $wt(\alpha) = \sum_{i=1}^{n} a_i$. For any $k \in \mathbb{N}$, $k \leq n$, the set of all n-dimensional binary vectors of weight k is called a *k-th layer* of the n-dimensional Boolean cube. It is denoted by $L_{n,k} = \{\alpha| \alpha \in \{0,1\}^n : wt(\alpha) = k\}$ and we have $|L_{n,k}| = \binom{n}{k}$, for $k = 0, 1, \ldots, n$. These numbers are the binomial coefficients from the n-th row of Pascal's triangle and so $\sum_{k=0}^{n} \binom{n}{k} = 2^n = |\{0,1\}^n|$. The family of all layers $L_n = \{L_{n,0}, L_{n,1}, \ldots, L_{n,n}\}$ is a *partition* of the n-dimensional Boolean cube into layers.

For arbitrary vectors $\alpha = (a_1, a_2, \ldots, a_n)$ and $\beta = (b_1, b_2, \ldots, b_n) \in \{0,1\}^n$, we say that "$\alpha$ *precedes lexicographically* β" and denote this by $\alpha \leq \beta$, if $\alpha = \beta$ or if $\exists k, 1 \leq k \leq n$, such that $a_k < b_k$ and $a_i = b_i$, for all $i < k$. The relation "$\leq$" is a *total* (unique) order in $\{0,1\}^n$, called *lexicographic order*. The vectors of $\{0,1\}^n$ are ordered lexicographically in the sequence $\alpha_0, \alpha_1, \ldots \alpha_k, \ldots, \alpha_{2^n-1}$ if and only if:

- $\alpha_l \leq \alpha_k, \forall l \leq k$ and $\alpha_k \leq \alpha_r, \forall k \leq r$;
- the sequence of their serial numbers $\#\alpha_0, \#\alpha_1, \ldots, \ \#\alpha_k, \ldots, \#\alpha_{2^n-1}$ is exactly $0, 1, \ldots, k, \ldots, 2^n - 1$.

A *Boolean function* of n variables (denoted usually by $x_1, x_2, \ldots, x_n$) is a mapping $f : \{0,1\}^n \rightarrow \{0,1\}$, i.e. f maps any binary input $x = (x_1, x_2, \ldots, x_n) \in \{0,1\}^n$ to a single binary output $y = f(x) \in \{0,1\}$. Any Boolean function f can be represented in a unique way by the vector of its functional values, called a *Truth Table* vector and denoted by $TT(f) = (f_0, f_1, \ldots f_{2^n-1})$, where $f_i = f(\alpha_i)$ and α_i is the i-th vector in the lexicographic order of $\{0,1\}^n$, for $i = 0, 1, \ldots, 2^n - 1$. The set of all Boolean functions of n variables is denoted by $\mathscr{B}_n$ and its size is $|\mathscr{B}_n| = 2^{2^n}$.

Another unique representation of the Boolean function $f \in \mathscr{B}_n$ is the *algebraic normal form* (ANF) of f, which is a multivariate polynomial

$$f(x_1, x_2, \ldots, x_n) = \bigoplus_{\gamma \in \{0,1\}^n} a_{\#\gamma} \, x^\gamma.$$

Here $\gamma = (c_1, c_2, \ldots, c_n) \in \{0,1\}^n$, the coefficient $a_{\#\gamma} \in \{0,1\}$, and x^γ means the monomial $x_1^{c_1} x_2^{c_2} \ldots x_n^{c_n} = \prod_{i=1}^{n} x_i^{c_i}$, where $x_i^0 = 1$ and $x_i^1 = x_i$, for $i = 1, 2, \ldots n$. A *degree* of the monomial $x = x_1^{c_1} x_2^{c_2} \ldots x_n^{c_n}$ is the integer $deg(x) = wt(\gamma)$—it is the number of variables of the type $x_i^1 = x_i$, or the essential variables for x^γ. The *algebraic degree* (or simply *degree*) of f is defined as $deg(f) = max\{deg(x^\gamma) \mid a_{\#\gamma} = 1\}$. When $f \in \mathscr{B}_n$ and the $TT(f)$ is given, the values of the coefficients $a_0, a_1, \ldots, a_{2^n-1}$ can be computed by a fast algorithm, usually called an *ANF transform* (ANFT)[1]. The ANFT is well studied, it is derived in different ways by many authors, for example [5,6,9]. Its byte-wise implementation has a time-complexity $\Theta(n.2^n)$. The vector $(a_0, a_1, \ldots, a_{2^n-1}) \in \{0,1\}^n$ obtained after the ANFT is denoted by A_f. When $f \in \mathscr{B}_n$ is the constant zero function (i.e., $TT(f) = (0, 0, \ldots, 0)$), its ANF is $A_f = (0, 0, \ldots, 0)$ and its algebraic degree is defined as $deg(f) = -\infty$. If f is the constant one function ($TT(f) = (1, 1, \ldots, 1)$), then $A_f = (1, 0, 0, \ldots, 0)$ and $deg(f) = 0$.

3 Some Preliminary Results

3.1 Distribution of Boolean Functions According to Their Algebraic Degrees

It is well-known that half of all Boolean functions of n variables have an algebraic degree equal to n, for $n \in \mathbb{N}^+$ [5,6,8,10]. Furthermore, in [6, p. 49] Carlet notes that when n tends to infinity, random Boolean functions have almost surely algebraic degrees at least $n-1$. We consider that the overall enumeration and distribution of all Boolean functions of n variables ($n \in \mathbb{N}^+$) according to their algebraic degrees is very important for our study. The paper where we explore them is still in review, but some results can be seen in OEIS [11], sequence A319511. We will briefly outline the results needed for further exposition.

Let $d(n, k)$ be the number of all Boolean functions $f \in \mathscr{B}_n$ such that $deg(f) = k$.

Theorem 1. *For arbitrary integers $n \in \mathbb{N}$ and $0 \leq k \leq n$, the number*

$$d(n, k) = (2^{\binom{n}{k}} - 1).2^{\sum_{i=0}^{k-1} \binom{n}{i}}.$$

---

[1] In dependence of the area of consideration, the same algorithm is called also (fast) Möbius Transform, Zhegalkin Transform, Positive polarity Reed-Muller Transform, etc.

Sketch of proof: let X be the set of n variables. There are $\binom{n}{k}$ monomials of degree $= k$ because so many are the ways to choose k variables from X. The first multiplier in the formula denotes the number of ways to choose at least one such monomial to participate in the ANF. The second multiplier is the number of ways to choose 0 or more monomials of degrees $< k$ and to add them to the ANF.

Corollary 1. *The number $d(n, n-1)$ tends to $\dfrac{1}{2} \cdot |\mathscr{B}_n|$ when $n \to \infty$.*

Let $p(n, k)$ be the discrete probability a random Boolean function $f \in \mathscr{B}_n$ to have an algebraic degree $= k$. It is defined as

$$p(n,k) = \frac{d(n,k)}{|\mathscr{B}_n|} = \frac{d(n,k)}{2^{2^n}},$$

for $n \geq 0$ and $0 \leq k \leq n$. The values of $p(n, k)$ obtained for a fixed n give the distribution of the functions from $\mathscr{B}_n$ according to their algebraic degrees. Table 1 represents this distribution, for $3 \leq n \leq 10$ and $n-3 \leq k \leq n$. The values of $p(n, k)$ in it are rounded up to 10 digits after the decimal point. Furthermore, $p(n, k) \approx 0$, for $0 \leq k < n-3$, and their values are not shown in the table.

Table 1. Distribution of the functions from $\mathscr{B}_n$ according to their algebraic degrees, for $n = 3, 4, \ldots, 10$

	The values of $p(n, k)$, for:			
n	$k = n - 3$	$k = n - 2$	$k = n - 1$	$k = n$
3	0.00390625	0.0546875	0.4375	0.5
4	0.0004577637	0.0307617187	0.46875	0.5
5	0.0000152439	0.0156097412	0.484375	0.5
6	0.0000002384	0.0078122616	0.4921875	0.5
7	0.0000000019	0.0039062481	0.49609375	0.5
8	0	0.0019531250	0.498046875	0.5
9	0	0.0009765625	0.4990234375	0.5
10	0	0.0004882812	0.4995117187	0.5

These results were used:

- To check for representativeness the files used to test all algorithms discussed here. These are 4 files containing $10^6, 10^7, 10^8$ and 10^9 randomly generated unsigned integers in 64-bit computer words. We used each of these files as an input for Boolean functions of $6, 8, 10, \ldots, 16$ variables (reading 2^{n-6} integers from the chosen file) and we computed the algebraic degrees of all these functions. The absolute value of the difference between the theoretical and computed distribution is less than 0.88% (it exceeds 0.1% in only a few cases), for all tests. So we consider that the algorithms work with samples of Boolean functions which are representative enough.

– When creating the algorithms represented in the following sections. The distribution shows why the WLO has been studied in detail and what to expect for the running time of algorithms that use WLO.

3.2   WLO of the Vectors of n-dimensional Boolean Cube

The simplest algorithm for computing the algebraic degree of a Boolean function is an *Exhaustive Search* (we refer to it as **ES algorithm**): if $f \in \mathscr{B}_n$ and $A_f = (a_0, a_1, \ldots, a_{2^n-1})$ is given, it checks consecutively whether $a_i = 1$, for $i = 0, 1, \ldots, 2^n - 1$. The algorithm selects the vector of maximal weight among all vectors $\alpha_i \in \{0,1\}^n$ such that $a_i = 1$. The algorithm checks exhaustively all values in A_f (which correspond to the lexicographic order of the vectors of $\{0,1\}^n$) and so it performs $\Theta(2^n)$ checks.

The basic parts of a faster way for the same computing are considered in [1,2]. Here they are given in short, but all related notions, proofs, illustrations, algorithms and programming codes, details, etc., can be seen in [2].

The *sequence of layers* $L_{n,0}, L_{n,1}, \ldots, L_{n,n}$ gives an *order* of the vectors of $\{0,1\}^n$ in accordance with their weights. When $\alpha, \beta \in \{0,1\}^n$ and $wt(\alpha) < wt(\beta)$, then α *precedes* β in the sequence of layers, and if $wt(\alpha) = wt(\beta) = k$, then $\alpha, \beta \in L_{n,k}$ and there is no precedence between them. We define the corresponding relation $R_{<wt}$ as follows: for arbitrary $\alpha, \beta \in \{0,1\}^n$, $(\alpha, \beta) \in R_{<wt}$ if $wt(\alpha) < wt(\beta)$ or if $\alpha = \beta$. When $(\alpha, \beta) \in R_{<wt}$ we say that "α *precedes by weight* β" and write also $\alpha <_{wt} \beta$. Thus $R_{<wt}$ is a partial order in $\{0,1\}^n$ and we refer to it (and to the order determined by it) as a *Weight-Order* (WO). To develop an algorithm we use the serial numbers of the vectors in the sequence of layers instead of the vectors themselves. For an arbitrary layer $L_{n,k} = \{\alpha_0, \alpha_1, \ldots, \alpha_m\}$ of $\{0,1\}^n$, we define the *sequence of serial numbers* of the vectors of $L_{n,k}$ and denote it by $l_{n,k} = \#\alpha_0, \#\alpha_1, \ldots, \#\alpha_m$. Let $l_n = l_{n,0}, l_{n,1}, \ldots, l_{n,n}$ be the *sequence of all serial numbers* corresponding to the vectors in the sequence of layers $L_{n,0}, L_{n,1}, \ldots, L_{n,n}$. Thus l_n represents a WO of the vectors of $\{0,1\}^n$ and we call l_n a *WO sequence* of $\{0,1\}^n$. One of all possible $\prod_{k=0}^{n} \binom{n}{k}!$ WO sequences[2] deserves a special attention. Firstly, we define the operation *addition of the natural number to a sequence* as follows: if $n, m \in \mathbb{N}^+$ and $s = a_1, a_2, \ldots, a_n$ is a sequence of integers, then $s + m = a_1 + m, a_2 + m, \ldots, a_n + m$. Following Definition 1, we obtain:

Definition 2. (1) The WO sequence of the one-dimensional Boolean cube is
$l_1 = 0, 1$.
(2) Let $l_{n-1} = l_{n-1,0}, l_{n-1,1}, \ldots, l_{n-1,n-1}$ be the WO sequence of the $(n-1)$-dimensional Boolean cube.
(3) The WO sequence of n-dimensional Boolean cube $l_n = l_{n,0}, l_{n,1}, \ldots, l_{n,n}$ is defined as follows:
  • $l_{n,0} = 0$ and it corresponds to the layer $L_{n,0} = \{\tilde{0}_n\}$, where $\tilde{0}_n$ is the zero vector of n coordinates;

---

[2] You can see the sequence A051459 in the OEIS [11] for details.

- $l_{n,n} = 2^n - 1$ and it corresponds to the layer $L_{n,n} = \{\tilde{1}_n\}$, where $\tilde{1}_n$ is the all-ones vector of n coordinates;
- $l_{n,k} = l_{n-1,k}, l_{n-1,k-1} + 2^{n-1}$, for $k = 1, 2, \ldots, n - 1$. Here $l_{n,k}$ is a concatenation of two sequences: the sequence $l_{n-1,k}$ is taken (or copied) firstly, and the sequence $l_{n-1,k-1} + 2^{n-1}$ follows after it. The sequence $l_{n,k}$ corresponds to the layer $L_{n,k}$.

Theorem 2. *Let $n \in \mathbb{N}^+$ and $l_n = l_{n,0}, l_{n,1}, \ldots, l_{n,n}$ be the WO sequence, obtained in accordance with Definition 2. Then, the serial numbers in the sequence $l_{n,k}$ determine a lexicographic order of the vectors of the corresponding layer $L_{n,k}$, for $k = 0, 1, \ldots, n$.*

Theorem 2 is proven by mathematical induction in [2]. It states that Definition 2 determines a *second criterion* for ordering the vectors within the existing WO of the Boolean cube—this is the *lexicographic order*. Since it is a total order for each subsequence $l_{n,k}$, $0 \leq k \leq n$, a total weight order for the sequence l_n is obtained. We call it a *Weight-Lexicographic Order* (WLO).

The *WLO algorithm* is based on Definition 2 and Theorem 2, and so they imply its correctness. For a given input $n \in \mathbb{N}^+$, it starts from l_1 and computes consecutively the sequences $l_2, l_3, \ldots, l_n$. Some results computed by the algorithm are given in Table 2. More results can be seen in OEIS [11], sequence A294648.

The time complexity of the WLO algorithm is $\Theta(2^n)$, it is exponential with respect to the size of the input n. Furthermore, it is linear with respect to the size of the output. The space complexity of the algorithm is of the same type. We note that the running time for precomputation of the sequence l_n in a lookup table is negligible ($\approx 0\,\mathrm{s}$).

4   Computing the Algebraic Degree of Boolean Functions by WLO

4.1   Byte-Wise Approach

The terms of the WLO sequence l_n form a permutation of the numbers $0, 1, \ldots, 2^n - 1$ and we denote this permutation as $l_n = (i_0, i_1, \ldots, i_{2^n-1})$. We use the sequence l_n to compute the algebraic degree of a given Boolean function

Table 2. Results obtained by the WLO algorithm for $n = 1, 2, \ldots, 5$

n	l_n
1	0, 1
2	0, 1, 2, 3
3	0, 1, 2, 4, 3, 5, 6, 7
4	0, 1, 2, 4, 8, 3, 5, 6, 9, 10, 12, 7, 11, 13, 14, 15
5	0, 1, 2, 4, 8, 16, 3, 5, 6, 9, 10, 12, 17, 18, 20, 24, 7, 11, 13, 14, 19, 21, 22, 25, 26, 28, 15, ...

$f \in \mathscr{B}_n$. The proposed algorithm is similar to the ES algorithm, but it checks the coordinates of $A_f = (a_0, a_1, \ldots, a_{2^n-1})$ in accordance with the values of l_n, from right to left. It starts with the i_{2^n-1}-th coordinate of A_f. If it is equal to zero the algorithm checks the i_{2^n-2}-th coordinate of A_f, and so on, looking for the first coordinate of A_f which is equal to one and then it stops. If there is not such a coordinate, then f is the constant zero function. Otherwise, if the algorithm stops the searching on the i_j-th coordinate ($0 \leq j < 2^n$) of A_f, it returns the number of the subsequence that contains the number i_j as an output. If i_j is a term of $l_{n,k}$, $0 \leq k \leq n$, then the layer $L_{n,k}$ contains a vector which serial number is i_j and therefore $deg(f) = k$. The algorithm is correct, since it follows the WLO and stops at the right place—if it continues with the checks, it will find possible monomials of degree $\leq k$. Thus the algorithm performs $O(2^n)$ checks and this is its *time complexity*. This general estimation concerns a very small number of functions $f \in \mathscr{B}_n$ because the computing will finish after $O(n)$ checks at almost 100% of all such functions (especially when n grows)—as it is shown in Sect. 3.1. Since this algorithm works in a byte-wise manner and after the byte-wise ANFT, we call it ***Byte-wise WLO algorithm***.

4.2 Bitwise Approach

In [3] we represented a comprehensive study of the bitwise implementation of the ANFT. When 64-bit computer words are used, the obtained algorithm has a time-complexity $\Theta((9n-2).2^{n-7})$ and a space complexity $\Theta(2^{n-6})$, i.e., both are of the type $\Theta(2^n)$. But the experimental results show that the bitwise version of the algorithm is about 25 times faster in comparison to the byte-wise version[3]. Analogous research concerning the parallel bitwise implementation of the ANFT is represented in [4] and similar results about its efficiency are obtained.

After these results it is natural to think about a bitwise implementation of the last algorithm. Otherwise, bitwise computing an ANFT seems unnecessary, since computing the other cryptographic parameters of Boolean functions needs a byte-wise representation (see Fig. 1). Our first idea is to check all vectors in the same layer in one (or several) step(s). For this purpose we use $n+1$ masks $m_{n,0}, m_{n,1}, \ldots, m_{n,n}$ corresponding to the vectors in the layers $L_{n,0}, L_{n,1}, \ldots, L_{n,n}$. The mask $m_{n,i}$ is a binary vector of the same length as A_f and $m_{n,i}$ contains units only in these bits, whose coordinates correspond to the numbers in the subsequence $l_{n,i}$, for $i = 0, 1, \ldots, n$. So we need to repeat bitwise conjunctions between A_f and $m_{n,i}$, for $i = n, n-1, \ldots, 0$, until $A_f \wedge m_{n,i} = 0$. If this equality holds for all values of i, then f is the constant zero function. Otherwise, if k is the first value of i (when i decreases from n to 0) such that $A_f \wedge m_{n,k} > 0$, then k is the algebraic degree of f. So the algorithm stops and returns k. We call it ***Bitwise WLO algorithm*** accepting that it always uses masks.

---

[3] Both algorithms have been implemented as C++ programs in Code::Blocks 13.12 IDE, built as 32-bit applications in Release mode and tested with the largest file of 10^9 integers.

When A_f occupies one computer word, the algorithm performs at most $n+1$ steps and so its time complexity is $O(n)$, i.e., it is of logarithmic type ($n = \log_2 2^n$) with respect to the size of the input. If the size of the computer word is $64 = 2^6$ bits and f is a function of $n > 6$ variables, then $TT(f)$ and A_f occupy $s = 2^n/64 = 2^{n-6}$ computer words. So $m_{n,i}$ will occupy s computer words too and the computing $A_f \wedge m_{n,i}$ will be done in s steps, for $i = n, n-1, \ldots, 0$. If on some of these steps the conjunction between the corresponding computer words of A_f and $m_{n,i}$ is greater than zero, the algorithm returns i and stops. Therefore, in the general case, the bitwise WLO algorithm has a *time complexity* $O(n+1).O(s) = O(n.2^{n-6})$. This estimation concerns a very small number of functions $f \in \mathscr{B}_n$ again—the computing will finish after $O(1+s) = O(2^{n-6})$ checks at almost 100% of all such functions.

Let us consider the masks' generating. For arbitrary $i, 0 \le i \le n$, it is easy to put units in all these bits of $m_{n,i}$ that correspond to the numbers in the subsequence $l_{n,i}$. We note that we use the serial numbers of the masks, stored in the necessary number of 64-bit computer words, as well as the vectors $TT(f)$ and A_f. Furthermore, we generate them in accordance with the following definition.

Definition 3. (1) For $n = 1$, the serial numbers of the masks corresponding to the subsequences $l_{1,0}$ and $l_{1,1}$ are $\#m_{1,0} = 2$ and $\#m_{1,1} = 1$.

(2) Let $\#m_{n-1,0}, \#m_{n-1,1}, \ldots, \#m_{n-1,n-1}$ be the serial numbers of the masks corresponding to the subsequences $l_{n-1,0}, l_{n-1,1}, \ldots, l_{n-1,n-1}$.

(3) The serial number of the mask $m_{n,i}$ corresponding to the subsequence $l_{n,i}$ is:

$$\#m_{n,i} = \begin{cases} 2^{2^{n-1}}.\#m_{n-1,0} = 2^{2^n-1}, & \text{if } i = 0, \\ 1, & \text{if } i = n, \\ 2^{2^{n-1}}.\#m_{n-1,i} + \#m_{n-1,i-1}, & \text{if } 0 < i < n, \end{cases}$$

for $i = 0, 1, \ldots, n$.

Definition 3 corresponds to Definitions 1 and 2. Its correctness can be proven strictly by mathematical induction on n. The running time for generating (pre-computation of) the masks in accordance with Definition 3 is negligible ($\approx 0\,\mathrm{s}$). We note that when $n > 6$, the generating algorithm has some particularities because it works with $s = 2^{n-6}$ computer words for each mask. The serial numbers of masks grow exponentially—see Table 3, as well as the sequence A305860 in OEIS [11].

Example 1. Let us consider $f \in \mathscr{B}_4$ whose ANF, the coordinates' (or bits') numbers (these which are greater than 9 are represented by their last digit) and the masks (for $n = 4$) are given in Table 4. When we use the byte-wise WLO Algorithm, it checks consecutively the coordinates of A_f, from right to left, i.e., 15, 14, 13, 11, 7, 12—see the WLO sequence l_4 in Table 2. A_f contains zeros in all coordinates before 12-th, in this coordinate A_f contains one and so the algorithm stops after **6 checks**. Since 12 is a term of $l_{4,2}$, hence $deg(f) = 2$. When the bitwise WLO algorithm is used, it computes the conjunctions: $A_f \wedge m_{4,4} = 0$, $A_f \wedge m_{4,3} = 0$, $A_f \wedge m_{4,2} > 0$ and thereafter it stops. So $deg(f) = 2$ and it is computed in **3 steps**.

Table 3. Serial numbers of the masks, for $n = 1, \ldots, 5$

n	$\#m_{n,0}$	$\#m_{n,1}$	$\#m_{n,2}$	$\#m_{n,3}$	$\#m_{n,4}$	$\#m_{n,5}$
1	2	1	–	–	–	–
2	8	6	1	–	–	–
3	128	104	22	1	–	–
4	32768	26752	5736	278	1	–
5	2147483648	1753251840	375941248	18224744	65814	1

Table 4. The data used in Example 1

Coordinates' numbers	0 1 2 3 4 5 6 7 8 9 0 1 2 3 4 5
$A_f =$	1 0 0 1 0 1 1 0 1 0 1 0 1 0 0 0
$\#m_{4,0} = 32768,\ m_{4,0} =$	1 0 0 0 0 0 0 0 0 0 0 0 0 0 0 0
$\#m_{4,1} = 26752,\ m_{4,1} =$	0 1 1 0 1 0 0 0 1 0 0 0 0 0 0 0
$\#m_{4,2} = 5736,\ \ \ m_{4,2} =$	0 0 0 1 0 1 1 0 0 1 1 0 1 0 0 0
$\#m_{4,3} = 278,\ \ \ \ m_{4,3} =$	0 0 0 0 0 0 0 1 0 0 0 1 0 1 1 0
$\#m_{4,4} = 1,\ \ \ \ \ \ \ \ m_{4,4} =$	0 0 0 0 0 0 0 0 0 0 0 0 0 0 0 1

The second idea for a new bitwise algorithm is to check the bits of A_f in accordance with the WLO sequence. This algorithm will be similar to the byte-wise WLO algorithm and it will have a time complexity of the same type: $O(2^n)$. We discarded this idea because the time complexity of the bitwise WLO algorithm is $O(n.2^{n-6})$ and $n.2^{n-6} < 2^n$ when $6 < n < 64$. But during the revision of this paper, we noticed that for almost 100% of all $f \in \mathscr{B}_n$, the bitwise WLO algorithm performs $O(2^{n-6})$ checks, whereas the byte-wise WLO algorithm (as well as the new bitwise algorithm) performs $O(n)$ checks. Furthermore, the check of a serial bit of A_f (in accordance with the WLO sequence) needs no more than 5 bitwise operations. Hence the new bitwise algorithm will have a small constant hidden in the O-notation. For example, the bitwise WLO algorithm will be better for small n (say $n \leq 8$). But for $n = 16$ the bitwise WLO algorithm will perform quite more operations than the new bitwise algorithm. The forthcoming tests will show when and how faster is the new algorithm.

5   Experimental Results

We return to the main problem of this study—fast computing the algebraic degree of a Boolean function $f \in \mathscr{B}_n$ given by its $TT(f)$. A scheme of the computations and used algorithms is shown in Fig. 1.

In accordance with this scheme, the time complexities of the algorithms considered are summarized as follows:

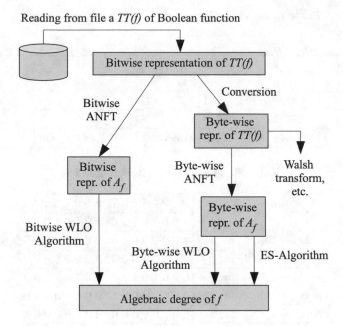

Fig. 1. A scheme for computing the algebraic degree of Boolean functions

1. The byte-wise ANFT algorithm followed by the ES algorithm are referred as **Byte-wise ANFT&ES** further. So, their time complexity is a sum of $\Theta(n.2^n) + \Theta(2^n) = \Theta(n.2^n)$.
2. The byte-wise ANFT algorithm followed by the byte-wise WLO algorithm are referred as **Byte-wise ANFT&WLO**. Their time complexity is $\Theta(n.2^n) + O(2^n) = \Theta(n.2^n)$.
3. The bitwise ANFT algorithm followed by the bitwise WLO algorithm are referred as **Bitwise algorithms**. When 64-bit computer words are used, the time complexity of the bitwise algorithms is $\Theta((9n-2).2^{n-7}) + O(n.2^{n-6}) = \Theta((9n-2).2^{n-7}) = \Theta(n.2^n)$.

It has to be noted that these time complexities are:

- dominated by the time complexity of the corresponding ANFT—the cost of search is relatively small and it is absorbed into the cost of ANFT;
- of the same type $\Theta(n.2^n)$, and the differences between them are in the constants hidden in the Θ-notation.

To understand what these theoretical time complexities mean in practice, we have done a lot of tests. Some more important tests' parameters are:

1. Hardware parameters: Intel Pentium CPU G4400, 3.3 GHz, 4 GB RAM, Samsung SSD 650 120 GB.
2. Software parameters: Windows 10 OS and MVS Express 2015 for Windows Desktop. The algorithms are written in C++. All programs were built in

Table 5. Experimental results about all 2^{32} Boolean functions of 5 variables

Tested algorithms	Pure running time in seconds for:	
	32-bit application	64-bit application
Byte-wise ANFT& ES	540.824	507.407
Byte-wise ANFT& WLO	450.521	378.374
Bitwise algorithms	6.470	6.512

Table 6. Experimental results for 32-bit applications

32-bit implementation of:	Pure running time in seconds for Boolean functions of:				
	6 vars, 10^8 BFs	8 vars, $10^8/4$ BFs	10 vars, $10^8/16$ BFs	12 vars, $10^8/64$ BFs	16 vars, 97 656 BFs
Byte-wise ANFT& ES	38.834	42.400	42.664	43.466	44.740
Byte-wise ANFT& WLO	22.003	20.022	18.758	18.230	18.808
Bitwise algorithms	1.078	1.958	1.560	1.563	1.431

Table 7. Experimental results for 64-bit applications

64-bit implementation of:	Pure running time in seconds for Boolean functions of:				
	6 vars, 10^8 BFs	8 vars, $10^8/4$ BFs	10 vars, $10^8/16$ BFs	12 vars, $10^8/64$ BFs	16 vars, 97 656 BFs
Byte-wise ANFT& ES	37.429	39.178	37.699	38.789	40.350
Byte-wise ANFT& WLO	17.443	15.880	14.224	14.243	14.454
Bitwise algorithms	0.861	0.819	0.709	0.640	0.718

Release mode as 32-bit and 64-bit console applications and executed without Internet connection.

3. Methodology of testing: all tests were executed 3 times, on the same computer, under the same conditions. The running times are taken in average. All results were checked for coincidence. The time for reading from file and conversion to byte-wise representation is excluded.

Table 5 shows the obtained running times of the compared algorithms for all 2^{32} Boolean functions of 5 variables.

Functions of 6 and more variables have been tested with the file of 10^8 integers. Depending on the number of variables, 2^{n-6} integers are read from the file and so they form the serial Boolean function. Tables 6 and 7 show the results for Boolean functions (BFs) of 6 and more variables (vars).

6   Conclusions

We hope that the obtained results show convincingly the advantages of the WLO approaches in computing the algebraic degree of Boolean functions. The bitwise implementations of the considered algorithms are dozens of times faster than the byte-wise implementations. Their usage economizes valuable time, especially in generating S-boxes. The natural continuation of the topic under consideration includes an experimental study of:

- The second bitwise algorithm proposed at the end of Sect. 4.2.
- Combination of both approaches discussed in Sect. 1 as follows. First, compute the weight of $TT(f)$. If it is an odd number, then f is of maximal degree. Otherwise, continue with the bitwise algorithms. Some tests with the largest file (of 10^9 integers) have already begun. The first results show that due to this modification, the bitwise algorithms run about two times faster.
- More appropriate software environment (for example, Linux) in order to minimize the effects of background processes running during the executions of the tests. Afterward, repeat all tests since some running times in the last two tables are less than one second and they might not been precise enough.
- Application of the bitwise algorithms in computing the algebraic degree of true examples of S-boxes.
- Parallel implementations of the bitwise algorithms.

References

1. Bakoev, V.: Ordinances of the vectors of the n-dimensional Boolean cube in accordance with their weights, (Presented Conference Paper style). In: Book of Abstracts, XIV Serbian Mathematical Congress, Kragujevac, Serbia, 16–19 May 2018, p. 103 (2018)
2. Bakoev, V.: About the ordinances of the vectors of the n-dimensional Boolean cube in accordance with their weights (2018). https://arxiv.org/abs/1811.04421
3. Bakoev, V.: Fast bitwise implementation of the algebraic normal form transform. Serdica J. Comput. **11**(1), 45–57 (2017)
4. Bikov, D., Bouyukliev, I.: Parallel fast Möbius (Reed-Muller) transform and its implementation with CUDA on GPUs. In: Proceedings of the International Workshop on Parallel Symbolic Computation, PASCO 2017, 23–24 July 2017, Kaiserslautern, Germany, pp. 5:1–5:6 (2017)
5. Canteaut, A.: Lecture notes on Cryptographic Boolean Functions, Inria, Paris, France (2016)
6. Carlet, C.: Boolean functions for cryptography and error correcting codes. In: Crama, Y., Hammer, P.L. (eds.) Boolean Models and Methods in Mathematics, Computer Science, and Engineering, pp. 257–397. Cambridge University Press, Cambridge (2010)

7. Carlet, C.: Vectorial Boolean functions for cryptography. In: Crama, Y., Hammer, P.L. (eds.) Boolean Models and Methods in Mathematics, Computer Science, and Engineering, pp. 398–469. Cambridge University Press, Cambridge (2010)

8. Climent, J.-J., García, F., Requena, V.: The degree of a Boolean function and some algebraic properties of its support. In: Data Management and Security, pp. 25–36. WIT Press (2013)

9. Joux, A.: Algorithmic Cryptanalysis. Chapman & Hall/CRC Cryptography and Network Security (2012)

10. MacWilliams, F.J., Sloane, N.J.A.: The Theory of Error-Correcting Codes, North-Holland, Amsterdam (1978)

11. OEIS Foundation Inc.: The On-line Encyclopedia of Integer Sequences. https://oeis.org/

On the Scalar Complexity of Chudnovsky^2 Multiplication Algorithm in Finite Fields

Stéphane Ballet[1,2], Alexis Bonnecaze[1,2], and Thanh-Hung Dang[1,2(✉)]

[1] Aix Marseille Univ, CNRS, Centrale Marseille, I2M, Marseille, France
{stephane.ballet,alexis.bonnecaze}@univ-amu.fr
[2] Institut de Mathématiques de Marseille, UMR 7373, CNRS, Aix-Marseille Université, case 930, 13288 Marseille cedex 9, France
thanh-hung.dang@etu.univ-amu.fr

Abstract. We propose a new construction for the multiplication algorithm of D.V. and G.V. Chudnovsky in order to improve scalar algebraic complexity. In particular, we improve the Baum-Shokrollahi construction for multiplication in $\mathbb{F}_{256}/\mathbb{F}_4$ based on the elliptic Fermat curve $x^3 + y^3 = 1$.

Keywords: Finite field · Algebraic function field ·
Algebraic complexity

1 Introduction

We are interested by the multiplicative complexity of multiplication in a finite field $\mathbb{F}_{q^n}$, i.e. by the number of multiplications required to multiply in the $\mathbb{F}_q$-vector space $\mathbb{F}_{q^n}$ of dimension n. There exist two types of multiplications in $\mathbb{F}_q$: the scalar multiplication and the bilinear one. The scalar multiplication is the multiplication by a constant (in $\mathbb{F}_q$). The bilinear multiplication is a multiplication that depends on the elements of $\mathbb{F}_{q^n}$ that are multiplied. The bilinear complexity is independent of the chosen representation of the finite field.

Definition 1. The total number of scalar multiplications in $\mathbb{F}_q$ used in an algorithm $\mathcal{U}$ of multiplication in $\mathbb{F}_{q^n}$ is called scalar complexity and denoted $\mu_s(\mathcal{U})$.

More precisely, the multiplication of two elements of $\mathbb{F}_{q^n}$ is an $\mathbb{F}_q$-bilinear map from $\mathbb{F}_{q^n} \times \mathbb{F}_{q^n}$ onto $\mathbb{F}_{q^n}$. Then, it can be considered as an $\mathbb{F}_q$-linear map from the tensor product $\mathbb{F}_{q^n} \otimes_{\mathbb{F}_q} \mathbb{F}_{q^n}$ onto $\mathbb{F}_{q^n}$. Therefore, it can also be considered as an element T of $(\mathbb{F}_{q^n})^\star \otimes_{\mathbb{F}_q} (\mathbb{F}_{q^n})^\star \otimes_{\mathbb{F}_q} \mathbb{F}_{q^n}$, where $\mathbb{F}_{q^n}^\star$ denotes the dual of $\mathbb{F}_{q^n}$.

Set $T = \sum_{i=1}^{r} x_i^\star \otimes y_i^\star \otimes c_i$, where $x_i^\star \in \mathbb{F}_{q^n}^\star$, $y_i^\star \in \mathbb{F}_{q^n}^\star$ and $c_i \in \mathbb{F}_{q^n}$. The following holds for any $x, y \in \mathbb{F}_{q^n}$:

$$x \cdot y = T(x \otimes y) = \sum_{i=1}^{r} x_i^\star(x) y_i^\star(y) c_i.$$

© Springer Nature Switzerland AG 2019
M. Ćirić et al. (Eds.): CAI 2019, LNCS 11545, pp. 64–75, 2019.
https://doi.org/10.1007/978-3-030-21363-3_6

Definition 2. A bilinear multiplication algorithm $\mathcal{U}$ is an expression

$$x \cdot y = \sum_{i=1}^{r} x_i^\star(x) y_i^\star(y) c_i,$$

where $x_i^\star, y_i^\star \in (\mathbb{F}_{q^n})^\star$, and $c_i \in \mathbb{F}_{q^n}$. Such an algorithm is said symmetric if $x_i^\star = y_i^\star$ for all i. The number r of summands in this expression is called the bilinear (resp. symmetric bilinear) complexity of the algorithm $\mathcal{U}$ and is denoted by $\mu(\mathcal{U})$ (resp. $\mu^{sym}(\mathcal{U})$).

Definition 3. The minimal number of summands in a decomposition of the tensor T of the multiplication is called the bilinear (resp. symmetric bilinear) complexity of the multiplication and is denoted by $\mu_q(n)$ (resp. $\mu_q^{sym}(n)$):

$$\mu_q(n)(\text{resp. } \mu_q^{sym}(n)) = \min_{\mathcal{U}} \mu(\mathcal{U})(\text{resp. } \mu^{sym}(\mathcal{U}))$$

where $\mathcal{U}$ is running over all bilinear (resp. symmetric bilinear) multiplication algorithms in $\mathbb{F}_{q^n}$ over $\mathbb{F}_q$.

In their seminal papers, Winograd [11] and De Groote [7] have shown that $\mu_q(n) \geq 2n - 1$, with equality holding if and only if $n \leq \frac{1}{2}q + 1$. Winograd has also proved [11] that optimal multiplication algorithms realizing the lower bound belong to the class of interpolation algorithms. Later, generalizing interpolation algorithms on the projective line over $\mathbb{F}_q$ to algebraic curves of higher genus over $\mathbb{F}_q$, D.V. and G.V. Chudnovsky provided a method [6] which enabled to prove the *linearity* [2] of the bilinear complexity of multiplication in finite extensions of a finite field. This is the so-called Chudnovsky^2 algorithm (or CCMA). Note that the original algorithm CCMA is naturally symmetric.

Several studies focused on the qualitative improvement of CCMA but the problem of its scalar complexity was only addressed in 2015 by Atighehchi, Ballet, Bonnecaze and Rolland [1]. They proposed a new construction which slightly improved the scalar complexity eventhough the main objective of this work was not to optimize scalar complexity. Thus, in the absence of a dedicated strategy to scalar optimization, the number of scalar multiplications has not been significantly reduced in finite distance. Therefore, we note that so far, practical implementations of multiplication algorithms of type Chudnovsky over finite fields have failed to simultaneously optimize the number of scalar multiplications and bilinear multiplications.

Our main goal is to seek an optimal construction of Chudnovsky^2 algorithm in order to optimize its multiplicative complexity. We will consider the elliptic case for which it has been proven that the bilinear complexity of the algorithm is optimal [9]. Therefore, we will focus on optimizing the scalar complexity of this algorithm.

The paper is arranged as follows. Section 2, describes CCMA in the general case. Section 3 proposes a new method of construction with an objective to reduce the scalar complexity of Chudnovsky^2 multiplication algorithms. An optimized basis representation of the Riemann-Roch space $\mathcal{L}(2D)$ is sought in order to minimize the number of scalar multiplications in the algorithm. Considering the multiplication in $\mathbb{F}_{256}/\mathbb{F}_4$, which is the case study of Baum and Shokrollahi in [4], our strategy leads to improve the scalar complexity of their algorithm.

2   The Chudnovsky² Multiplication Algorithm

2.1   Description and Construction of CCMA Algorithm

Let $F/\mathbb{F}_q$ be an algebraic function field over the finite field $\mathbb{F}_q$ of genus $g(F)$. We denote by $N_k(F/\mathbb{F}_q)$ the number of places of degree k of F over $\mathbb{F}_q$. If D is a divisor, $\mathcal{L}(D)$ denotes the Riemann-Roch space associated to D. We denote by $\mathcal{O}_Q$ the valuation ring of the place Q and by F_Q its residue class field $\mathcal{O}_Q/Q$ which is isomorphic to $\mathbb{F}_{q^{\deg Q}}$ where $\deg Q$ is the degree of the place Q. The order of a divisor $D = \sum_P a_P P$ in the place P is the number a_P, denoted $ord_P(D)$. The support of a divisor D is the set $supp\, D$ of the places P such that $ord_P(D) \neq 0$. The divisor D is called effective if $ord_P(D) \geq 0$ for any P. Let us define the following Hadamard product in $\mathbb{F}_{q^{l_1}} \times \mathbb{F}_{q^{l_2}} \times \cdots \times \mathbb{F}_{q^{l_N}}$ denoted by $\odot$, where the l_i's denote positive integers, by $(u_1, \ldots, u_N) \odot (v_1, \ldots, v_N) = (u_1 v_1, \ldots, u_N v_N)$. The following theorem describes the original multiplication algorithm of D.V. and G.V. Chudnovsky [6].

Theorem 1. *Let*

- *n be a positive integer,*
- *$F/\mathbb{F}_q$ be an algebraic function field,*
- *Q be a degree n place of $F/\mathbb{F}_q$,*
- *D be a divisor of $F/\mathbb{F}_q$,*
- *$\mathcal{P} = \{P_1, \ldots, P_N\}$ be an ordered set of places of degree one of $F/\mathbb{F}_q$.*

We suppose that $supp\, D \cap \{Q, P_1, ..., P_N\} = \emptyset$ and that

(i) *The evaluation map*

$$Ev_Q : \mathcal{L}(D) \rightarrow F_Q$$
$$f \mapsto f(Q)$$

is surjective

(ii) *The evaluation map*

$$Ev_\mathcal{P} : \mathcal{L}(2D) \rightarrow \mathbb{F}_q^N$$
$$f \mapsto \left(f(P_1), \ldots, f(P_N)\right)$$

is injective

Then

(1) *For any two elements x, y in $\mathbb{F}_{q^n}$, we have:*

$$xy = E_Q \circ Ev_\mathcal{P}|_{ImEv_\mathcal{P}}^{-1} \left(E_\mathcal{P} \circ Ev_Q^{-1}(x) \odot E_\mathcal{P} \circ Ev_Q^{-1}(y) \right), \qquad (1)$$

where E_Q denotes the canonical projection from the valuation ring $\mathcal{O}_Q$ of the place Q in its residue class field F_Q, $E_\mathcal{P}$ the extension of $Ev_\mathcal{P}$ on the valuation ring $\mathcal{O}_Q$ of the place Q, $Ev_\mathcal{P}|_{ImEv_\mathcal{P}}^{-1}$ the restriction of the inverse map of $Ev_\mathcal{P}$ on its image, and $\circ$ the standard composition map.

(2)

$$\mu_q^{sym}(n) \leq N.$$

Since Q is a place of degree n, the residue class field F_Q of place Q is an extension of degree n of $\mathbb{F}_q$ and it therefore can be identified to $\mathbb{F}_{q^n}$. Moreover, the evaluation map Ev_Q being onto, one can associate the elements $x, y \in \mathbb{F}_{q^n}$ with elements of $\mathbb{F}_q$-vector space $\mathcal{L}(D)$, denoted respectively f and g. We define $h := fg$ by

$$(h(P_1), ..., h(P_N)) = E_\mathcal{P}(f) \odot E_\mathcal{P}(g) = (f(P_1)g(P_1), ..., f(P_N)g(P_N)). \qquad (2)$$

We know that such an element h belongs to $\mathcal{L}(2D)$ since the functions f, g lie in $\mathcal{L}(D)$. Moreover, thanks to injectivity of $Ev_\mathcal{P}$, the function h is in $\mathcal{L}(2D)$ and is uniquely determined by (2). We have

$$xy = Ev_Q(f)Ev_Q(g) = E_Q(h)$$

where E_Q is the canonical projection from the valuation ring $\mathcal{O}_Q$ of the place Q in its residue class field F_Q, Ev_Q is the restriction of E_Q over the vector space $\mathcal{L}(D)$.

In order to make the study and the construction of this algorithm easier, we proceed in the following way. We choose a place Q of degree n and a divisor D of degree $n + g - 1$, such that Ev_Q and $Ev_\mathcal{P}$ are isomorphisms. In this aim in [2], S. Ballet introduces simple numerical conditions on algebraic curves of an arbitrary genus g giving a sufficient condition for the application of the algorithm CCMA (existence of places of certain degree, of non-special divisors of degree $g - 1$) generalizing the result of A. Shokrollahi [9] for the elliptic curves. Let us recall this result:

Theorem 2. *Let q be a prime power and let n be an integer > 1. If there exists an algebraic function field $F/\mathbb{F}_q$ of genus g satisfying the conditions*

1. *$N_n > 0$ (which is always the case if $2g + 1 \leq q^{\frac{n-1}{2}}(q^{\frac{1}{2}} - 1)$),*
2. *$N_1 > 2n + 2g - 2$,*

then there exists a divisor D of degree $n + g - 1$ and a place Q such that:

(i) *The evaluation map*

$$Ev_Q : \mathcal{L}(D) \to \frac{\mathcal{O}_Q}{Q}$$
$$f \mapsto f(Q)$$

is an isomorphism of vector spaces over $\mathbb{F}_q$.

(ii) *There exist places $P_1,...,P_N$ such that the evaluation map*

$$Ev_{\mathcal{P}} : \mathcal{L}(2D) \to \mathbb{F}_q^N$$
$$f \mapsto \left(f(P_1),\ldots,f(P_N)\right)$$

is an isomorphism of vector spaces over $\mathbb{F}_q$ with $N = 2n + g - 1$.

Remark 1. First, note that in the elliptic case, the condition (2) is a large inequality thanks to a result due to Chaumine [5]. Secondly, note also that the divisor D is not necessarily effective.

By this last remark, it is important to add the property of effectivity for the divisor D in a perspective of implemention. Indeed, it is easier to construct the algorithm CCMA with this assumption because in this case $\mathcal{L}(D) \subseteq \mathcal{L}(2D)$ and we can directly apply the evaluation map $Ev_{\mathcal{P}}$ instead of $E_{\mathcal{P}}$ in the algorithm (1), by means of a suitable representation of $\mathcal{L}(2D)$. Moreover, in this case we need to consider simultaneously the assumption that the support of the divisor D does not contain the rational places and the place Q of degree n and the assumption of effectivity of the divisor D. Indeed, it is known that the support moving technic (cf. [8, Lemma 1.1.4.11]), which is a direct consequence of Strong Approximation Theorem (cf. [10, Proof of Theorem I.6.4]), applied on an effective divisor generates the loss of effectivity of the initial divisor (cf. also [1, Remark 2.2]). So, let us suppose these two last assumptions.

Remark 2. As in [3], in practice, we take as a divisor D one place of degree $n + g - 1$. It has the advantage to solve the problem of the support of divisor D (cf. also [1, Remark 2.2]) as well as the problem of the effectivity of the divisor D. However, it is not required to be considered in the theoretical study, but, as we will see, it will have some importance in the strategy of optimization.

We can therefore consider the basis $\mathcal{B}_Q$ of the residue class field F_Q over $\mathbb{F}_q$ as the image of a basis of $\mathcal{L}(D)$ by Ev_Q or equivalently (which is sometimes useful following the considered situation) the basis of $\mathcal{L}(D)$ as the reciprocal image of a basis of the residue class field F_Q over $\mathbb{F}_q$ by Ev_Q^{-1}. Let

$$\mathcal{B}_D := (f_1, ..., f_n) \tag{3}$$

be a basis of $\mathcal{L}(D)$ and let us denote the basis of the supplementary space $\mathcal{M}$ of $\mathcal{L}(D)$ in $\mathcal{L}(2D)$ by

$$\mathcal{B}_D^c := (f_{n+1}, ..., f_N) \tag{4}$$

where $N := dim\mathcal{L}(2D) = 2n + g - 1$. Then, we choose

$$\mathcal{B}_{2D} := \mathcal{B}_D \cup \mathcal{B}_D^c \tag{5}$$

as the basis of $\mathcal{L}(2D)$.

We denote by T_{2D} the matrix of the isomorphism $Ev_{\mathcal{P}} : \mathcal{L}(2D) \to \mathbb{F}_q^N$ in the basis $\mathcal{B}_{2D}$ of $\mathcal{L}(2D)$ (the basis of $\mathbb{F}_q^N$ will always be the canonical basis). Then, we denote by T_D the matrix of the first n columns of the matrix T_{2D}.

Therefore, T_D is the matrix of the restriction of the evaluation map $Ev_{\mathcal{P}}$ on the Riemann-Roch vector space $\mathcal{L}(D)$, which is an injective morphism.

Note that the canonical surjection E_Q is the extension of the isomorphism Ev_Q since, as $Q \notin supp(D)$, we have $\mathcal{L}(D) \subseteq \mathcal{O}_Q$. Moreover, as $supp(2D) = supp(D)$, we also have $\mathcal{L}(2D) \subseteq \mathcal{O}_Q$. We can therefore consider the images of elements of the basis $\mathcal{B}_{2D}$ by E_Q and obtain a system of N linear equations as follows:

$$E_Q(f_r) = \sum_{m=1}^{n} c_r^m Ev_Q(f_i), \quad r = 1, ..., N$$

where E_Q denotes the canonical projection from the valuation ring $\mathcal{O}_Q$ of the place Q in its residue class field F_Q, Ev_Q is the restriction of E_Q over the vector space $\mathcal{L}(D)$ and $c_r^m \in \mathbb{F}_q$ for $r = 1, ..., N$. Let C be the matrix of the restriction of the map E_Q on the Riemann-Roch vector space $\mathcal{L}(2D)$, from the basis $\mathcal{B}_{2D}$ in the basis $\mathcal{B}_Q$. We obtain the product $z := xy$ of two elements $x, y \in \mathbb{F}_{q^n}$ by the algorithm (1) in Theorem 1, where M^t denotes the transposed matrix of the matrix M:

Algorithm 1. Multiplication algorithm in $\mathbb{F}_{q^n}$

Require: $x = \sum_{i=1}^{n} x_i Ev_Q(f_i)$, and $y = \sum_{i=1}^{n} y_i Ev_Q(f_i)$ // $x_i, y_i \in \mathbb{F}_q$

 1. $X := (X_1, ..., X_N) \leftarrow (x_1, ..., x_n) T_D^t$
    $Y := (Y_1, ..., Y_N) \leftarrow (y_1, ..., y_n) T_D^t$
 2. $Z := X \odot Y = (Z_1, ..., Z_N) \leftarrow (X_1 Y_1, ..., X_N Y_N)$
 3. $(z_1, ..., z_n) \leftarrow (Z_1, ..., Z_N)(T_{2D}^t)^{-1} C^t$.

Ensure: $z = xy = \sum_{i=1}^{n} z_i Ev_Q(f_i)$ // $z := xy$

Now, we present an initial setup algorithm which is only done once.

Algorithm 2. Setup algorithm

Require: $F/\mathbb{F}_q$, Q, D, $\mathcal{P} = \{P_1, ..., P_{2n+g-1}\}$.
Ensure: $\mathcal{B}_{2D}$, T_{2D} and CT_{2D}^{-1}.
  (i) Check the function field $F/\mathbb{F}_q$, the place Q, the divisors D are such that Conditions (i) and (ii) in Theorem 2 can be satisfied.
 (ii) Represent $\mathbb{F}_{q^n}$ as the residue class field of the place Q.
(iii) Construct a basis $\mathcal{B}_{2D} := (f_1, ..., f_n, f_{n+1}, ..., f_{2n+g-1})$ of $\mathcal{L}(2D)$, where $\mathcal{B}_D := (f_1, ..., f_n)$ is a basis of $\mathcal{L}(D)$, and $\mathcal{B}_D^c := (f_{n+1}, ..., f_{2n+g-1})$ a basis of the supplementary space $\mathcal{M}$ of $\mathcal{L}(D)$ in $\mathcal{L}(2D)$.
(iv) Compute the matrices T_{2D}, C and CT_{2D}^{-1}.

2.2 Complexity Analysis

The total complexity, in terms of number of multiplications in $\mathbb{F}_q$, is equal to $(3n + 1)(2n + g - 1)$, including $3n(2n + g - 1)$ scalar multiplications. Recall that the bilinear complexity of Chudnovsky[2] algorithms of type (1) in Theorem 1 satisfying assumptions of Theorem 2 is optimized. Therefore, we only focus on optimizing the scalar complexity of the algorithm. From Algorithm (1), we observe that the number of the scalar multiplications, denoted by N_s, depends directly on the number of zeros in the matrices T_D and $C.T_{2D}^{-1}$, respectively denoted by $N_{zero}(T_D)$ and $N_{zero}(C.T_{2D}^{-1})$. Indeed, all the involved matrices being constructed once, the multiplication by a coefficient zero in a matrix has not to be taken into account. Thus, we get the formula to compute the number of scalar multiplications of this algorithm with respect to the number of zeros of the involved matrices as follows:

$$N_s = 2\left(n(2n + g - 1) - N_{zero}(T_D) \right) + \left(n(2n + g - 1) - N_{zero}(C.T_{2D}^{-1}) \right) \quad (6)$$
$$= 3n(2n + g - 1) - N_{zero},$$

where

$$N_{zero} = 2N_{zero}(T_D) + N_{zero}(C.T_{2D}^{-1}). \quad (7)$$

3   Optimization of the Scalar Complexity

By Sect. 2.2, reducing the number of operations means finding an algebraic function field $F/\mathbb{F}_q$ having a genus g as small as possible and a suitable set of divisors and place $(D, Q, \mathcal{P})$ with a good representation of the associated Riemann-Roch spaces, namely such that the matrices T_D, T_{2D} and $C.T_{2D}^{-1}$ are as hollow as possible. Therefore, for a place Q and a suitable divisor D, we seek the best possible representations of Riemann-Roch spaces $\mathcal{L}(D)$ and $\mathcal{L}(2D)$ to maximize both parameters $N_{zero}(T_D)$ and $N_{zero}(C.T_{2D}^{-1})$.

3.1   Different Types of Strategy

With Fixed Divisor and Places. In this section, we consider the optimization for a fixed suitable set of divisor and places $(D, Q, \mathcal{P})$ for a given algebraic function field $F/\mathbb{F}_q$ of genus g. So, let us give the following definition:

Definition 4. We call $\mathcal{U}_{D,Q,\mathcal{P}}^{F,n} := (\mathcal{U}_{D,Q,\mathcal{P}}^A, \mathcal{U}_{D,Q,\mathcal{P}}^R)$ a Chudnovsky[2] multiplication algorithm of type (1) where $\mathcal{U}_{D,Q,\mathcal{P}}^A := E_\mathcal{P} \circ Ev_Q^{-1}$ and $\mathcal{U}_{D,Q,\mathcal{P}}^R := E_Q \circ Ev_\mathcal{P}|_{ImEv_\mathcal{P}}^{-1}$, satisfying the assumptions of Theorem 1. We will say that two algorithms are equal, and we will note: $\mathcal{U}_{D,Q,\mathcal{P}}^{F,n} = \mathcal{U}_{D',Q',\mathcal{P}'}^{F,n}$, if $\mathcal{U}_{D,Q,\mathcal{P}}^A = \mathcal{U}_{D',Q',\mathcal{P}'}^A$ and $\mathcal{U}_{D,Q,\mathcal{P}}^R = \mathcal{U}_{D',Q',\mathcal{P}'}^R$.

Note that in this case, this definition makes sense only if the bases of implied vector-spaces are fixed. So, we denote respectively by $\mathcal{B}_Q$, $\mathcal{B}_D$, and $\mathcal{B}_{2D}$ the basis of the residue class field F_Q, and of Riemann-Roch vector-spaces $\mathcal{L}(D)$, and $\mathcal{L}(2D)$ associated to $\mathcal{U}_{D,Q,\mathcal{P}}^{F,n}$. Note that the basis of the $\mathbb{F}_q$-vector space $\mathbb{F}_q^N$ is always the canonical basis. Then, we obtain the following result:

Proposition 1. *Let us consider an algorithm $\mathcal{U}_{D,Q,\mathcal{P}}^{F,n}$ such that the divisor D is an effective divisor, $D-Q$ a non-special divisor of degree $g-1$, and such that the cardinal of the set $\mathcal{P}$ is equal to the dimension of the Riemann-Roch space $\mathcal{L}(2D)$. Then we can choose the basis $\mathcal{B}_{2D}$ as (5) and for any σ in $GL_{\mathbb{F}_q}(2n+g-1)$, where $GL_{\mathbb{F}_q}(2n+g-1)$ denotes the linear group, we have*

$$\mathcal{U}_{\sigma(D),Q,\mathcal{P}}^{F,n} = \mathcal{U}_{D,Q,\mathcal{P}}^{F,n},$$

where $\sigma(D)$ denotes the action of σ on the basis $\mathcal{B}_{2D}$ of $\mathcal{L}(2D)$ in $\mathcal{U}_{D,Q,\mathcal{P}}^{F,n}$, with a fixed basis $\mathcal{B}_Q$ of the residue class field of the place Q and $\mathcal{B}_c$ the canonical basis of $\mathbb{F}_q^{2n+g-1}$. In particular, the quantity $N_{zero}(C.T_{2D}^{-1})$ is constant under this action.

Proof. Let E, F and H be three vector spaces of finite dimension on a field K respectively equipped with the basis $\mathcal{B}_E$, $\mathcal{B}_F$ and $\mathcal{B}_H$. Consider two morphisms f and h respectively defined from E into F and from F into H and consider respectively their associated matrix $M_f(\mathcal{B}_E, \mathcal{B}_F)$ and $M_h(\mathcal{B}_F, \mathcal{B}_H)$. Then it is obvious that the matrix $M_{h \circ f}(\mathcal{B}_E, \mathcal{B}_H)$ of the morphism $h \circ f$ is independant from the choice of the basis $\mathcal{B}_F$ of F. As the divisor D is effective, we have $\mathcal{L}(D) \subset \mathcal{L}(2D)$ and then $\mathcal{U}_{D,Q,\mathcal{P}}^{A} := E_{\mathcal{P}} \circ Ev_Q^{-1} = Ev_{\mathcal{P}} \circ Ev_Q^{-1}$ and as $D-Q$ a non-special divisor of degree $g-1$, Ev_Q is an isomorphism from $\mathcal{L}(D)$ into F_Q and we have $\mathcal{U}_{D,Q,\mathcal{P}}^{A} = Ev_{\mathcal{P}}|_{\mathcal{L}(D)} \circ Ev_Q^{-1}$. Moreover, as the cardinal of the set $\mathcal{P}$ is equal to the dimension of the Riemann-Roch space $\mathcal{L}(2D)$, $Ev_{\mathcal{P}}$ is an isomorphism from $\mathcal{L}(2D)$ into $\mathbb{F}_q^{2n+g-1}$ equipped with the canonical basis $\mathcal{B}_c$. Thus, $\mathcal{U}_{D,Q,\mathcal{P}}^{R} := E_Q \circ Ev_{\mathcal{P}}^{-1}|_{ImEv_{\mathcal{P}}} = E_Q|_{\mathcal{L}(2D)} \circ Ev_{\mathcal{P}}^{-1}$. Then, the matrix of $\mathcal{U}_{D,Q,\mathcal{P}}^{A}$ (resp. $\mathcal{U}_{D,Q,\mathcal{P}}^{R}$) is invariant under the action of σ in $GL_{\mathbb{F}_q}(n)$ (resp. in $GL_{\mathbb{F}_q}(2n+g-1)$) on the basis $\mathcal{B}_D$ (resp. $\mathcal{B}_{2D}$) since the set (E, F, H) is equal to $(F_Q, \mathcal{L}(D), \mathcal{B}_c)$ (resp. $(\mathbb{F}_q^{2n+g-1}, \mathcal{L}(2D), \mathcal{B}_Q)$) for $h \circ f := Ev_{\mathcal{P}}|_{\mathcal{L}(D)} \circ Ev_Q^{-1}$ (resp. $E_Q|_{\mathcal{L}(2D)} \circ Ev_{\mathcal{P}}^{-1}$). $\qquad\square$

Remark 3. Note that a priori for any permutation τ of the set $\mathcal{P}$, we have $\mathcal{U}_{\sigma(D),Q,\tau(\mathcal{P})}^{F,n}$ different from $\mathcal{U}_{D,Q,\mathcal{P}}^{F,n}$, where $\sigma(D)$ denotes the action of σ on the basis $\mathcal{B}_{2D}$ of $\mathcal{L}(2D)$ in $\mathcal{U}_{D,Q,\mathcal{P}}^{F,n}$, with a fixed basis $\mathcal{B}_Q$ of the residue class field of the place Q. Indeed, the action of τ corresponds to a permutation of the canonical basis $\mathcal{B}_c$ of $\mathbb{F}_q^{2n+g-1}$. It corresponds to a permutation of the lines of the matrix T_{2D}. In this case, $N_{zero}(T_{2D})$ is obviously constant under the action of τ but nothing enables us to claim that $N_{zero}(C.T_{2D}^{-1})$ is constant.

Proposition 2. *Let $\mathcal{U}_{D,Q,\mathcal{P}}^{F,n}$ be a Chudnovsky^2 multiplication algorithm in a finite field $\mathbb{F}_{q^n}$, satisfying the assumptions of Proposition 1. The optimal scalar complexity $\mu_{s,o}(\mathcal{U}_{D,Q,\mathcal{P}}^{F,n})$ of $\mathcal{U}_{D,Q,\mathcal{P}}^{F,n}$ is reached for the set $\{\mathcal{B}_{D,max}, \mathcal{B}_Q\}$ such that $\mathcal{B}_{D,max}$ is the basis of $\mathcal{L}(D)$ satisfying*

$$N_{zero}(T_{D,max}) = \max_{\sigma \in GL_{\mathbb{F}_q}(n)} N_{zero}(T_{\sigma(D)}),$$

where $\sigma(D)$ denotes the action of σ on the basis $\mathcal{B}_D$ of $\mathcal{L}(D)$ in $\mathcal{U}_{D,Q,\mathcal{P}}^{F,n}$, $T_{D,max}$ the matrix of the restriction of the evaluation map $Ev_{\mathcal{P}}$ on the Riemann-Roch vector space $\mathcal{L}(D)$ equipped with the bases $\mathcal{B}_{D,max}$ and $\mathcal{B}_Q = Ev_Q(\mathcal{B}_{D,max})$.

In particular,

$$\mu_{s,o}(\mathcal{U}_{D,Q,\mathcal{P}}^{F,n}) = \min_{\sigma \in GL_{\mathbb{F}_q}(n)} \{\mu_s(\mathcal{U}_{\sigma(D),Q,\mathcal{P}}^{F,n}) \mid \sigma(\mathcal{B}_D) \text{ is the basis of } \mathcal{L}(D)$$

$$\text{and } \mathcal{B}_Q = Ev_Q(\mathcal{B}_D)\})$$

$$= 3n(2n + g - 1) - (2N_{zero}(T_{D,max}) + N_{zero}(T_{2D,n}^{-1})),$$

where matrices C and T_{2D} are defined with respect to the basis $\mathcal{B}_Q = Ev_Q(\mathcal{B}_{D,max})$, and $\mathcal{B}_{2D} = \mathcal{B}_{D,max} \cup \mathcal{B}_D^c$ with $\mathcal{B}_D^c$ a basis of the kernel of $E_Q|_{\mathcal{L}(2D)}$, and $T_{2D,n}^{-1}$ denotes the matrix constituted of the n first lines of the matrix T_{2D}^{-1}.

Proof. It follows directly from Proposition 1 and formulae (6) and (7). Note that since the quantity $N_{zero}(C.T_{2D}^{-1})$ is constant for any basis $\mathcal{B}_{2D}$ of $\mathcal{L}(2D)$, we can take the matrix $C.T_{2D}^{-1} = T_{2D,n}^{-1}$ if $\mathcal{B}_D^c$ is a basis of the kernel of $E_Q|_{\mathcal{L}(2D)}$.   □

Other Strategies of Optimization. In the view of a complete optimization (with respect to scalar complexity i.e. with fixed bilinear complexity) of the multiplication in a finite field $\mathbb{F}_{q^n}$ by a Chudnovsky[2] type multiplication algorithm, we have to vary the eligible sets $(F, D, Q, \mathcal{P})$. As an example, for a fixed integer n, a given algebraic function field $F/\mathbb{F}_q$, and a couple divisor and place (D, Q) satisfying the conditions of Proposition 1, we must apply the optimization strategy studied in Sect. 3.1 on each suitable ordered subset $\mathcal{P}$ (of cardinal $2n + g - 1$) of the set of rational places (i.e. each suitable subset $\mathcal{P}$ and all their associated permutations $\tau(\mathcal{P})$). Then we have to vary the couples (D, Q) and apply the previous step: for example, we can start by fixing the place Q and then vary the suitable divisors D. We can then look for a fixed suitable algebraic function field of genus g, up to isomorphism, and repeat all the previous steps. Finally, it is still possible to look at the trade-off between scalar complexity and bilinear complexity by increasing the genus and then re-conducting all the previous optimizations.

3.2   Optimization of Scalar Complexity in the Elliptic Case

Now, we study a specialisation of the Chudnovsky[2] multiplication algorithm of type (1) in the case of the elliptic curves. In particular, we improve the effective algorithm constructed in the article of U. Baum and M.A. Shokrollahi [4] which presented an optimal algorithm from the point of view of the bilinear complexity in the case of the multiplication in $\mathbb{F}_{256}/\mathbb{F}_4$ based on Chudnovsky[2] multiplication algorithm applied on the Fermat curve $x^3 + y^3 = 1$ defined over $\mathbb{F}_4$. Our method of construction leads to a multiplication algorithm in $\mathbb{F}_{256}/\mathbb{F}_4$ having a lower scalar complexity with an optimal bilinear complexity.

Experiment of Baum-Shokrollahi. The article [4] presents Chudnovsky^2 multiplication in $\mathbb{F}_{4^4}$, for the case $q = 4$ and $n = 4$. The elements of $\mathbb{F}_4$ are denoted by $0, 1, \omega$ and ω^2. The algorithm construction requires the use of an elliptic curve over $\mathbb{F}_4$ with at least 9 $\mathbb{F}_4$-rational points (which is the maximum possible number by Hasse-Weil Bound). Note that in this case, Conditions 1) and 2) of Theorem 2 are well satisfied. It is well known that the Fermat curve $u^3 + v^3 = 1$ satisfies this condition. By the substitutions $x = 1/(u + v)$ and $y = u/(u + v)$, we get the isomorphic curve $y^2 + y = x^3 + 1$. From now on, $F/\mathbb{F}_q$ denotes the algebraic function field associated to the elliptic curve $\mathcal{C}$ with plane model $y^2 + y = x^3 + 1$, of genus one. The projective coordinates $(x : y : z)$ of $\mathbb{F}_4$-rational points of this elliptic curve are:

$$P_\infty = (0 : 1 : 0), P_1 = (0 : \omega : 1), P_2 = (0 : \omega^2 : 1), P_3 = (1 : 0 : 1),$$
$$P_4 = (1 : 1 : 1), P_5 = (\omega : 0 : 1), P_6 = (\omega : 1 : 1), P_7 = (\omega^2 : 0 : 1), P_8 = (\omega^2 : 1 : 1).$$

Now, we represent $\mathbb{F}_{256}$ as $\mathbb{F}_4[x]/\mathcal{Q}(x)$ with primitive root α, where $\mathcal{Q}(x) = x^4 + x^3 + \omega x^2 + \omega x + \omega$.

- For the place Q of degree 4, the authors considered $Q = \sum_{i=1}^{4} \mathfrak{p}_i$ where $\mathfrak{p}_1$ corresponds to the $\mathbb{F}_{4^4}$-rational point with projective coordinates $(\alpha^{16} : \alpha^{174} : 1)$ and $\mathfrak{p}_2, \mathfrak{p}_3, \mathfrak{p}_4$ are its conjugates under the Frobenius map. We see that α^{16} is a root of the irreducible polynomial $\mathcal{Q}(x) = x^4 + x^3 + \omega x^2 + \omega x + \omega$. Thus, the place Q is a place lying over the place $(\mathcal{Q}(x))$ of $\mathbb{F}_4(x)/\mathbb{F}_4$. Note also that the place $((\mathcal{Q}(x))$ of $\mathbb{F}_4(x)/\mathbb{F}_4$ is totally splitted in the algebraic function field $F/\mathbb{F}_4$, which means that there exist two places of degree n in $F/\mathbb{F}_4$ lying over the place $(\mathcal{Q}(x))$ of $\mathbb{F}_4(x)/\mathbb{F}_4$, since the function field $F/\mathbb{F}_q$ is an extension of degree 2 of the rational function field $\mathbb{F}_4(x)/\mathbb{F}_q$. The place Q is one of the two places in $F/\mathbb{F}_4$ lying over the place $(\mathcal{Q}(x))$. Notice that the second place is given by the orbit of the conjugated point $(\alpha^{16} : \alpha^{174} + 1 : 1)$. Therefore, we can represent $\mathbb{F}_{256} = \mathbb{F}_{4^4} = \mathbb{F}_4[x]/\mathcal{Q}(x)$ as the residue class field F_Q of the place Q in $F/\mathbb{F}_4$.
- For the divisor D, we choose the place described as $\sum_{i=1}^{4} \mathfrak{d}_i$ where $\mathfrak{d}_1$ corresponds to the $\mathbb{F}_{4^4}$-rational point $(\alpha^{17} : \alpha^{14} : 1)$ and $\mathfrak{d}_2, \mathfrak{d}_3, \mathfrak{d}_4$ are its conjugates under the Frobenius map. By computation we see that α^{17} is a root of irreducible polynomial $\mathcal{D}(x) = x^2 + x + \omega$ and $\deg D = 4$ because $\mathfrak{d}_1, \mathfrak{d}_2, \mathfrak{d}_3, \mathfrak{d}_4$ are all distinct. Therefore, D is the only place in $F/\mathbb{F}_4$ lying over the place $(\mathcal{D}(x))$ of $\mathbb{F}_4(x)$ since the residue class field F_D of the place D is a quadratic extension of the residue class field $F_{\mathcal{D}}$ of the place $\mathcal{D}$, which is an inert place of $\mathbb{F}_4(x)$ in $F/\mathbb{F}_4$.

The matrix T_{2D} obtained in the basis of Riemann-Roch space $L(2D)$:
$\mathcal{B}_{2D} = \{f_1 = 1/f, f_2 = x/f, f_3 = y/f, f_4 = x^2/f, f_5 = 1/f^2, f_6 = xy/f^2, f_7 = y/f^2, f_8 = x/f^2\}$, with $f = x^2 + x + \omega$ is the following:

$$T_{2D} = \begin{pmatrix} 0 & 0 & 0 & 1 & 0 & 0 & 0 & 0 \\ \omega^2 & 0 & 1 & 0 & \omega & 0 & \omega^2 & 0 \\ \omega^2 & 0 & \omega & 0 & \omega & 0 & 1 & 0 \\ \omega^2 & \omega^2 & 0 & \omega^2 & \omega & 0 & 0 & \omega \\ \omega^2 & \omega^2 & \omega^2 & \omega^2 & \omega & \omega & \omega & \omega \\ \omega & \omega^2 & 0 & 1 & \omega^2 & 0 & 0 & 1 \\ \omega & \omega^2 & \omega & 1 & \omega^2 & 1 & \omega^2 & 1 \\ \omega & 1 & 0 & \omega^2 & \omega^2 & 0 & 0 & \omega \end{pmatrix}.$$

Then, computation gives:

$$C = \begin{pmatrix} 1 & 0 & 0 & 0 & \omega & 0 & \omega^2 & \omega \\ 0 & 1 & 0 & 0 & 0 & \omega^2 & \omega & 0 \\ 0 & 0 & 1 & 0 & 1 & 0 & 0 & 1 \\ 0 & 0 & 0 & 1 & 1 & \omega & 0 & \omega \end{pmatrix} \quad \text{and} \quad CT_{2D}^{-1} = \begin{pmatrix} 1 & \omega & 1 & \omega & 1 & 1 & \omega & 0 \\ 1 & 0 & \omega^2 & \omega & 1 & \omega^2 & 1 & \omega \\ 1 & \omega & \omega & \omega^2 & 1 & \omega^2 & \omega & \omega \\ 0 & \omega & \omega^2 & \omega & 1 & \omega^2 & 0 & 0 \end{pmatrix}.$$

Consequently, we obtain:

$$N_{zero}(T_D) = 10, \quad N_{zero}(CT_{2D}^{-1}) = 5.$$

Thus, the total number N_s of scalar multiplications in the algorithm constructed by Baum and Shokrollahi in [4] is $N_s = 71$ by the formula (6). In the next section, we follow the approach described in Sect. 3, and we improve the Chudnovsky^2 multiplication algorithm in $\mathbb{F}_{4^4}$ constructed by Baum and Shokrollahi in [4]. By using the same elliptic curve and the same set $\{D, Q, \mathcal{P}\}$, we obtain an algorithm with the same bilinear complexity and lower scalar complexity.

New Design of the Baum-Shokrollahi Construction. The new construction of Chudnovsky^2 algorithm for the multiplication in $\mathbb{F}_{256}/\mathbb{F}_4$ using strategy given in Proposition 2 of Sect. 3.1 gives the following matrices T_{2D} with a better basis $\mathcal{B}_{2D} = (f_1, f_2, ..., f_8)$ of $\mathcal{L}(2D)$ space, where

$f_1 = (\omega x^2 + x)/(x^2 + x + \omega),$

$f_2 = (\omega^2 x^2 + \omega^2 x + \omega^2)/(x^2 + x + \omega),$

$f_3 = \omega^2/(x^2 + x + \omega)c + (\omega^2 x + 1)/(x^2 + x + \omega),$

$f_4 = \omega^2/(x^2 + x + \omega)c + (\omega^2 x + \omega)/(x^2 + x + \omega),$

$f_5 = (x^2 + x)/(x^4 + x^2 + \omega^2)c + (x^4 + \omega x^3 + \omega x^2 + \omega x)/(x^4 + x^2 + \omega^2),$

$f_6 = \omega^2 x/(x^4 + x^2 + \omega^2)c + (\omega x^4 + x^2 + \omega x + 1)/(x^4 + x^2 + \omega^2),$

$f_7 = (\omega^2 x + 1)/(x^4 + x^2 + \omega^2)c + (\omega^2 x^4 + \omega^2 x^3 + \omega x^2 + \omega)/(x^4 + x^2 + \omega^2),$

$f_8 = (x^2 + \omega x + 1)/(x^4 + x^2 + \omega^2)c + (x^4 + \omega x^3 + x^2 + \omega^2 x + \omega^2)/(x^4 + x^2 + \omega^2).$

$$T_{2D} = \begin{pmatrix} \omega & \omega^2 & 0 & 0 & 1 & \omega & \omega^2 & 1 \\ 0 & \omega & 0 & \omega & 0 & \omega & 0 & \omega \\ 0 & \omega & \omega & 0 & 0 & \omega & \omega & 0 \\ 1 & 0 & 0 & 1 & 1 & 1 & \omega^2 & \omega^2 \\ 1 & 0 & 1 & 0 & \omega & \omega & \omega^2 & 0 \\ 0 & 0 & 1 & 0 & \omega & \omega & 0 & 1 \\ 0 & 0 & 0 & 1 & 1 & \omega^2 & \omega & 0 \\ \omega & \omega & 1 & \omega^2 & 1 & 0 & 0 & \omega^2 \end{pmatrix} \text{ and } T_{2D,4}^{-1} = \begin{pmatrix} 0 & \omega & 1 & 0 & 0 & 1 & 1 & \omega^2 \\ 0 & 0 & 0 & 0 & 1 & \omega & \omega & \omega^2 \\ \omega^2 & \omega & \omega^2 & \omega^2 & \omega & \omega & 0 & 0 \\ 1 & \omega^2 & \omega & \omega^2 & 0 & 0 & 1 & \omega^2 \end{pmatrix}$$

Therefore, $N_{zero}(T_D) = 16$ and $N_{zero}(T_{2D,4}^{-1}) = 11$. By the formula (6), we obtain $N_s = 53$, a gain of 25% over Baum and Shokrollahi's method.

References

1. Atighehchi, K., Ballet, S., Bonnecaze, A., Rolland, R.: Arithmetic in finite fields based on Chudnovsky's multiplication algorithm. Math. Comput. **86**(308), 297–3000 (2017)
2. Ballet, S.: Curves with many points and multiplication complexity in any extension of $\mathbb{F}_q$. Finite Fields Their Appl. **5**, 364–377 (1999)
3. Ballet, S.: Quasi-optimal algorithms for multiplication in the extensions of $\mathbb{F}_{16}$ of degree 13, 14, and 15. J. Pure Appl. Algebra **171**, 149–164 (2002)
4. Baum, U., Shokrollahi, A.: An optimal algorithm for multiplication in $\mathbb{F}_{256}/\mathbb{F}_4$. Appl. Algebra Eng. Commun. Comput. **2**(1), 15–20 (1991)
5. Chaumine, J.: On the bilinear complexity of multiplication in small finite fields. Comptes Rendus de l'Académie des Sciences, Série **I**(343), 265–266 (2006)
6. Chudnovsky, D., Chudnovsky, G.: Algebraic complexities and algebraic curves over finite fields. J. Complex. **4**, 285–316 (1988)
7. De Groote, H.: Characterization of division algebras of minimal rank and the structure of their algorithm varieties. SIAM J. Comput. **12**(1), 101–117 (1983)
8. Pieltant, J.: Tours de corps de fonctions algébriques et rang de tenseur de la multiplication dans les corps finis. Ph.D. thesis, Université d'Aix-Marseille, Institut de Mathématiques de Luminy (2012)
9. Shokhrollahi, A.: Optimal algorithms for multiplication in certain finite fields using algebraic curves. SIAM J. Comput. **21**(6), 1193–1198 (1992)
10. Stichtenoth, H.: Algebraic Function Fields and Codes. Lectures Notes in Mathematics, vol. 314. Springer, Heidelberg (1993)
11. Winograd, S.: On multiplication in algebraic extension fields. Theor. Comput. Sci. **8**, 359–377 (1979)

Maximal Diameter on a Class
of Circulant Graphs

Milan Bašić[1]([✉]), Aleksandar Ilić[2], and Aleksandar Stamenković[1]

[1] Faculty of Sciences and Mathematics, University of Niš, Niš, Serbia
basic_milan@yahoo.com, aca@pmf.ni.ac.rs
[2] Facebook Inc., Menlo Park, California, USA
aleksandari@gmail.com

Abstract. Integral circulant graphs are proposed as models for quantum spin networks. Specifically, it is important to know how far information can potentially be transferred between nodes of the quantum networks modeled by integral circulant graphs and this task is related to calculating the maximal diameter of a graph. The integral circulant graph $\mathrm{ICG}_n(D)$ has the vertex set $Z_n = \{0, 1, 2, \ldots, n-1\}$ and vertices a and b are adjacent if $\gcd(a-b, n) \in D$, where $D \subseteq \{d : d \mid n, \ 1 \le d < n\}$. Motivated by the result on the upper bound of the diameter of $\mathrm{ICG}_n(D)$ given in [N. Saxena, S. Severini, I. Shparlinski, *Parameters of integral circulant graphs and periodic quantum dynamics*, International Journal of Quantum Information 5 (2007), 417–430], which is equal to $2|D|+1$, in this paper we prove that the maximal value of the diameter of the integral circulant graph $\mathrm{ICG}_n(D)$ of a given order n with its prime factorization $p_1^{\alpha_1} \cdots p_k^{\alpha_k}$ and $|D| = k$, is equal to $k + |\{i \mid \alpha_i > 1, \ 1 \le i \le k\}|$. This way we further improve the upper bound of Saxena, Severini and Shparlinski. Moreover, we characterize all such extremal graphs. We also show that the upper bound is attainable for integral circulant graphs $\mathrm{ICG}_n(D)$ such that $|D| \le k$.

Keywords: Integral circulant graphs · Diameter ·
Chinese remainder theorem · Quantum networks

1 Introduction

Circulant graphs are Cayley graphs over a cyclic group. A graph is called integral if all the eigenvalues of its adjacency matrix are integers. In other words, the corresponding adjacency matrix of a circulant graph is the circulant matrix (a special kind of a Toeplitz matrix where each row vector is rotated one element to the right relative to the preceding row vector). Integral graphs are extensively studied in the literature and there has been a vast research on some types of classes of graphs with integral spectrum. The interest for circulant graphs in

This work was supported by Research Grant 174013 of Serbian Ministry of Science and Technological Development.

M. Ćirić et al. (Eds.): CAI 2019, LNCS 11545, pp. 76–87, 2019.
https://doi.org/10.1007/978-3-030-21363-3_7

graph theory and applications has grown during the last two decades. They appear in coding theory, telecommunication network, VLSI design, parallel and distributed computing (see [8] and references therein).

Since they possess many interesting properties (such as vertex transitivity called mirror symmetry), circulants are applied in quantum information transmission and proposed as models for quantum spin networks that permit the quantum phenomenon called perfect state transfer [1, 2, 15]. In the quantum communication scenario, the important feature of this kind of quantum graphs (especially those with integral spectrum) is the ability of faithfully transferring quantum states without modifying the network topology. Integral circulant graphs have found important applications in molecular chemistry for modeling energy-like quantities such as the heat of formation of a hydrocarbon [9, 11, 14, 16]. Recently there has been a vast research on the interconnection schemes based on circulant topology – circulant graphs represent an important class of interconnection networks in parallel and distributed computing (see [8]). Recursive circulants are proposed as an interconnection structure for multicomputer networks [13]. While retaining the attractive properties of hypercubes such as node-symmetry, recursive structure, connectivity etc., these graphs achieve noticeable improvements in diameter.

Various properties of integral circulant graphs were recently investigated especially having in mind the study of certain parameters useful for modeling a good quantum (or in general complex) network that allows periodic dynamics. It was observed that integral circulant graphs represent very reliable networks, meaning that the vertex connectivity of these graphs is equal to the degree of their regularity. Moreover, for even orders they are bipartite [15] – note that many of the proposed networks mainly derived from the hypercube structure by twisting some pairs of edges (twisted cube, crossed cub, multiply twisted cube, Möbius cube, generalized twisted cube) are nonbipartite. Other important network metrics of integral circulant graphs are analyzed as well, such as the degree, chromatic number, the clique number, the size of the automorphism group, the size of the longest cycle, the number of induced cycles of prescribed order ([3–6, 10, 12]).

In this paper we continue the study of properties of circulant networks relevant for the purposes of information transfer. Specifically, it would be interesting to know how far information can potentially be transferred between nodes of the networks modeled by the graph. So, it is important to know the maximum length of all shortest paths between any pair of nodes. Moreover, for a fixed number of nodes in the network, a larger diameter potentially implies a larger maximum distance between nodes for which (perfect) transfer is possible (communication distance). Therefore, for a given order of a circulant graph the basic question is to find the circulant graph which maximizes the diameter. The sharp upper bound of the diameter of a graph of a given order is important for estimating the degradation of performance of the network obtained by deleting a set of a certain number of vertices.

Throughout the paper we let $\mathrm{ICG}_n(D)$ be an arbitrary integral circulant graph of order n and set of divisors D. In [15], the following sharp bounds on the diameter of integral circulant graphs are obtained.

Theorem 1. *Let t be the size of the smallest set of additive generators of Z_n contained in D. Then*

$$t \leq diam(\mathrm{ICG}_n(D)) \leq 2t + 1.$$

Direct consequences of the previous assertion are the following bounds

$$2 \leq diam(\mathrm{ICG}_n(D)) \leq 2|D| + 1. \tag{1}$$

In this paper we find the maximal diameter of all integral circulant graphs $\mathrm{ICG}_n(D)$ of a given order n and the cardinality of the divisor set D equal to k, for $n = p_1^{\alpha_1} \cdots p_k^{\alpha_k}$. We actually show that the maximal diameter of this class of graphs is equal to $r(n)$, where

$$r(n) = k + |\{i \mid \alpha_i > 1, \ 1 \leq i \leq k\}|.$$

As $r(n) \leq 2t + 1$, for $t = k$ we conclude that the diameter of these graphs can not attain the upper bound given by (1). Moreover, we characterize all $\mathrm{ICG}_n(D)$ such that the maximal diameter is attained by describing all divisor sets D such that $diam(\mathrm{ICG}_n(D)) = r(n)$.

2   Preliminaries

In this section we introduce some basic notations and definitions.

A *circulant graph* $G(n; S)$ is a graph on vertices $\mathbb{Z}_n = \{0, 1, \ldots, n-1\}$ such that vertices i and j are adjacent if and only if $i - j \equiv s \pmod{n}$ for some $s \in S$. The set S is called the *symbol* of graph $G(n; S)$. As we will consider undirected graphs without loops, we assume that $S = n - S = \{n - s \mid s \in S\}$ and $0 \notin S$. Note that the degree of the graph $G(n; S)$ is $|S|$. A graph is *integral* if all its eigenvalues are integers. A circulant graph $G(n; S)$ is integral if and only if

$$S = \bigcup_{d \in D} G_n(d),$$

for some set of divisors $D \subseteq D_n$ [17]. Here $G_n(d) = \{k \ : \ \gcd(k, n) = d, \ 1 \leq k \leq n - 1\}$, and D_n is the set of all divisors of n, different from n. Therefore an *integral circulant graph* (in further text ICG) $G(n; S)$ is defined by its order n and the set of divisors D. An integral circulant graph with n vertices, defined by the set of divisors $D \subseteq D_n$ will be denoted by $\mathrm{ICG}_n(D)$. The term 'integral circulant graph' was first introduced in the work of So, where the characterization of the class of circulant graphs with integral spectra was given. The class of integral circulant graphs is also known as 'gcd-graphs' and arises as a generalization of unitary Cayley graphs [3,12]. From the above characterization of integral circulant graphs we have that the degree of an integral circulant graph is $\deg \mathrm{ICG}_n(D) = \sum_{d \in D} \varphi(n/d)$. Here $\varphi(n)$ denotes the Euler-phi function [7]. If $D = \{d_1, \ldots, d_t\}$, it can be seen that $\mathrm{ICG}_n(D)$ is connected if and only if $\gcd(d_1, \ldots, d_t) = 1$, given that $G(n; s)$ is connected if and only if $\gcd(n, S) = 1$.

Recall that the greatest distance between any pair of vertices in a graph is graph diameter. The distance $d(u, v)$ between two vertices u and v of a graph is the minimum length of the paths connecting them. Throughout this paper we let that the order of $\mathrm{ICG}_n(D)$ have the following prime factorization $n = p_1^{\alpha_1} \cdots p_k^{\alpha_k}$. Also, for a given prime number p and an integer n, denote by $S_p(n)$ the maximal number α such that $p^\alpha \mid n$. If $S_p(n) = 1$ we write $p \| n$.

3   Main Result

Let $G = \mathrm{ICG}_n(D)$ be a connected graph with maximal diameter in the class of all integral circulant graphs of order n, and $D = \{d_1, d_2, \ldots, d_t\}$.

Let D' be an arbitrary subset of D, such that $\gcd(\{d \mid d \in D'\}) = 1$. The graph $\mathrm{ICG}_n(D')$ is connected and clearly a subgraph of $\mathrm{ICG}_n(D)$, so it follows that

$$diam(\mathrm{ICG}_n(D)) \leq diam(\mathrm{ICG}_n(D')).$$

Since $\mathrm{ICG}_n(D)$ has maximal diameter, we have $diam(\mathrm{ICG}_n(D')) = diam(\mathrm{ICG}_n(D))$. Therefore, we will find maximal diameter among all $\mathrm{ICG}_n(D)$ such that for every subset of divisors $D' \subset D$ it holds that $\gcd(\{d \mid d \in D'\}) > 1$ (the graphs $\mathrm{ICG}_n(D')$ are unconnected). Furthermore, from the last assumption it follows that $\gcd(d_1, \ldots, d_{s-1}, d_{s+1}, \ldots, d_t) > 1$ for every $1 \leq s \leq t$ and from the connectedness of $\mathrm{ICG}_n(D)$ we have $\gcd(d_1, d_2, \ldots, d_t) = 1$. Thus, we conclude that for each s there exists a prime divisor p_{i_s} of n such that $p_{i_s} \nmid d_s$ and $p_{i_s} \mid d_j$ for all $1 \leq j \neq s \leq t$.

Therefore, we may define a bijective mapping $f : \{d_1, \ldots, d_t\} \rightarrow \{p_{i_1}, \ldots, p_{i_t}\}$, since $d_{s_1} \neq d_{s_2}$ implies $p_{i_{s_1}} \neq p_{i_{s_2}}$. Finally, we conclude that for every divisor d_s, $1 \leq s \leq t$, it holds that

$$p_{i_s} \nmid d_s \tag{2}$$

$$p_{i_1}, \ldots, p_{i_{s-1}}, p_{i_{s+1}}, \ldots, p_{i_t} \mid d_s. \tag{3}$$

So, in the rest of the paper we will assume that the divisors of the integral circulant graph $\mathrm{ICG}_n(D)$ have the property described above, unless it is stated otherwise.

In the following example we illustrate how to narrow the search for the finding maximal diameter among all integral circulanat with order $n = 12$ and divisor set D' that has the properties (2) and (3). It is easy to see that we have only two such examples $D' = \{2, 3\}$ and $D' = \{3, 4\}$ and the diameters of the graphs $\mathrm{ICG}_{12}(\{2, 3\})$ and $\mathrm{ICG}_{12}(\{3, 4\})$ are 2 and 3, respectively. However, the graph $\mathrm{ICG}_{12}(\{3, 4, 6\})$ does not have the properties (2) and (3) and its diameter can not exceed the diameters of the previous graphs (Fig. 1).

 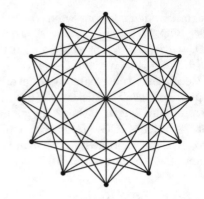

Fig. 1. Integral circulant graphs $\mathrm{ICG}_{12}(\{3,4\})$ on the left and $\mathrm{ICG}_{12}(\{3,4,6\})$ on the right whose diameters are equal to 3 and 2, respectively

Recall that two vertices a and b of $\mathrm{ICG}_n(d_1, d_2, \ldots, d_t)$ are adjacent if and only if $\gcd(a - b, n) = d_i$ for some $1 \leq i \leq t$. Given that the graph $\mathrm{ICG}_n(D)$ is vertex-transitive, we are going to focus on the vertex 0 and construct shortest paths from the vertex 0 to any other vertex $0 \leq l = |a - b| \leq n - 1$.

If we manage to find a solution $x = (x_1, x_2, \ldots, x_q)$ to the following system of equations

$$x_1 + x_2 + \cdots + x_q \equiv l \pmod{n}$$

and

$$\gcd(x_i, n) = d_{h(i)}$$

for $1 \leq i \leq q$ and some function $h : \{1, \ldots q\} \mapsto \{1, \ldots, t\}$, then we can construct the path from 0 to l of length q passing through the vertices

$$0, x_1, x_1 + x_2, \ldots, x_1 + x_2 + \cdots + x_q.$$

We are going to look for solutions of the above system of congruence equations and greatest common divisor equalities, in the following representation

$$d_1 y_1 + d_2 y_2 + \cdots + d_t y_t \equiv l \pmod{n} \tag{4}$$

with the constraints

$$\gcd(d_j y_j, n) = d_j \qquad \text{for} \qquad 1 \leq j \leq t, \tag{5}$$

where each summand $d_j y_j$ corresponds to one or more x_i's. Notice that some of the y_j, $1 \leq j \leq t$, can be equal to zero and in that case we do not consider the constraint (5).

In the following lemma we prove that an upper bound of the diameter of $\mathrm{ICG}_n(d_1, d_2, \ldots, d_t)$, for $t = k$, is equal to $k + |\{i \,|\alpha_i > 1, \ 1 \leq i \leq k\}|$ and we denote this value by $r(n)$.

Lemma 1. *The diameter of the integral circulant graph* $\mathrm{ICG}_n(d_1, d_2, \ldots, d_k)$, *where the set of divisors* $D = \{d_1, d_2, \ldots, d_k\}$ *satisfies the properties (2) and (3), is less or equal to* $r(n)$.

Proof. Since the order of the divisors in $|D| = k$ is arbitrary and divisor set satisfies (2) and (3), we may suppose with out loss of generality that $i_s = s$, for $1 \le s \le k$. We are going to use the representation (4) for $t = k$, which requires solving the following system

$$d_1 y_1 + d_2 y_2 + \cdots + d_k y_k \equiv l \pmod{p_j^{\alpha_j}} \quad 1 \le j \le k, \tag{6}$$

with the constraints

$$\gcd(d_j y_j, n) = d_j \quad \text{for} \quad 1 \le j \le k.$$

The last equation is equivalent to

$$\gcd(y_j, \frac{n}{d_j}) = 1 \quad \Leftrightarrow \quad p_i \nmid y_j, \text{ if } S_{p_i}(d_j) < \alpha_i, 1 \le i \le k. \tag{7}$$

From (2) we have $S_{p_j}(d_j) = 0 < \alpha_j$ and it directly holds that $p_j \nmid y_j$.

Furthermore, as $\gcd(d_j, p_j^{\alpha_j}) = 1$ it follows

$$y_j \equiv (l - p_j t) \cdot d_j^{-1} \pmod{p_j^{\alpha_j}},$$

where by $p_j t$ we denote $d_1 y_1 + \cdots + d_{j-1} y_{j-1} + d_{j+1} y_{j+1} + \cdots + d_k y_k$. Note that the only constraint is that d_j cannot be divisible by p_j as we already discuss above.

If p_j does not divide l, we can directly compute y_j which is not divisible by p_j. Assume now that p_j divides l.

For $\alpha_j = 1$, as $p_j \mid l$ we have

$$y_j \equiv (l - p_j t) \cdot d_j^{-1} \equiv 0 \pmod{p_j^{\alpha_j}}$$

and we can decide not to include the summand $d_j y_j$ in the above summation (otherwise, we obtain a contradiction with the constraint $p_j \nmid y_j$). This basically means that we are going to put $y_j = 0$ (ignore that summand) and in this case we do not consider the constraint $\gcd(d_j y_j, n) = d_j$.

For $\alpha_j > 1$, assume that

$$y_j \equiv (l - p_j t) \cdot d_j^{-1} \equiv p_j^{\beta_j} \cdot s \pmod{p_j^{\alpha_j}}, \tag{8}$$

where $p_j \nmid s$. For $\beta_j = \alpha_j$, we can similarly drop the summand $d_j y_j$ from the summation.

If $\beta_j < \alpha_j$, then $y_j = 0$ obviously is not a solution of the congruence equation. The trick now is to split y_j into two summands $y_j' + y_j''$ which are both coprime with p_j and the sum is equal to $p_j^{\beta_j} \cdot s$ modulo $p_j^{\alpha_j}$. This can be easily done by taking $y_j' = 1$ and $y_j'' = p_j^{\beta_j} \cdot s - 1$. Therefore, we split the summand $d_j y_j$ into two summands that satisfy all the conditions

$$d_j \cdot 1 + d_j \cdot (p_j^{\beta_i} \cdot s - 1).$$

This means that for the prime factors with $\alpha_j > 1$ we need two edges in order to construct a path from 0 to l when $p_j^{\beta_j} \| l - p_j t$, $\beta_j < \alpha_j$.

Finally, we conclude that if we want to construct a path from 0 to l, for an arbitrary l, we need at most one edge that corresponds the modulo p_j, if $\alpha_j = 1$, and at most two edges that correspond the modulo p_j, if $\alpha_j > 1$, for every $1 \leq j \leq k$. This means that $diam(\mathrm{ICG}_n(d_1, \ldots, d_k)) \leq r(n)$.

In the following theorem we show that the maximal diameter of $\mathrm{ICG}_n(d_1, d_2, \ldots, d_t)$, for $t = k$, is equal to $r(n)$ and characterize all extremal graphs.

Theorem 2. *The maximal diameter of the integral circulant graph* $\mathrm{ICG}_n(d_1, d_2, \ldots, d_k)$, *where the set of divisors* $D = \{d_1, d_2, \ldots, d_k\}$ *satisfies the properties (2) and (3), is equal to* $r(n)$. *The equality holds in two following cases*

(i) *if* $\alpha_j > 1$ *then* $S_{p_j}(d_i) > 1$, *for* $1 \leq i \neq j \leq k$

(ii) *if* $n \in 4\mathbb{N} + 2$ *and* $d_1 \in 2\mathbb{N} + 1$ *then there exists exactly one* $2 \leq j \leq k$ *such that* $p_j \| d_1$, $S_{p_j}(d_i) > 1$, *for* $2 \leq i \leq k$ *and other prime factors* p_l *($1 \leq l \neq j \leq k$) satisfy* $S_{p_l}(d_i) > 1$, *if* $\alpha_l > 1$, *for* $1 \leq i \neq l \leq k$.

Proof. Since the diameter of $\mathrm{ICG}_n(d_1, d_2, \ldots, d_k)$ is less or equal to $r(n)$, where the set of divisors $D = \{d_1, d_2, \ldots, d_k\}$ satisfies the properties (2) and (3), according to Lemma 1, in the first part of the proof we analyze when the graph has diameter less than $r(n)$. From (8) we conclude that this is the case when the condition $l - p_j t \equiv 0 \pmod{p_j^{\alpha_j}}$ is satisfied, for $\alpha_j > 1$ and $p_j \mid l$.

Observe that the sum $p_j t$ can be rewritten in the following form

$$p_j(t_1 + \cdots + t_{u_j}) + p_j^2(t_{u_j+1} + \cdots + t_{k-1}), \tag{9}$$

where $u_j = |\{d_i \mid i \in \{1, \ldots, j-1, j+1, \ldots, k\}, \ p_j \| d_i, \ d_i y_i = p_j t_i\}|$ and $p_j \nmid t_1, \ldots, t_{u_j}$. From (7) we conclude that this form is indeed always possible since it holds that $p_j \nmid y_1, \ldots, y_{u_j}$ as $1 = S_{p_j}(d_i) < \alpha_j$, $1 \leq i \leq u_j$. Furthermore, since we take into consideration the values $d_i y_i$ modulo $p_j^{\alpha_j}$, we can assume, without loss of generality, that $S_{p_j}(p_j^2 t_i) < \alpha_j$ for $u_j + 1 \leq i \leq k$, where $d_i y_i = p_j^2 t_i$. In that case, according to (7) it must be that $p_j \nmid y_i$, for all $1 \leq i \leq k$.

Assume first that $u_j \geq 2$. We see that $l - p_j t \equiv 0 \pmod{p_j^{\alpha_j}}$ if and only if

$$t_1 + \cdots + t_{u_j} + p_j(t_{u_j+1} + \cdots + t_{k-1}) \equiv \frac{l}{p_j} \pmod{p_j^{\alpha_j-1}}. \tag{10}$$

Furthermore, since $p_j \nmid t_1$ we see that t_1 must satisfy the following system $t_1 \equiv \frac{l}{p_j} - t_2 - \cdots - t_{u_j} - p_j(t_{u_j+1} + \cdots + t_{k-1}) \pmod{p_j^{\alpha_j-1}}$ and $t_1 \not\equiv p_j s_1$

(mod $p_j^{\alpha_j-1}$), for some $0 \le s_1 < p_j^{\alpha_j-2}$. Again, since $p_j \nmid t_2$ too, we obtain that $t_2 \not\equiv \{\frac{l}{p_j}-t_3-\cdots-t_{u_j}-p_j(t_{u_j+1}+\cdots+t_{k-1})-p_js_1, p_js_2\}$ (mod $p_j^{\alpha_j-1}$), for some $0 \le s_2 < p_j^{\alpha_j-2}$. Therefore, if $p_j > 2$ we can find the value t_2 modulo $p_j^{\alpha_j-1}$ which is a solution of the system if we take arbitrary values $t_3, \ldots, t_{k-1}$ modulo $p_j^{\alpha_j-1}$ (the value t_1 modulo $p_j^{\alpha_j-1}$ can be obviously computed thereafter using the values $t_2, \ldots, t_{k-1}$). If $p_j = 2$ and $\frac{l}{p_j} - t_3 - \cdots - t_{u_j} \in 2\mathbb{N}+1$, we can not find an odd t_2 which would be a solution of the system. For such l, in the same way we can prove that the system $d_1(y_1^{(1)}+y_1^{(2)})+d_2y_2+\cdots+d_{j-1}y_{j-1}+d_{j+1}y_{j+1}+\cdots+d_ky_k \equiv l$ (mod $p_j^{\alpha_j}$), $p_j \nmid y_1^{(1)}, y_1^{(2)}, y_2, \ldots, y_k$, has a solution by reducing it to the form $t_1^1+t_1^2+t_2+\cdots+t_{u_j}+p_j(t_{u_j+1}+\cdots+t_{k-1}) \equiv \frac{l}{p_j}$ (mod $p_j^{\alpha_j-1}$). So, in this case we need at most two edges for the parts of the path from 0 to l corresponding to the moduli p_1 and p_j, since $l - p_jt \equiv 0$ (mod $p_j^{\alpha_j}$) and $y_j = 0$. Finally, we conclude that the diameter of this graph can not reach the value $r(n)$, since $r(p_1^{\alpha_1}) + r(p_j^{\alpha_j}) > 2$.

Now, let $u_j = 1$. If $p_j > 2$, similarly to the previous case by examining the system $d_1(y_1^{(1)}+y_1^{(2)})+d_2y_2+\cdots+d_{j-1}y_{j-1}+d_{j+1}y_{j+1}+\cdots+d_ky_k \equiv l$ (mod $p_j^{\alpha_j}$) (which corresponds to the equation $t_1^1+t_1^2+p_j(t_2+\cdots+t_{k-1}) \equiv \frac{l}{p_j}$ (mod $p_j^{\alpha_j-1}$)) and $p_j \nmid y_1^{(1)}, y_1^{(2)}, y_2, \ldots, y_k$, we see that we can find $t_1^1, t_1^2, t_2, \ldots, t_{k-1}$ such that $l - p_jt \equiv 0$ (mod $p_j^{\alpha_j}$) and therefore we can set $y_j = 0$. Furthermore, as $p_1 \mid d_2, \ldots, d_k$ it holds that $d_1(y_1^{(1)}+y_1^{(2)}) \equiv l$ (mod p_1) and if both $y_1^{(1)}, y_1^{(2)} \neq 0$ then we conclude that for $p_1 = 2$ and $l \in 2\mathbb{N}+1$ the parity of left and right hand sides of the equation is violated ($p_1 \nmid y_1^{(1)}, y_1^{(2)}$). So, in this case we need one extra edge and therefore at most three edges for the parts of the diameter path corresponding to moduli p_1 and p_j. Finally, we conclude that maximal diameter of this graph can be attained if the value $r(p_1^{\alpha_1}) + r(p_j^{\alpha_j})$ is equal to 3 and this is the case if only if $\alpha_1 = 1$. If $p_j = 2$, we have already proved that for $\frac{l}{p_j} \in 2\mathbb{N}$ the system $t_1^1 + t_1^2 + p_j(t_2 + \cdots + t_{k-1}) \equiv \frac{l}{p_j}$ (mod $p_j^{\alpha_j-1}$), $p_j \nmid t_1^1, t_1^2$ has a solution (case $u_j = 2$) and therefore $l - p_jt \equiv 0$ (mod $p_j^{\alpha_j}$) ($y_j = 0$). Considering the modulus p_1, we can find $y_1^{(1)}, y_1^{(2)}$ not divisible by p_1 and $y_1^{(1)} + y_1^{(2)} = (l - p_1t')d_1^{-1}$, where $p_1t' = d_2y_2 + \ldots + d_{j-1}y_{j-1} + d_{j+1}y_{j+1} + \cdots + d_ky_k$ as $p_1 > 2$. In this case we need at most two edges for the parts of the path from 0 to l corresponding to the moduli p_1 and p_j (the diameter of this graph can not reach the value $r(n)$). If $\frac{l}{p_j} \in 2\mathbb{N} + 1$, assume that $p_1 \nmid l$, then by examining the system $d_1y_1 + d_2y_2 + \cdots + d_{j-1}y_{j-1} + d_{j+1}y_{j+1} + \cdots + d_ky_k \equiv l$ (mod $p_j^{\alpha_j}$) (which corresponds to the equation $t_1 + p_j(t_2 + \cdots + t_{k-1}) \equiv \frac{l}{p_j}$ (mod $p_j^{\alpha_j-1}$)) and $p_j \nmid y_1, y_2, \ldots, y_k$, we see that $t_1 \in 2\mathbb{N} + 1$ satisfies the following system $t_1 \equiv \frac{l}{p_j} - p_j(t_{u_j+1} + \cdots + t_{k-1})$ (mod $p_j^{\alpha_j-1}$) and $t_1 \not\equiv p_js$ (mod $p_j^{\alpha_j-1}$), for some $0 \le s < p_j^{\alpha_j-2}$. Moreover, $y_1 \equiv (l - p_1t')d_1^{-1}$ (mod $p_1^{\alpha_1}$) is a solution of the congruence equation modulo p_1 and in this case we need one edge for the parts of the path from 0 to l corresponding to the moduli p_1 and p_j (the diameter of this graph can not reach the value $r(n)$). Finally, if $\frac{l}{p_j} \in 2\mathbb{N} + 1$ and $p_1 \mid l$, the following system can be similarly solved $d_2 y_2 + \cdots + d_j(y_j^1 + y_j^2) + \cdots + d_k y_k \equiv l$ $(\text{mod } p_j^{\alpha_j})$, $p_j \nmid y_j^1, y_j^2$.

We have proved that if $u_i = 1$, $n \in 2\mathbb{N}$ and $l, d_1 \in 2\mathbb{N} + 1$, we can find a solution of the congruence system $d_1(y_1^{(1)} + y_1^{(2)} + y_1^{(3)}) + d_2 y_2 + \cdots + d_{j-1} y_{j-1} + d_{j+1} y_{j+1} + \cdots + d_k y_k \equiv l$ $(\text{mod } p_i^{\alpha_j})$, $p_i \nmid y_1^{(1)}, y_1^{(2)}, y_1^{(3)}, y_2, \ldots, y_k$. Notice, that if $u_j = 1$ and $p_j \| d_1$ we can find a solution of the congruence system $d_1(y_1^{(1)} + y_1^{(2)} + y_1^{(3)}) + d_2 y_2 + \cdots + d_{j-1} y_{j-1} + d_{j+1} y_{j+1} + \cdots + d_k y_k \equiv l$ $(\text{mod } p_j^{\alpha_j})$, $p_j \nmid y_1^{(1)}, y_1^{(2)}, y_1^{(3)}, y_2, \ldots, y_k$. This implies that we can set $y_j = 0$ and conclude that we need three edges for the parts of the path corresponding to the moduli p_1, p_i and p_j and this path can not attain the value $r(n)$. The same conclusion holds for more than two prime divisors.

From the discussion above we conclude that the diameter path can attain the value $r(n)$, if $S_{p_j}(d_i) > 1$ for $\alpha_j > 1$ and $i \in \{1, \ldots, j-1, j+1, \ldots, k\}$ (the case $u_j = 0$). In yet another case the value $r(n)$ of the diameter can be attained if $S_{p_j}(d_1) = 1 < S_{p_j}(d_i)$, $2 \leq i \leq k$, and $p_j, d_1 \in 2\mathbb{N} + 1$, for $n \in 4\mathbb{N} + 2$ (the case $u_j = 1$). Now, we prove that there exists a vertex l_0 where the distance from 0 to l_0 is equal to $r(n)$ in both of the mentioned cases.

In the first case, we derive l_0 from the system of congruence equations composed by choosing exactly one of the following equation, for each $1 \leq j \leq k$ (which exists due to the Chinese remainder theorem)

$$l_0 \equiv -1 \quad (\text{mod } p_j) \qquad \text{if } \alpha_j = 1$$
$$l_0 \equiv p_j \quad (\text{mod } p_j^{\alpha_j}) \qquad \text{if } \alpha_j > 1.$$

For all $1 \leq j \leq k$ such that $\alpha_j = 1$, we need at least one summand $d_j y_j$ in the representation (6) of l_0 since $l_0 \not\equiv 0$ $(\text{mod } p_j^{\alpha_j})$ and all other d_i are divisible by $p_j^{\alpha_j}$ for $i \neq j$. On the other hand, for all $1 \leq j \leq k$ such that $\alpha_j > 1$ we cannot have exactly one such summand as otherwise we would have $d_j y_j \equiv l_0 - p_j^2 t' \equiv p_j s$ $(\text{mod } p_j^{\alpha_j})$ and $p_j \nmid s$, which would be a contradiction as $p_j \nmid d_j y_j$, where $p_j^2 t' = d_1 y_1 + \cdots + d_{j-1} y_{j-1} + d_{j+1} y_{j+1} + \cdots + d_k y_k$. Therefore, we need at least two summands for $\alpha_j > 1$.

In the second case, we derive l_0 from the following system of congruence equations

$$l_0 \equiv -1 \quad (\text{mod } p_i) \qquad \text{if } \alpha_i = 1$$
$$l_0 \equiv p_j^2 \quad (\text{mod } p_j^{\alpha_j}) \qquad \text{if } S_{p_j}(d_1) = 1 < S_{p_j}(d_l), \ 2 \leq l \leq k$$
$$l_0 \equiv p_i \quad (\text{mod } p_i^{\alpha_i}) \qquad \text{if } \alpha_i > 1, \ i \neq j.$$

It remains to prove that we need two edges by considering the modulus $p_j^{\alpha_j}$, for $p_j, d_1 \in 2\mathbb{N} + 1$ and $S_{p_j}(d_1) = 1 < S_{p_j}(d_l)$, $2 \leq l \leq k$. Indeed, we cannot have exactly one such summand as otherwise we would have $d_j y_j \equiv l_0 - p_j t \equiv p_j s$ $(\text{mod } p_j^{\alpha_j})$ and $p_j \nmid s$, which is a contradiction due to $p_j \nmid d_j y_j$.

This proves that the diameter of $\mathrm{ICG}_n(D)$ in these cases is greater or equal to $r(n)$ and therefore $diam(\mathrm{ICG}_n(D)) = r(n)$, which completes the proof of the theorem.

We provide several examples to illustrate how Theorem 2 is applied to determine whether the diameter of $\mathrm{ICG}_n(D)$ such that $|D| = k$ attains $r(n)$.

(i) If $n = 2^2 \cdot 3^3 \cdot 5$ and $D = \{3^2 \cdot 5, 2^2 \cdot 5, 2^2 \cdot 3^3\}$ then $diam(\mathrm{ICG}_n(D)) = 5$, according to the first part of Theorem 2 and the diameter of the graph attains $r(n)$.

We will also compute the diameter of $\mathrm{ICG}_n(d_1, d_2, d_3)$ for $n = 2^2 \cdot 3^3 \cdot 5$, $d_1 = 3^2 \cdot 5$, $d_2 = 2^2 \cdot 5$ and $d_3 = 2^2 \cdot 3^3$ to illustrate the methods from the proof of Theorem 2. Indeed, for any $0 \le l \le n - 1$ we will try to solve the equation in the following form $d_1 y_1 + d_2 y_2 + d_3 y_3 \equiv l \pmod{n}$, with the constraints $\gcd(d_j y_j, n) = d_j$, for $1 \le j \le 3$. This is equivalent with the following system

$$y_1 \equiv (l - d_2 y_2 + d_3 y_3) \cdot d_1^{-1} \pmod{5}$$
$$y_2 \equiv (l - d_1 y_1 + d_3 y_3) \cdot d_2^{-1} \pmod{3^3}$$
$$y_3 \equiv (l - d_1 y_1 + d_2 y_2) \cdot d_3^{-1} \pmod{2^2},$$

where $5 \nmid y_1$, $3 \nmid y_2$ and $2 \nmid y_3$. If $5 \nmid l$ then we can directly compute y_1. Similarly if $3 \nmid l$ or $2 \nmid l$, we can directly compute y_2 or y_3, respectively. In particular, if $5 \nmid l$, $3 \nmid l$ and $2 \nmid l$ we conclude that the distance between 0 and l is equal to 3.

If $5 \mid l - d_2 y_2 + d_3 y_3$ or $3^3 \mid l - d_1 y_1 + d_3 y_3$ or $2^2 \mid l - d_1 y_1 + d_2 y_2$ then we obtain that a solution of the corresponding equation is $y_1 = 0$ or $y_2 = 0$ or $y_3 = 0$, respectively. On the other hand, if $3^2 \| l - d_1 y_1 + d_3 y_3$ or $3 \| l - d_1 y_1 + d_3 y_3$ we can not find y_2 satisfying the above congruence equation such that $3 \nmid y_2$. However, for such l there exist y_2' and y_2'' such that $y_2' + y_2'' \equiv (l - d_1 y_1 + d_3 y_3) \cdot d_2^{-1}$ $\pmod{3^3}$ and $3 \nmid y_2', y_2''$. This means that we need two edges in order to construct a path from 0 to l when $3^2 \| l - d_1 y_1 + d_3 y_3$ or $3 \| l - d_1 y_1 + d_3 y_3$ corresponding to the modulus 3^3. Similarly, if we take $2 \| l - d_1 y_1 + d_2 y_2$ in order to construct a path from 0 to l we need two edges edges corresponding to the modulus 2^2. Finally, the distance between any two vertices is less or equal to 5. According to the above discussion we can, say, choose l such that

$$l \equiv -1 \pmod{5}$$
$$l \equiv 3 \pmod{3^3}$$
$$l \equiv 2 \pmod{2^2},$$

and calculate that $l = 354$ using the Chinese Remainder Theorem. Therefore, we conclude that the distance between 0 and l is equal to 5.

(ii) If $n = 2 \cdot 3^3 \cdot 5^3$ and $D = \{3 \cdot 5^2, 2 \cdot 5^3, 2 \cdot 3^2\}$ then $diam(\mathrm{ICG}_n(D)) = 5$, due to the part (ii) of Theorem 2 as there exists exactly one $p_j = 3 \mid d_1$ such that $S_{p_j}(d_1) = 1 < S_{p_j}(d_3)$ and $d_1 \in 2\mathbb{N} + 1$, for $n \in 4\mathbb{N} + 2$ and the diameter of the graph is equal to $r(n) = 5$.

(iii) If $n = 2^2 \cdot 3^2 \cdot 5 \cdot 7$ and $D = \{3 \cdot 5 \cdot 7, 2^2 \cdot 5 \cdot 7, 2^2 \cdot 3^2 \cdot 7, 2^2 \cdot 3^2 \cdot 5\}$ then $diam(\mathrm{ICG}_n(D)) = 5$ (less than $r(n)$), as $n \in 4\mathbb{N}$ even though the other conditions from the part (ii) of Theorem 2 are satisfied.

(iv) If $n = 2^2 \cdot 3 \cdot 5 \cdot 7$ and $D = \{3 \cdot 5 \cdot 7, 2 \cdot 5 \cdot 7, 2^2 \cdot 3 \cdot 7, 2^2 \cdot 3 \cdot 5\}$ then $diam(\mathrm{ICG}_n(D)) = 4$ (less than $r(n)$), as the conditions from the part (ii) of Theorem 2 are satisfied only for $p_j = 2$.

(v) If $n = 2 \cdot 3^2 \cdot 5^2 \cdot 7^2$ and $D = \{3 \cdot 5 \cdot 7, 2 \cdot 5^2 \cdot 7^2, 2 \cdot 3^2 \cdot 7^2, 2 \cdot 3^2 \cdot 5^2\}$ then $diam(\mathrm{ICG}_n(D)) = 5$ (less than $r(n)$), as the conditions from the part (ii) of Theorem 2 are satisfied for more than one divisor (in this case the three divisors $p_j \in \{3, 5, 7\}$).

We also directly conclude that in the class of all integral circulant graphs with a given order n and $k-$element divisor set D we can not find an example for which $diam(\mathrm{ICG}_n(d_1, \ldots, d_k)) = 2k + 1$, since $diam(\mathrm{ICG}_n(d_1, \ldots, d_k)) \le r(n) \le 2k$. So, in this case the upper bound from (1) is not attainable.

4   Conclusion

In this paper we find the maximal diameter of integral circulant graphs of a given order n and cardinality of the divisor set k. Moreover, we characterize all $\mathrm{ICG}_n(D)$ such that the maximal diameter is attained. We have already mentioned that the maximal diameter of some class of graphs of a given order plays an important role if that class is used for modeling a quantum network that allows quantum dynamics, having in mind applications like perfect state transfer. Therefore, it would be very important to find the maximal diameter of integral circulant graphs of a given order n and any cardinality of the divisor set D. Moreover, a possible challenging direction in future research would be examining the existence of an integral circulant graph of a given order allowing perfect state transfer that attains maximal diameter.

On the other hand, we may notice that we can not improve the lower bound in the inequality (1) for prescribed n and any prescribed cardinality of the divisor set D. Indeed, according to Theorem 9 from [12] we observe that $\mathrm{ICG}_n(1) = 2$ if and only if n is a power of 2 or n is odd (in both of the cases n is not prime). This implies that $diam(\mathrm{ICG}_n(1, d_2, \ldots, d_t)) = 2$, for any t and the mentioned values of n (as long as $\{1, d_2, \ldots, d_t\} \ne D_n$).

In the remaining case, for $n = 2^{\alpha_1} m$, where $m > 1$ is odd and $\alpha \ge 1$, we can prove that $diam(\mathrm{ICG}_n(1, 2^{\alpha_1})) = 2$. Indeed, for every $0 \le l \le n - 1$, such that l is even we will prove the existence of l in the form $s_1 + s_2 \equiv l \pmod{n}$ such that $\gcd(s_1, n) = 1$, $\gcd(s_2, n) = 1$. Now, let p_i be an arbitrary divisor of n such that $S_{p_i}(n) = \alpha_i$. By solving the above congruence equation system modulo $p_i^{\alpha_i}$ we get that $s_1 \equiv l - s_2 \not\equiv p_i u \pmod{p_i^{\alpha_i}}$ and $s_2 \not\equiv 0 \pmod{p_i}$ (which is equivalent to $s_2 \not\equiv p_i v \pmod{p_i^{\alpha_i}}$), for $0 \le u, v < p_i^{\alpha_i - 1} - 1$. Therefore, we obtain $s_2 \not\equiv \{p_i v, l - p_i u\} \pmod{p_i^{\alpha_i}}$ and it can be concluded that the maximal number of values that s_2 modulo $p_i^{\alpha_i}$ can not take is equal to $2p_i^{\alpha_i - 1}$. This number is less than the number of residues modulo $p_i^{\alpha_i}$ any p_i (as l is even), so this system has a solution in this case.

Now, suppose that $l \in 2\mathbb{N} + 1$. We find s_1 and s_2 such that $s_1 + s_2 \equiv l \pmod{n}$ such that $\gcd(s_1, n) = 1$, $\gcd(s_2, n) = 2^{\alpha_1}$. Similarly to the previous discussion, we can conclude that the above conditions can be reduced to the following system

$s_2 \not\equiv \{p_i v, l-p_i u\}$ (mod $p_i^{\alpha_i}$), for $2 \le i \le k$ and $s_2 \equiv 0$ (mod 2^{α_1}). As $p_i > 2$ this system has a solution. We finally conclude that $diam(\mathrm{ICG}_n(1, 2^{\alpha_1}, d_3, \ldots, d_t)) = 2$, for any t and the mentioned values of n (as long as $\{1, 2^{\alpha_1}, d_3, \ldots, d_t\} \ne D_n$).

References

1. Bašić, M.: Characterization of circulant graphs having perfect state transfer. Quantum Inf. Process. **12**, 345–364 (2013)
2. Bašić, M.: Which weighted circulant networks have perfect state transfer? Inf. Sci. **257**, 193–209 (2014)
3. Bašić, M., Ilić, A.: On the clique number of integral circulant graphs. Appl. Math. Lett. **22**, 1406–1411 (2009)
4. Bašić, M., Ilić, A.: On the automorphism group of integral circulant graphs. Electron. J. Comb. **18** (2011). #P68
5. Berrizbeitia, P., Giudic, R.E.: On cycles in the sequence of unitary Cayley graphs. Discrete Math. **282**, 239–243 (2004)
6. Fuchs, E.: Longest induced cycles in circulant graphs. Electron. J. Comb. **12**, 1–12 (2005)
7. Hardy, G.H., Wright, E.M.: An Introduction to the Theory of Numbers, 5th edn. Clarendon Press, Oxford University Press, New York (1979)
8. Hwang, F.K.: A survey on multi-loop networks. Theor. Comput. Sci. **299**, 107–121 (2003)
9. Ilić, A.: Distance spectra and distance energy of integral circulant graphs. Linear Algebra Appl. **433**, 1005–1014 (2010)
10. Ilić, A., Bašić, M.: On the chromatic number of integral circulant graphs. Comput. Math. Appl. **60**, 144–150 (2010)
11. Ilić, A., Bašić, M.: New results on the energy of integral circulant graphs. Appl. Math. Comput. **218**, 3470–3482 (2011)
12. Klotz, W., Sander, T.: Some properties of unitary Cayley graphs. Electron. J. Comb. **14** (2007). #R45
13. Park, J.H., Chwa, K.Y.: Recursive circulants and their embeddings among hypercubes. Theor. Comput. Sci. **244**, 35–62 (2000)
14. Ramaswamy, H.N., Veena, C.R.: On the energy of unitary Cayley graphs. Electro. J. Comb. **16** (2007). #N24
15. Saxena, N., Severini, S., Shparlinski, I.: Parameters of integral circulant graphs and periodic quantum dynamics. Int. J. Quant. Inf. **5**, 417–430 (2007)
16. Sander, J.W., Sander, T.: The maximal energy of classes of integral circulant graphs. Discrete Appl. Math. **160**, 2015–2029 (2012)
17. So, W.: Integral circulant graphs. Discrete Math. **306**, 153–158 (2006)

Parallelisms of PG(3, 4) Invariant Under Cyclic Groups of Order 4

Anton Betten[1] , Svetlana Topalova[2] , and Stela Zhelezova[2(✉)]

[1] Colorado State University, Fort Collins, USA
betten@math.colostate.edu
[2] Institute of Mathematics and Informatics, Bulgarian Academy of Sciences,
Sofia, Bulgaria
{svetlana,stela}@math.bas.bg

Abstract. A *spread* in $\mathrm{PG}(n,q)$ is a set of mutually skew lines which partition the point set. A *parallelism* is a partition of the set of lines by spreads. The classification of parallelisms in small finite projective spaces is of interest for problems from projective geometry, design theory, network coding, error-correcting codes, cryptography, etc. All parallelisms of $\mathrm{PG}(3,2)$ and $\mathrm{PG}(3,3)$ are known and parallelisms of $\mathrm{PG}(3,4)$ which are invariant under automorphisms of odd prime orders and under the Baer involution have already been classified. In the present paper, we classify all (we establish that their number is 252738) parallelisms in $\mathrm{PG}(3,4)$ that are invariant under cyclic automorphism groups of order 4. We compute the order of their automorphism groups and obtain invariants based on the type of their spreads and duality.

Keywords: Finite projective space · Parallelism · Automorphism

1 Introduction

For background material concerning projective spaces, spreads and parallelisms, refer, for instance, to [8] or [17].

Let $\mathrm{PG}(n,q)$ be the n-dimensional projective space over the finite field $\mathbb{F}_q$. For a positive integer t such that $t + 1$ divides $n + 1$, a partition of the set of points of $\mathrm{PG}(n,q)$ into subspaces $\mathrm{PG}(t,q)$ is possible. Such a partition is called a t-spread. Two t-spreads are isomorphic if there is a collineation of $\mathrm{PG}(n,q)$ which takes one to the other.

If $t + 1$ divides $n + 1$, a t-parallelism of $\mathrm{PG}(n,q)$ is a partition of the set of all t-dimensional subspaces of $\mathrm{PG}(n,q)$ into spreads. A 1-spread is called a spread or a line-spread and a 1-parallelism a parallelism or a line-parallelism.

Let $V = \{P_i\}_{i=1}^{v}$ be a finite set of *points*, and $\mathcal{B} = \{B_j\}_{j=1}^{b}$ a finite collection of k-element subsets of V, called *blocks*. $D = (V, \mathcal{B})$ is a *2-design* with parameters

The research of the second and the third author is partially supported by the Bulgarian National Science Fund under Contract No. DH 02/2, 13.12.2016.

2-(v, k, λ) if any 2-subset of V is contained in exactly λ blocks of $\mathcal{B}$. A *parallel class* is a partition of the point set by blocks. A *resolution* of the design is a partition of the collection of blocks by parallel classes. The incidence of the points and t-dimensional subspaces of PG(d, q) defines a 2-design [27, 2.35–2.36]. The points of this design correspond to the points of the projective space, and the blocks to the t-dimensional subspaces. There is a one-to-one correspondence between parallelisms and the resolutions of the related point-line design.

Parallelisms are also related to translation planes [11], network coding [12], error-correcting codes [14], cryptography [25], etc.

Two parallelisms are isomorphic if there is a collineation of PG(n, q) which takes the spreads of one parallelism to the spreads of the other. The classification problem for parallelisms is the problem of determining a set of representatives for the isomorphism classes of parallelisms.

An automorphism of a parallelism is a collineation which fixes the parallelism. Assuming non-trivial automorphisms is a popular approach, because the search space is reduced since the object must be a union of complete orbits of the assumed group.

A spread of PG$(3, q)$ has $q^2 + 1$ lines and a parallelism has $q^2 + q + 1$ spreads.

A *regulus* of PG$(3, q)$ is a set R of $q + 1$ mutually skew lines such that any line intersecting three elements of R intersects all elements of R. A spread S of PG$(3, q)$ is *regular* if for any three distinct elements of S, the unique regulus determined by them is contained in S. A parallelism is *regular* if all its spreads are regular. A spread is called *aregular* if it contains no regulus. A spread is called *subregular* if it can be obtained from a regular spread by successive replacements of some reguli by their opposites [27]. These three spreads are also known as Desarguesian, Hall and semified spread. Up to isomorphism, there is only one regular spread. The number of subregular and aregular spreads grows quickly in q and in n. A parallelism is *uniform* if it consists entirely of isomorphic spreads.

The dual space of PG(n, q) is the space whose points are the hyperplanes of PG(n, q), with the co-dimension two subspaces considered as lines and with reversed incidence. This is a PG(n, q) too. A dual spread in PG$(3, q)$ is a set of lines which have the property that each plane contains exactly one line. The lines of a spread in PG$(3, q)$ are lines of a dual spread too. Consequently each parallelism defines a dual parallelism. Dual parallelisms in PG$(3, q)$ are also parallelisms of the dual space (obtained by interchanging the points and the planes of the initial space). There exists a dual transformation (a permutation of the points and a permutation of the lines of the dual space) which maps the dual space to the initial space, and the dual parallelism P_D (dual of the parallelism P) to a parallelism P_d of the considered initial projective space. The parallelism P is self-dual if it is isomorphic to P_d.

A construction of parallelisms of PG$(n, 2)$ is presented in [1, 33]. A construction of parallelisms of PG$(2^n - 1, q)$ is given in [7]. Constructions for PG$(3, q)$ can be found in [10, 15, 16, 20].

All parallelisms of PG$(3, 2)$ and PG$(3, 3)$ are known [3]. For larger projective spaces, the classification problem for parallelisms is beyond reach at the moment. The most promising case is that of PG$(3, 4)$, where there are 5096448 spreads in exactly three isomorphism classes: regular, subregular and aregular [9].

Parallelism classification results which rely on an assumed group of symmetries can be found in [21, 23, 24, 26, 28–30], for instance. Bamberg [2] in an unpublished note claims that there are no regular parallelisms in PG(3, 4).

By considering the order of $P\Gamma L(4, 4)$, it follows that if p is the prime order of an automorphism of a parallelism of PG(3, 4), then p must be one of $2, 3, 5, 7$ and 17.

All parallelisms of PG(3, 4) with automorphisms of odd prime orders are known. Some of the parallelisms with automorphisms of order 2 have been classified too, namely those which are invariant under the Baer involution (Table 1). The problem of the classification of the remaining parallelisms with automorphisms of order 2, and of those with the trivial automorphism is still open.

Table 1. The known parallelisms of PG(3, 4) with nontrivial automorphisms

Autom. group order	Isomorphism classes	References
2	$\geq$270088	[6]
$2^m, m > 1$	$\geq$0	
3	8 115 559	[31]
5	31 830	[5, 29]
6	4 488	[31]
7	482	[32]
10	76	[5, 29]
12	52	[31]
15	40	[5, 29, 31]
17	0	[29]
20	52	[5, 29]
24	14	[31]
30	38	[5, 29, 31]
48	12	[31]
60	8	[5, 29, 31]
96	2	[31]
960	4	[5, 29, 31]
Total	$\geq$8422745	

The present paper considers parallelisms of PG(3, 4) invariant under cyclic groups of order 4. The investigation is computer-aided. We use our own C++ programs written for this purpose, as well as Orbiter [4], GAP [13], Magma [18] and Nauty [19]. As a result, 252738 nonisomorphic parallelisms are constructed, and 252620 of them are new.

2   Construction Method

2.1 $P\Gamma L(4, 4)$ and Its Sylow Subgroup of Order 2

The projective space PG(3, 4) has 85 points and 357 lines. We denote by G its full automorphism group, where $G \cong P\Gamma L(4, 4)$ and $|G| = 2^{13}.3^4.5^2.7.17$. The full automorphism group of all the related designs is G too. Each spread contains 17 lines which partition the point set and each parallelism has 21 spreads.

Consider $GF(4)$ with generating polynomial $x^2 = x + 1$ and the points of $PG(3, 4)$ as all the 4-dimensional vectors (v_1, v_2, v_3, v_4) over $GF(4)$ such that if $v_k = 0$ for all $k > i$ then $v_i = 1$. The second and third authors sort in lexicographic order these 85 vectors and assign them numbers, such that $(1, 0, 0, 0)$ is number 1, and $(3, 3, 3, 1)$ number 85. This lexicographic order on the points is used to define a lexicographic order on the lines and on the constructed parallelisms. The latter is necessary for the minimality test described in Sect. 2.6.

Each invertible matrix $(a_{i,j})_{4 \times 4}$ over $GF(4)$ defines an automorphism of the considered projective space by the map $v_i' = \sum_j a_{i,j} v_j$.

Denote by G_2 a Sylow 2-subgroup of G. It is of order 2^{13} and has elements of orders $a = 2, 4$ and 8 which are partitioned in 9 conjugacy classes (Table 2) and generate cyclic groups G_{a_f}, where f is the number of lines which are fixed by G_{a_f}. We denote by $N(G_{a_f})$ the normalizer of G_{a_f} in G, which is defined as $N(G_{a_f}) = \{g \in G \mid g G_{a_f} g^{-1} = G_{a_f}\}$.

Table 2. Conjugacy classes of elements of G_2 and the cyclic groups they generate

Element	1000	3000	1000	1000	1000	3000	3000	1000	3000
	0100	0300	0100	1100	0100	0300	0300	1100	0300
	1010	0030	0010	0110	1010	1030	0030	0110	2030
	0101	0003	1001	0011	0011	0103	1003	1011	1023
G_{a_f}	$G_{2_{21}}$	$G_{2_{35}}$	$G_{2_{37}}$	G_{4_1}	G_{4_5}	G_{4_7}	$G_{4_{11}}$	G_{8_1}	G_{8_3}
$|N(G_{a_f})|$	30720	40320	368640	256	3072	384	768	64	128
subgroups	$G_{2_{21}}$	$G_{2_{35}}$	$G_{2_{37}}$	$G_{2_{21}}$	$G_{2_{37}}$	$G_{2_{21}}$	$G_{2_{37}}$	$G_{2_{21}}, G_{4_1}$	$G_{2_{37}}, G_{4_5}$

There are three conjugacy classes of elements of order 2. We denote the groups generated by their representatives by $G_{2_{21}}$, $G_{2_{35}}$ and $G_{2_{37}}$ respectively.

The group $G_{2_{35}}$ contains a Baer involution, namely a collineation of $PG(3, 4)$ which fixes a subspace $PG(3, 2)$ pointwise. Up to conjugacy, there is a unique Baer involution. All parallelisms invariant under it are classified in [6].

Parallelisms invariant under $G_{2_{37}}$ cannot be constructed, because the group fixes 37 lines and therefore some of the fixed spreads should contain orbits of length 2 with mutually disjoint lines. There are, however, no nontrivial orbits under $G_{2_{37}}$ with mutually disjoint lines.

The group $G_{2_{21}}$ fixes 5 points and 21 lines and partitions the remaining lines in 168 orbits of length 2. Our investigations show that there are millions of parallelisms which are invariant under $G_{2_{21}}$. In the present paper we classify those of them which possess cyclic automorphism groups of order 4. Each such group is generated by a cyclic collineation of order 4.

There are 4 conjugacy classes of elements of order 4 in G_2. Consider the cyclic group of order 4 which such an element generates. It consists of two permutations of order 4, one of order 2 and the identity. We denote by G_{4_1}, G_{4_5}, G_{4_7} and $G_{4_{11}}$ the cyclic groups of order 4 generated by representatives of the four conjugacy classes. A collineation of order 2 that generates $G_{2_{37}}$ is contained in G_{4_5} and $G_{4_{11}}$. Hence they cannot be the automorphism group of a parallelism.

The other two subgroups of order 4 contain $G_{2_{21}}$. We construct here all parallelisms which are invariant under G_{4_1} or G_{4_7} and find the orders of their full automorphism groups.

As a check of the correctness of the obtained results, we also consider cyclic groups of order 8. The group G_2 has two conjugacy classes of elements of order 8. The cyclic group generated by an element of order 8 is made of four collineations of order 8, two of order 4, one of order 2 and the identity. The group G_{8_3} cannot be the automorphism group of a parallelism because its collineation of order 2 generates $G_{2_{37}}$. Hence it suffices to consider only G_{8_1} whose collineation of order 2 generates $G_{2_{21}}$.

2.2   Types of Line and Spread Orbits

Consider the action of $G_{a_f} \in \{G_{4_1}, G_{4_7}, G_{8_1}\}$ on the set of lines. There are f line orbits of length 1 (f fixed lines, or f trivial orbits), s orbits of length $a/2$ (short orbits) and l orbits of length a (full orbits), where $a \in \{4, 8\}$ is the order of the group. In general, a group of order $a = 8$ might have line orbits of length $a/4$ too, but G_{8_1} has no line orbits of length 2.

The nontrivial line orbits can be of two different types with respect to the way they can be used in the construction of parallelisms with assumed automorphisms. Orbits whose lines are mutually disjoint (the orbit lines contain each point at most once) are of the first type. We call them *spread-like* line orbits. A whole orbit of this type can be part of a spread which is fixed by the assumed group. Lines from orbits of the second type (*non-spread-like*) can only be used in nonfixed spreads.

A spread in PG$(3, 4)$ consists of 17 mutually disjoint lines. To construct a parallelism invariant under G_{a_f} we need to consider the action of this group on the spreads. The length of a spread orbit under G_{a_f} can be $1, a/2$ or a and there are 3 different types of spreads (and respectively spread orbits) according to the action of the different subgroups of G_{a_f}. The parallelisms invariant under this group can have spreads which are fixed only by some of its subgroups.

Type F: Such a spread is fixed by G_{a_f} (and therefore by all of its subgroups too). It is made of f_F fixed lines and several spread-like orbits. Among them we denote by l_F the number of orbits of length a, and by s_F the number of those of length $a/2$.

$$f_F + a l_F + \frac{a}{2} s_F = 17.$$

We further use the notation $\boldsymbol{F}_{f_F, s_F}$ for a spread of type F with f_F fixed lines and s_F orbits of length $a/2$.

Type S: The spread is fixed only by the subgroup $G_{2_{21}}$ of order 2. Its spread orbit under G_{a_f} is of length $a/2$. Such a spread contains s_S lines from short orbits under G_{a_f} and d_S whole spread-like orbits under $G_{2_{21}}$ which belong to line orbits of length a under G_{a_f}, such that the collineations of order a map each of these d_S orbits to $a/2 - 1$ other spread-like orbits under $G_{2_{21}}$.

$$s_S + 2d_S = 17.$$

We denote by $\boldsymbol{S_{ss}}$ a spread of type S with s_S lines from short orbits.

Type $\boldsymbol{L}$: This spread contains lines from 17 different line orbits (under G_{a_f}) of one and the same length l_L. To specify the orbit length we use the notation $\boldsymbol{L_{l_L}}$.

Remark: For cyclic groups of order 8, spreads fixed only by the subgroup of order 4 might also be possible, but this does not concern the case we consider. If $G_{a_f} = G_{8_1}$ we cannot construct a spread which is not fixed by G_{8_1}, but fixed by its subgroup of order 4, because its spread orbit under G_{8_1} should be of order 2, and (since the number of lines is 17) the spread must contain at least one line from a line orbit of length 2 under G_{8_1}. There are, however, no line orbits of length 2 under G_{8_1}.

2.3   The Automorphism Group G_{4_1}

It fixes a point and a line. There are $s = 10$ line orbits of length 2 and $l = 84$ of length 4. The number of spread-like orbits of length two is 8, and of length four 48. There is one fixed line and all the lines from short orbits share a point with it. Therefore the fixed spread should be of type $F_{1,0}$. It contains the fixed line a_1 and four spread-like orbits of length 4 ($c_i, c_{i+1}, c_{i+2}, c_{i+3}$, $i = 1, 5, 9, 13$).

$$\boxed{F_{1,0}} : \quad \boxed{a_1 \; c_1 \; c_2 \; c_3 \; c_4 \; \big| \; c_5 \; c_6 \; c_7 \; c_8 \; \big| \; c_9 \; c_{10} \; c_{11} \; c_{12} \; \big| \; c_{13} \; c_{14} \; c_{15} \; c_{16}}$$

There are 32 different spread orbits of type $F_{1,0}$ (Table 3).

Table 3. Spread orbits under the assumed automorphism groups

G_{4_1}		G_{4_7}		G_{8_1}	
type	number	type	number	type	number
$F_{1,0}$	32	$F_{1,0}$	40	$F_{1,0}$	4
S_5	1 216	$F_{3,1}$	64	S_5	100
L_4	159 936	S_5	144	L_8	37472
		L_4	193 600		

Since the number of short line orbits is less than 17, spreads of type L_2 are not possible. The short line orbits under G_{4_1} can only take part in spreads of type S. Two short line orbits contain the first point, so there should be at least two spreads of type S. Our computer-aided investigations show that the parallelisms invariant under G_{4_1} have two spreads of type S_5. They have 5 lines $b_1, b_2, \ldots, b_5$ from orbits $\{b_i, b_i'\}$, $i = 1, 2, \ldots, 5$ of length 2 under G_{4_1} and 6 whole orbits $\{c_i, c_{i+1}\}$, $i = 1, 3, \ldots, 11$ of length 2 under $G_{2_{21}}$ which are part of orbits $\{c_i, c_i', c_{i+1}, c_{i+1}'\}$, $i = 1, 3, \ldots, 11$ of length 4 under G_{4_1}, where a permutation of order 4 permutes them as $(c_i, c_i', c_{i+1}, c_{i+1}')$. This way the spread orbit under G_{4_1} contains the two spreads:

$\boxed{S_5\ S_5}$:

b_1	b_2	b_3	b_4	b_5	c_1	c_2	c_3	c_4	c_5	c_6	c_7	c_8	c_9	c_{10}	c_{11}	c_{12}
b_1'	b_2'	b_3'	b_4'	b_5'	c_1'	c_2'	c_3'	c_4'	c_5'	c_6'	c_7'	c_8'	c_9'	c_{10}'	c_{11}'	c_{12}'

There are 1216 different spread orbits of type S_5 (Table 3).

There are spreads of type L_4 too. They contain 17 lines $c_i, i = 1, 2, \ldots, 17$ of orbits $\{c_i, c_i', c_i'', c_i'''\}$ of length 4 under G_{4_1}. Their number is 159936 (Table 3) and their spread orbit looks like:

$\boxed{L_4\ L_4\ L_4\ L_4}$:

c_1	c_2	c_3	c_4	c_5	c_6	c_7	c_8	c_9	c_{10}	c_{11}	c_{12}	c_{13}	c_{14}	c_{15}	c_{16}	c_{17}
c_1'	c_2'	c_3'	c_4'	c_5'	c_6'	c_7'	c_8'	c_9'	c_{10}'	c_{11}'	c_{12}'	c_{13}'	c_{14}'	c_{15}'	c_{16}'	c_{17}'
c_1''	c_2''	c_3''	c_4''	c_5''	c_6''	c_7''	c_8''	c_9''	c_{10}''	c_{11}''	c_{12}''	c_{13}''	c_{14}''	c_{15}''	c_{16}''	c_{17}''
c_1'''	c_2'''	c_3'''	c_4'''	c_5'''	c_6'''	c_7'''	c_8'''	c_9'''	c_{10}'''	c_{11}'''	c_{12}'''	c_{13}'''	c_{14}'''	c_{15}'''	c_{16}'''	c_{17}'''

A parallelism invariant under G_{4_1} has 21 spreads among which one fixed spread, two spread orbits of length 2 with spreads of type S_5 and four spread orbits of length 4 with spreads of type L_4:

$\boxed{F_{1,0}\ \big|\ S_5\ S_5\ \big|\ S_5\ S_5\ \big|\ L_4\ L_4\ L_4\ L_4\ \big|\ L_4\ L_4\ L_4\ L_4\ \big|\ L_4\ L_4\ L_4\ L_4\ \big|\ L_4\ L_4\ L_4\ L_4}$

Using the normalizer of G_{4_1} which is of order 256 we take away most of the isomorphic solutions and obtain 3 possibilities for the first spread, 292 for the first 3 spread orbits and 107320 parallelisms. It takes several hours on a 3 GHz PC. The subsequent application of a full test for isomorphism shows that the number of nonisomorphic parallelisms is 107030.

2.4   The Automorphism Group G_{4_7}

It fixes 3 points and 7 lines. There are $s = 7$ line orbits of length 2 (4 of them spread-like), and $l = 84$ line orbits of length 4 (among them 48 spread-like). The maximum set of disjoint fixed lines is of size 3. That is why a spread of type F can have either 3 or 1 fixed line. There are only 4 short spread-like line orbits under G_{4_7} and they have a common point. Therefore a spread of type F may contain at most one of them, and there are spreads of types $F_{1,0}$ and $F_{3,1}$.

The structure of a spread of type $F_{1,0}$ is the same for G_{4_7} and G_{4_1} and was considered in the previous subsection. We construct 40 different spreads of type $F_{1,0}$ in $PG(3, 4)$, which are fixed by G_{4_7} (Table 3).

A spread of type $F_{3,1}$ contains three fixed lines a_1, a_2, a_3, a spread-like short line orbit $\{b_1, b_2\}$ and three spread-like orbits of length 4 $\{c_i, c_{i+1}, c_{i+2}, c_{i+3}\}$, $i = 1, 5, 9$. There are 64 spreads of type $F_{3,1}$ fixed under G_{4_7} (Table 3).

$\boxed{F_{3,1}}$:

a_1	a_2	a_3	b_1	b_2	c_1	c_2	c_3	c_4	c_5	c_6	c_7	c_8	c_9	c_{10}	c_{11}	c_{12}

We further establish that the structure of the spreads of types S and L is the same for G_{4_7} and G_{4_1}. It was presented in the previous subsection. For G_{4_7} we obtain 144 spread orbits of type S_5 and 193600 of type L_4 (Table 3).

A parallelism invariant under G_{4_7} has one spread of type $F_{1,0}$, two spreads of type $F_{3,1}$, one spread orbit of length 2 (under G_{4_7}) with spreads of type S_5, and four spread orbits of length 4 with spreads of type L_4:

$F_{1,0}$	$F_{3,1}$	$F_{3,1}$	S_5	S_5	L_4	L_4	L_4	L_4	L_4	L_4	L_4	L_4	L_4	L_4	L_4	L_4	L_4	L_4	L_4	L_4

Using the normalizer of G_{4_7} (of order 384) we take away most isomorphic solutions, and obtain 3 possibilities for the first spread, 49 for the first 3 fixed spreads and 146024 parallelisms. The search takes several hours. We next establish that 145780 parallelisms are nonisomorphic.

2.5   The Automorphism Group G_{8_1}

It fixes a point and a line. There are $s = 5$ line orbits of length 4 and $l = 42$ of length 8. The number of spread-like orbits of length four is 4, and of length eight 8. The fixed spread is of type $F_{1,0}$. It contains the fixed line a_1 and two spread-like orbits of length 8 $\{c_i, c_{i+1}, \ldots, c_{i+7}\}$, $i = 1, 9$.

$F_{1,0}$:		a_1	c_1 c_2 c_3 c_4 c_5 c_6 c_7 c_8	c_9 c_{10} c_{11} c_{12} c_{13} c_{14} c_{15} c_{16}

There are 4 different spread orbits of type $F_{1,0}$ (Table 3) constructed according to this pattern.

Parallelisms invariant under G_{8_1} contain spreads of type S_5. They have 5 lines $b_1, b_2, \ldots, b_5$ from orbits $\{b_i, b_i', b_i'', b_i'''\}$, $i = 1, 2, \ldots, 5$ of length 4 under G_{8_1} and 6 whole orbits $\{c_i, c_{i+1}\}$, $i = 1, 3, \ldots, 11$ of length 2 under $G_{2_{21}}$ which are part of orbits $\{c_i, c_i', c_i'', c_i''', c_{i+1}, c_{i+1}', c_{i+1}'', c_{i+1}'''\}$, $i = 1, 3, \ldots, 11$ of length 8 under G_{8_1}, where a permutation of order 8 permutes the lines as $(c_i, c_i', c_i'', c_i''', c_{i+1}, c_{i+1}', c_{i+1}'', c_{i+1}''')$. This way the spread orbit under G_{8_1} contains the following four spreads:

S_5 S_5 S_5 S_5 :		b_1	b_2	b_3	b_4	b_5	c_1	c_2	c_3	c_4	c_5	c_6	c_7	c_8	c_9	c_{10}	c_{11}	c_{12}
		b_1'	b_2'	b_3'	b_4'	b_5'	c_1'	c_2'	c_3'	c_4'	c_5'	c_6'	c_7'	c_8'	c_9'	c_{10}'	c_{11}'	c_{12}'
		b_1''	b_2''	b_3''	b_4''	b_5''	c_1''	c_2''	c_3''	c_4''	c_5''	c_6''	c_7''	c_8''	c_9''	c_{10}''	c_{11}''	c_{12}''
		b_1'''	b_2'''	b_3'''	b_4'''	b_5'''	c_1'''	c_2'''	c_3'''	c_4'''	c_5'''	c_6'''	c_7'''	c_8'''	c_9'''	c_{10}'''	c_{11}'''	c_{12}'''

There are 100 different spread orbits of type S_5 (Table 3).

The parallelisms have spreads of type L_8 too. They contain lines of 17 different orbits of length 8 under G_{8_1}. Their number is 37472 (Table 3).

A parallelism invariant under G_{8_1} has one fixed spread, one spread orbit of length 4 with spreads of type S_5 and two spread orbits of length 8 with spreads of type L_8:

$F_{1,0}$	S_5	S_5	S_5	S_5	L_8	L_8	L_8	L_8	L_8	L_8	L_8	L_8	L_8	L_8	L_8	L_8	L_8	L_8	L_8	L_8

Using the normalizer of G_{8_1} (which is of order 64), we take away part of the isomorphic solutions and obtain 248 parallelisms. The computation takes about an hour. We next find out that the number of nonisomorphic parallelisms is 220.

2.6   Computer Search

We construct the parallelisms by backtrack search with rejection of equivalent partial solutions at several stages. Since each spread orbit is defined by any one of its spreads, we find only one spread from each orbit and call it *spread orbit leader*. This way, instead of all 21 spreads, we construct 7 orbit leaders for the parallelisms with an assumed group G_{4_1}, 8 for G_{4_7}, and 4 for G_{8_1}.

The three authors offer slightly differing construction methods. The approach of the first and third authors implies construction of all the possible spread orbit leaders in advance. Their number is presented in Table 3. They are obtained by backtrack search on the lines of the projective space. A spread consists of disjoint lines. If we have chosen part of its lines, we add the line l if it is disjoint to all of them, and if it meets the requirements of the spread type (for instance, if the spread is of type F, the line orbit of l under G_{a_f} must contain mutually disjoint lines, all of them disjoint to each of the already chosen spread lines). The parallelisms are constructed next by backtrack search on these possible spreads. The spread which we add must be disjoint to all the spreads from the current partial solution.

The second author does not construct the possible spreads first. The search is on the lines. The line l is added to the current spread if it meets the requirements of the spread type, and has not been used in the already constructed spreads of the parallelism. The latter requirement is quite restrictive and that is why the computation times needed by the two approaches are comparable.

All the authors apply isomorphism testing at several stages within the search. For that purpose the first author uses the graph theory package Nauty [19], while the other authors apply a normalizer-based **minimality test**. It checks if there is an element of the normalizer $N(G_{a_f})$ (of G_{a_f} in G) which takes the constructed partial parallelism to a lexicographically smaller partial solution. If so, the current partial solution is discarded.

3   Results

All the constructed parallelisms are available online. They can be downloaded from http://www.moi.math.bas.bg/moiuser/~stela. A summary of the obtained results is presented in Table 4, where the order of the full automorphism group $|G_{\mathcal{P}}|$ of the parallelisms is given in the first column and the number of solutions obtained with the assumed groups G_{4_1} and G_{4_7} in the next two columns. There are parallelisms which are invariant both under G_{4_1} and G_{4_7}, so that the total number is less than the sum of the numbers in the previous two columns. In that case the row background is gray. You can see from the fifth column that the number of self-dual parallelisms is relatively small.

We determine the type of each spread in the parallelism. The number of subregular, regular and aregular spreads form an invariant which partitions the constructed parallelisms to 129 classes. The number of such classes for the parallelisms with each order of the full automorphism group is presented in the last column of Table 4.

Table 4. The order of the full automorphism group of the constructed parallelisms

$|G_{\mathcal{P}}|$	G_{4_1}	G_{4_7}	Total	Selfdual	Invariants
4	106464	145372	251836	522	129
8	420	232	596	14	14
12		40	40	–	1
16	122	48	170	6	5
20		52	52	–	8
32	14	8	14	2	2
48		12	12	–	1
60		8	8	–	1
64	4	4	4	–	1
96	2	–	2	2	1
960	4	4	4	–	1
Total	107030	145780	252738	546	129

Table 5. Types of spreads of the parallelisms with a full automorphism group of order greater than 4

$|G_{\mathcal{P}}|$	S R A	G_{4_1}	G_{4_7}	G_{8_1}	Total
8	0 1 20	56		56	56
8	0 5 16	2		2	2
8	4 0 17	16	16		16
8	4 1 16	122	136	40	258
8	4 9 8	8		2	8
8	6 1 14	8			8
8	8 1 12	30		30	30
8	8 5 8	4		4	4
8	10 1 10	16			16
8	12 0 9	40	40		40
8	12 1 8	72	8	38	80
8	12 9 0	10		2	10
8	16 1 4	4		4	4
8	20 1 0	32	32	6	64
12	20 1 0		40		40
16	4 1 16	28	20	24	48
16	4 9 8	12			12
16	12 1 8	16			16

$|G_{\mathcal{P}}|$	S R A	G_{4_1}	G_{4_7}	G_{8_1}	Total
16	12 9 0	12			12
16	20 1 0	54	28		82
20	6 0 15		4		4
20	10 1 10		4		4
20	11 0 10		4		4
20	11 5 5		4		4
20	15 1 5		10		10
20	16 0 5		10		10
20	20 1 0		8		8
20	21 0 0		8		8
32	4 1 16	8	4	4	8
32	20 1 0	6	4		6
48	20 1 0		12		12
60	20 1 0		8		8
64	20 1 0	4	4	4	4
96	20 1 0	2			2
960	20 1 0	4	4	4	4

Table 6. Uniform parallelisms invariant under cyclic groups of order 4

Type	$|G_{\mathcal{P}}| = 4$	$|G_{\mathcal{P}}| = 20$	Assumed automorphism group
Subregular	244	8	G_{4_7}
Aregular	4816		G_{4_1}

To find out if there are parallelisms invariant both under G_{4_1} and G_{4_7} (and thus constructed with each of these assumed groups) we test for isomorphism all parallelisms which have the same invariants and an automorphism group of order more than 4. The results are presented in Table 5, where columns S (subregular), R (regular) and A (aregular) present the invariants. The gray rows mark the parallelisms whose automorphism group contains both G_{4_1} and G_{4_7}.

We use our results for G_{8_1} to partially check the correctness of our computations. All the constructed parallelisms with automorphism group G_{8_1} are isomorphic to parallelisms obtained with an assumed group G_{4_1}. The data about G_{8_1} is presented in Table 5.

Uniform parallelisms are of particular interest [22]. We obtain many aregular and subregular parallelisms and no regular ones (Table 6). The parallelisms we construct have at most 10 regular spreads.

The number of parallelisms with automorphisms of prime orders 5 and 3 is in consistence with the classifications in [29, 31]. Only 118 of the parallelisms constructed here were known before the present work.

References

1. Baker, R.D.: Partitioning the planes of $AG_{2m}(2)$ into 2-designs. Discret. Math. **15**, 205–211 (1976)
2. Bamberg, J.: There are no regular packings of PG(3, 3) or PG(3, 4). https://symomega.wordpress.com/2012/12/01/. Accessed 24 Jan 2019
3. Betten, A.: The packings of PG(3, 3). Des. Codes Cryptogr. **79**(3), 583–595 (2016)
4. Betten, A.: Orbiter - a program to classify discrete objects, 2016–2018. https://github.com/abetten/orbiter. Accessed 24 Jan 2019
5. Betten, A.: Spreads and packings - an update. In: Booklet of Combinatorics 2016, Maratea (PZ), Italy, 29th May–5th June, p. 45 (2016)
6. Betten, A., Topalova, S., Zhelezova, S.: Parallelisms of PG(3, 4) invariant under a baer involution. In: Proceedings of the 16th International Workshop on Algebraic and Combinatorial Coding Theory, Svetlogorsk, Russia, pp. 57–61 (2018). http://acct2018.skoltech.ru/
7. Beutelspacher, A.: On parallelisms in finite projective spaces. Geom. Dedicata **3**(1), 35–40 (1974)
8. Colbourn, C., Dinitz, J. (eds.) Handbook of combinatorial designs. In: Rosen, K. (ed.) Discrete Mathematics and its Applications, 2nd edn. CRC Press, Boca Raton (2007)
9. Dempwolff, U., Reifart, A.: The classification of the translation planes of order 16. I. Geom. Dedicata **15**(2), 137–153 (1983)
10. Denniston, R.H.E.: Some packings of projective spaces. Atti Accad. Naz. Lincei Rend. Cl. Sci. Fis. Mat. Natur. **52**(8), 36–40 (1972)
11. Eisfeld, J., Storme L.: (Partial) t-spreads and minimal t-covers in finite projective spaces. Lecture notes from the Socrates Intensive Course on Finite Geometry and its Applications, Ghent, April 2000
12. Etzion, T., Silberstein, N.: Codes and designs related to lifted MRD codes. IEEE Trans. Inform. Theory **59**(2), 1004–1017 (2013)
13. GAP - Groups, algorithms, programming - a system for computational discrete algebra. http://www.gap-system.org/. Accessed 24 Jan 2019

14. Gruner, A., Huber, M.: New combinatorial construction techniques for low-density parity-check codes and systematic repeat-accumulate codes. IEEE Trans. Commun. **60**(9), 2387–2395 (2012)
15. Fuji-Hara, R.: Mutually 2-orthogonal resolutions of finite projective space. Ars Combin. **21**, 163–166 (1986)
16. Johnson, N.L.: Some new classes of finite parallelisms. Note Mat. **20**(2), 77–88 (2000)
17. Johnson, N.L.: Combinatorics of Spreads and Parallelisms. Chapman & Hall Pure and Applied Mathematics. CRC Press, Boca Raton (2010)
18. Magma. The Computational Algebra Group within the School of Mathematics and Statistics of the University of Sydney (2004)
19. McKay, B.: Nauty User's Guide (Version 2.4). Australian National University (2009)
20. Penttila, T., Williams, B.: Regular packings of PG(3, q). Eur. J. Comb. **19**(6), 713–720 (1998)
21. Prince, A.R.: Parallelisms of $PG(3,3)$ invariant under a collineation of order 5. In: Johnson, N.L. (ed.) Mostly Finite Geometries. Lecture Notes in Pure and Applied Mathematics, vol. 190, pp. 383–390. Marcel Dekker, New York (1997)
22. Prince, A.R.: Uniform parallelisms of PG(3,3). In: Hirschfeld, J., Magliveras, S., Resmini, M. (eds.) Geometry, Combinatorial Designs and Related Structures. London Mathematical Society Lecture Note Series, vol. 245, pp. 193–200. Cambridge University Press, Cambridge (1997)
23. Prince, A.R.: The cyclic parallelisms of $PG(3,5)$. Eur. J. Comb. **19**(5), 613–616 (1998)
24. Sarmiento, J.: Resolutions of $PG(5,2)$ with point-cyclic automorphism group. J. Comb. Des. **8**(1), 2–14 (2000)
25. Stinson, D.R.: Combinatorial Designs: Constructions and Analysis. Springer, New York (2004). https://doi.org/10.1007/b97564
26. Stinson, D.R., Vanstone, S.A.: Orthogonal packings in $PG(5,2)$. Aequationes Math. **31**(1), 159–168 (1986)
27. Storme, L.: Finite geometry. In: Colbourn, C., Dinitz, J., Handbook of Combinatorial Designs, Rosen, K. (eds.) Discrete Mathematics and its Applications, 2nd edn, pp. 702–729. CRC Press, Boca Raton (2007)
28. Topalova, S., Zhelezova, S.: On transitive parallelisms of PG(3, 4). Appl. Algebra Engrg. Comm. Comput. **24**(3–4), 159–164 (2013)
29. Topalova, S., Zhelezova, S.: On point-transitive and transitive deficiency one parallelisms of PG(3, 4). Des. Codes Cryptogr. **75**(1), 9–19 (2015)
30. Topalova, S., Zhelezova, S.: New regular parallelisms of $PG(3,5)$. J. Comb. Des. **24**, 473–482 (2016)
31. Topalova, S., Zhelezova, S.: New parallelisms of PG(3, 4). Electron. Notes Discret. Math. **57**, 193–198 (2017)
32. Topalova, S., Zhelezova, S.: Types of spreads and duality of the parallelisms of $PG(3,5)$ with automorphisms of order 13. Des. Codes Cryptogr. **87**(2–3), 495–507 (2019)
33. Zaicev, G., Zinoviev, V., Semakov, N.: Interrelation of Preparata and Hamming codes and extension of Hamming codes to new double-error-correcting codes. In: 1971 Proceedings of Second International Symposium on Information Theory, Armenia, USSR, pp. 257–263. Academiai Kiado, Budapest (1973)

Bounds on Covering Codes in RT Spaces Using Ordered Covering Arrays

André Guerino Castoldi[1], Emerson Luiz do Monte Carmelo[2],
Lucia Moura[3]([⊠]), Daniel Panario[4], and Brett Stevens[4]

[1] Departamento Acadêmico de Matemática,
Universidade Tecnológica Federal do Paraná, Pato Branco, Brazil
andrecastoldi@utfpr.edu.br
[2] Departamento de Matemática,
Universidade Estadual de Maringá, Maringá, Brazil
elmcarmelo@uem.br
[3] School of Electrical Engineering and Computer Science,
University of Ottawa, Ottawa, Canada
lmoura@uottawa.ca
[4] School of Mathematics and Statistics, Carleton University, Ottawa, Canada
{daniel,brett}@math.carleton.ca

Abstract. In this work, constructions of ordered covering arrays are discussed and applied to obtain new upper bounds on covering codes in Rosenbloom-Tsfasman spaces (RT spaces), improving or extending some previous results.

Keywords: Rosenbloom-Tsfasman metric · Covering codes ·
Bounds on codes · Ordered covering arrays

1 Introduction

Roughly speaking, covering codes deal with the following problem: Given a metric space, how many balls are enough to cover all the space? Several applications, such as data transmission, cellular telecommunications, decoding of errors, football pool problem, have motivated the study of covering codes in Hamming spaces. Covering codes also have connections with other branches of mathematics and computer science, such as finite fields, linear algebra, graph theory, combinatorial optimization, mathematical programming, and metaheuristic search. We refer the reader to the book by Cohen et al. [7] for an overview of the topic.

Rosenbloom and Tsfasman [18] introduced the RT metric on linear spaces over finite fields, motivated by possible applications to interference in parallel channels of communication systems. Since the RT metric generalizes the Hamming metric, central concepts on codes in Hamming spaces have been investigated in RT space, like perfect codes, MDS codes, linear codes, distribution, packing and covering problems.

M. Ćirić et al. (Eds.): CAI 2019, LNCS 11545, pp. 100–111, 2019.
https://doi.org/10.1007/978-3-030-21363-3_9

Most research on codes using the RT metric focuses on packing codes; covering codes in RT spaces have not been much explored. Brualdi et al. [2] implicitly investigated such codes when the space is induced by a chain. The minimum covering code for a chain is computed in [19]. An extension to an arbitrary RT space is proposed in [4], which deals mainly with upper bounds, inductive relations and some sharp bounds as well as relations with MDS codes. More recently, [5] improved the sphere covering bound in RT spaces under some conditions by generalizing the excess counting method. In this work, we explore upper bounds and inductive relations for covering codes in RT spaces by using ordered covering arrays (OCA), as briefly described below.

Ordered covering arrays (OCAs) are a generalization of ordered orthogonal arrays (OOA) and covering arrays (CA). Orthogonal arrays are classical combinatorial designs with close connections to coding theory; see the book on orthogonal arrays by Hedayat et al. [11]. Covering arrays generalize orthogonal arrays and have been given a lot of attention due to their use in software testing and interesting connections with other combinatorial designs; see the survey paper by Colbourn [8]. Ordered orthogonal arrays are a generalization of orthogonal arrays introduced independently by Lawrence [14] and Mullen and Schmid [15], and are used in numerical integration. OCAs have been introduced more recently by Krikorian [13], generalizing several of the mentioned designs; their definition is given in Sect. 3. In [13], Krikorian gives recursive and Roux-type constructions of OCAs as well as other constructions using the columns of a covering array and discusses an application of OCAs to numerical integration (evaluating multi-dimensional integrals).

In this paper, we apply OCAs to obtain new upper bounds on covering codes in RT spaces. We review the basics on RT metric and covering codes in Sect. 2. CAs and OCAs are defined in Sect. 3. Section 4 is devoted to recursive relations on the parameters of OCAs. Finally, in Sect. 5, we obtain upper bounds on covering codes in RT spaces from OCAs.

2   Preliminaries: RT Metric and Covering Codes

We review the RT metric based on [2]. Let P be a finite partial ordered set (poset) and denote its partial order relation by $\preceq$. A poset is a *chain* when any two elements are comparable; a poset is an *anti-chain* when no two distinct elements are comparable. A subset I of P is an *ideal* of P when the following property holds: if $b \in I$ and $a \preceq b$, then $a \in I$. The *ideal generated* by a subset A of P is the ideal of the smallest cardinality which contains A, denoted by $\langle A \rangle$. An element $a \in I$ is *maximal in* I if $a \preceq b$ implies that $b = a$. Analogously, an element $a \in I$ is *minimal in* I if $b \preceq a$ implies that $b = a$. A subset J of P is an *anti-ideal* of P when it is the complement of an ideal of P. If an ideal I has t elements, then its corresponding anti-ideal has $n - t$ elements, where n denotes the number of elements in P.

Let m and s be positive integers and $\Omega[m, s]$ be a set of ms elements partitioned into m blocks B_i having s elements each, where $B_i = \{b_{is}, \ldots, b_{(i+1)s-1}\}$ for $i = 0, \ldots, m - 1$ and the elements of each block are ordered as $b_{is} \preceq$

$b_{is+1} \preceq \cdots \preceq b_{(i+1)s-1}$. The set $\Omega[m, s]$ has a structure of a poset: it is the union of m disjoint chains, each one having s elements, which is known as the *Rosenbloom-Tsfasman poset* $\Omega[m, s]$, or briefly an RT poset $\Omega[m, s]$. When $\Omega[m, s] = [m \times s] := \{1, \ldots, ms\}$, the RT poset $\Omega[m, s]$ is denoted by RT poset $[m \times s]$ and its blocks are $B_i = \{i + 1, \ldots, (i + 1)s\}$, for $i = 0, \ldots, m - 1$.

For $1 \leq i \leq ms$ and $1 \leq j \leq \min\{m, i\}$, the parameter $\Omega_j(i)$ denotes the number of ideals of the RT poset $[m \times s]$ whose cardinality is i with exactly j maximal elements. In [5, Proposition 1], it is shown that $\Omega_j(i) = \binom{m}{j}\binom{i-1}{j-1}$ if $j \leq i \leq s$.

The *RT distance* between $x = (x_1, \ldots, x_{ms})$ and $y = (y_1, \ldots, y_{ms})$ in $\mathbb{Z}_q^{ms}$ is defined as [2]

$$d_{RT}(x, y) = |\langle supp(x - y)\rangle| = |\langle\{i : x_i \neq y_i\}\rangle|.$$

A set $\mathbb{Z}_q^{ms}$ endowed with the RT distance is a *Rosenbloom-Tsfasman space*, or simply, an *RT space*.

The RT sphere centered at x of radius R, denoted by $B^{RT}(x, R) = \{y \in \mathbb{Z}_q^{ms} : d_{RT}(x, y) \leq R\}$, has cardinality given by the formula

$$V_q^{RT}(m, s, R) = 1 + \sum_{i=1}^{R} \sum_{j=1}^{\min\{m,i\}} q^{i-j}(q - 1)^j \Omega_j(i). \tag{1}$$

As expected, the case $s = 1$ corresponds to the classical Hamming sphere. Indeed, each subset produces an ideal formed by minimal elements of the anti-chain $[m \times 1]$, thus the parameters $\Omega_i(i) = \binom{m}{i}$ and $\Omega_j(i) = 0$ for $j < i$ yield

$$V_q(m, R) = V_q^{RT}(m, 1, R) = 1 + \sum_{i=1}^{R}(q - 1)^i \binom{m}{i}. \tag{2}$$

In contrast with the Hamming space, the computation of the sum in Eq. (1) is not a feasible procedure for a general RT space. In addition to the well studied case $s = 1$, it is known that $V_q^{RT}(1, s, R) = q^R$ for a space induced by a chain $[1 \times s]$, see [2, Theorem 2.1] and [19]. Also, it is proved in [5, Corollary 1] that, for $R \leq s$,

$$V_q^{RT}(m, s, R) = 1 + \sum_{i=1}^{R} \sum_{j=1}^{\min\{m,i\}} q^{i-j}(q - 1)^j \binom{m}{j}\binom{i-1}{j-1}.$$

We now define covering codes in an arbitrary RT space, and refer the reader to [4] for an overview.

Definition 1. Given an RT poset $[m \times s]$, let C be a subset of $\mathbb{Z}_q^{ms}$. The code C is an *R-covering* of the RT space $\mathbb{Z}_q^{ms}$ if for every $x \in \mathbb{Z}_q^{ms}$ there is a codeword $c \in C$ such that $d_{RT}(x, c) \leq R$, or equivalently,

$$\bigcup_{c \in C} B^{RT}(c, R) = \mathbb{Z}_q^{ms}.$$

The number $K_q^{RT}(m, s, R)$ denotes the smallest cardinality of an R-covering of the RT space $\mathbb{Z}_q^{ms}$.

In particular, $K_q^{RT}(m, 1, R) = K_q(m, R)$. The sphere covering bound and a general upper bound are stated below.

Proposition 1. *([4, Propositions 6 and 7]) For every $q \geq 2$ and R such that $0 < R < ms$,*

$$\frac{q^{ms}}{V_q^{RT}(m, s, R)} \leq K_q^{RT}(m, s, R) \leq q^{ms-R}.$$

3   Ordered Covering Arrays

In this section, we define an important combinatorial object for this paper. We start recalling two classical combinatorial structures.

Definition 2. Let t, v, λ, n, N be positive integers and $N \geq \lambda v^t$. Let A be an $N \times n$ array over an alphabet V of size v. An $N \times t$ subarray of A is λ-**covered** if it has each t-tuple over V as a row at least λ times. A set of t columns of A is λ-**covered** if the $N \times t$ subarray of A formed by them is λ-covered; when $\lambda = 1$ we simply say it is covered.

In what follows, whenever $\lambda = 1$, we omit λ from the notation.

Definition 3. *(CA and OA)* Let N, n, v and λ be positive integers such that $2 \leq t \leq n$. A **covering array** $CA_\lambda(N; t, n, v)$ is an $N \times n$ array A with entries from a set V of size v such that any t-set of columns of A is λ-covered. The parameter t is the **strength** of the covering array. The **covering array number** $CAN_\lambda(t, n, v)$ is the smallest positive integer N such that a $CA_\lambda(N; t, n, v)$ exists. An **orthogonal array** is a covering array with $N = \lambda v^t$.

We are now ready to introduce ordered covering arrays.

Definition 4. *(OCA and OOA)* Let t, m, s, v and λ be positive integers such that $2 \leq t \leq ms$. An **ordered covering array** $OCA_\lambda(N; t, m, s, v)$ is an $N \times ms$ array A with entries from an alphabet V of size v, whose columns are labeled by an RT poset $\Omega[m, s]$, satisfying the property: for each anti-ideal J of the RT poset $\Omega[m, s]$ with $|J| = t$, the set of columns of A labeled by J is λ-covered. The parameter t is the **strength** of the ordered covering array. The **ordered covering array number** $OCAN_\lambda(t, m, s, v)$ is the smallest positive integer N such that there exists an $OCA_\lambda(N; t, m, s, v)$. An **ordered orthogonal array** is an ordered covering array with $N = \lambda v^t$.

Remark 1. Ordered covering arrays are special cases of variable strength covering arrays [16,17]. Ordered covering arrays were first studied by Krikorian [13].

Example 1. The following array is an OCA of strength 2 with 5 rows:

$$OCA(5; 2, 4, 2, 2) = \begin{array}{c} \begin{array}{cc|cc|cc|cc} 1 & 2 & 3 & 4 & 5 & 6 & 7 & 8 \end{array} \\ \left[\begin{array}{cc|cc|cc|cc} 0 & 1 & 0 & 1 & 0 & 1 & 0 & 1 \\ 1 & 1 & 1 & 0 & 0 & 0 & 0 & 0 \\ 0 & 0 & 1 & 1 & 1 & 0 & 1 & 0 \\ 1 & 0 & 0 & 0 & 1 & 1 & 0 & 0 \\ 0 & 0 & 0 & 0 & 0 & 0 & 1 & 1 \end{array} \right] \end{array}.$$

The columns of this array are labeled by $[4 \times 2] = \{1, \ldots, 8\}$ and the blocks of the RT poset $[4 \times 2]$ are $B_0 = \{1,2\}$, $B_1 = \{3,4\}$, $B_2 = \{5,6\}$ are $B_3 = \{7,8\}$. We have ten anti-ideals of size 2, namely,

$$\{1,2\}, \{3,4\}, \{5,6\}, \{7,8\}, \{2,4\}, \{2,6\}, \{2,8\}, \{4,6\}, \{4,8\}, \{6,8\}.$$

The 5×2 subarray constructed from each one of theses anti-ideals covers all the pairs $(0,0)$, $(0,1)$, $(1,0)$ and $(1,1)$ at least once.

In an $OCA_\lambda(N; t, m, s, v)$ such that $s > t$, each one of the first $s - t$ elements of a block in the RT poset $\Omega[m, s]$ is not an element of any anti-ideal of size t. Therefore, a column labeled by one of these elements will not be part of any $N \times t$ subarray that must be λ-covered in an OCA. So we assume $s \leq t$ from now on.

Two trivial relationships between the ordered covering array number and the covering array number $CAN_\lambda(t, n, v)$ are:

(1) $\lambda v^t \leq OCAN_\lambda(t, m, s, v) \leq CAN_\lambda(t, ms, v)$;
(2) For $t \leq m$, $CAN_\lambda(t, m, v) \leq OCAN_\lambda(t, m, t, v)$.

If $\lambda = 1$, we just write $OCA(N; t, m, s, v)$. We observe that if $N = \lambda v^t$, an $OCA_\lambda(\lambda v^t; t, m, s, v)$ is an ordered orthogonal array $OOA_\lambda(\lambda v^t; t, m, s, v)$. When $s = 1$, an $OCA_\lambda(N; t, m, 1, v)$ is the well-known covering array $CA_\lambda(N; t, m, v)$.

4   Recursive Relations for Ordered Covering Arrays

In this section, we show recursive relations for ordered coverings arrays.

Proposition 2. (1) *If there exists an $OCA_\lambda(N; t, m, t - 1, v)$, then there exists an $OCA_\lambda(N; t, m, t, v)$.*
(2) *If there exists an $OCA_\lambda(N; t, m, s, v)$, then there exists an $OCA_\lambda(N; t, m, s - 1, v)$.*
(3) *If there exists an $OCA_\lambda(N; t, m, s, v)$, then there exists an $OCA_\lambda(N; t, m - 1, s, v)$.*

Proof. Parts (2) and (3) are easily obtained by deletion of appropriate columns of the $OCA_\lambda(N; t, m, s, v)$. We prove part (1). Let P be an RT poset $\Omega[m, t]$, and let $B_0, B_1, \ldots, B_{m-1}$ be the blocks of P. Each B_i, $0 \leq i < m$, is a chain, and we denote by $\min(B_i)$ and $\max(B_i)$ the minimum and maximum elements of B_i, respectively. Let $M = \{\max(B_0), \ldots, \max(B_{m-1})\}$ and let π be a derangement of M. Let $P' = P \setminus \{\min(B_0), \ldots, \min(B_{m-1})\}$, an RT poset $\Omega[m, (t - 1)]$, and let A' be an $OCA_\lambda(N; t, m, t - 1, v)$ with columns labeled by P'. We take a map $f \colon P \to P'$ given by $f(x) = x$ if $x \in P'$ and $f(x) = \pi(\max(B_i))$, if $x = \min(B_i)$. In Fig. 1, we depict P' (above) and P (below) where $\min(B_i)$ is labeled by $\bar{a}$ where $a = \pi(\max(B_i))$. Construct an array A with columns labeled by elements of P by taking the column of A labeled by x to be the column of A' labeled by $f(x)$. Let J be any anti-ideal of P of size t and let $J' = f(J)$.

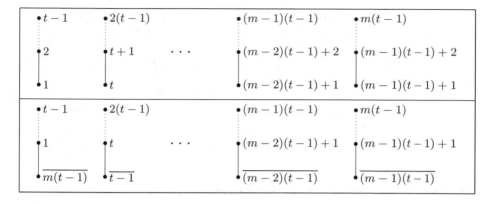

Fig. 1. Above, P' is an RT poset $\Omega[m, t-1]$; below, $P \supset P'$ is an RT poset $\Omega[m, t]$.

Then either $J = J' \subseteq P'$, or $\min(B_i) \in J$ for some i, which implies $J = B_i$ and $J' = (B_i \setminus \{\min(B_i)\}) \cup \max(B_j)$ for some $j \neq i$. In either case, J' is an anti-ideal of P', and so the set of t columns of A' corresponding to J' is λ-covered. Therefore, the set of t columns of A corresponding to J is λ-covered, for any anti-ideal J of P of size t, and A is an $OCA_\lambda(N; t, m, t, v)$. ☐

As a straightforward consequence of Proposition 2 items (1) and (2), we have the following result.

Corollary 1. *There exists an $OCA_\lambda(N; t, m, t, v)$ if and only if there exists an $OCA_\lambda(N; t, m, t-1, v)$.*

By the above corollary, when $t = 2$, the right hand side OCA has $s = t - 1 = 1$ which corresponds to a covering array; therefore, in this case, the ordered covering array number is equal to the covering array number. This also shows that we need $t > 2$ in order to have ordering covering arrays essentially different than covering arrays.

Example 2. $OCAN_\lambda(2, m, 2, v) = CAN_\lambda(2, m, v)$.

Let us label the columns of a $CA_\lambda(N; 2, m, v)$ by the elements of $[m] = \{1, \ldots, m\}$. Choose the columns of $CA_\lambda(N; 2, m, v)$ labeled by the elements of $[m]$ given in Fig. 2. We use again the notation $\bar{a}$ to duplicate $a \in [m]$ in the RT poset $\Omega[m, 2]$ in such a way that a and $\bar{a}$ are not comparable, but the columns of $CA_\lambda(N; 2, m, v)$ labeled by a and $\bar{a}$ are equal.

Fig. 2. Blocks of the RT poset $\Omega[m, 2]$.

It is known from [12] that $CAN(2, m, 2)$ is the smallest positive integer N such that $m \leq \binom{N-1}{\lfloor \frac{N}{2} \rfloor - 1}$. As a consequence, we obtain the following result.

Corollary 2. *The ordered covering array number $OCAN(2, m, 2, 2)$ is the smallest positive integer N such that $m \leq \binom{N-1}{\lfloor \frac{N}{2} \rfloor - 1}$.*

In the next result, we show an upper bound for the ordered covering array number over an alphabet of size v from the ordered covering array number over an alphabet of size $v + 1$. It generalizes [9, Lemma 3.1] and part of [10, Lemma 3.1].

Theorem 1. *(Fusion Theorem) $OCAN_\lambda(t, m, s, v) \leq OCAN_\lambda(t, m, s, v+1) - 2$.*

Proof. Let $V = \{1, \ldots, v + 1\}$ be an alphabet of size $v + 1$ and consider an $OCA_\lambda(N; t, m, s, v + 1)$ over V. The permutation of the entries of any column of the $OCA_\lambda(N; t, m, s, v + 1)$ still produces an ordered covering array with the same parameters. If necessary, applying a permutation in each of the columns, we can guarantee that there exists a row in $OCA_\lambda(N; t, m, s, v + 1)$ such that all the entries are $v + 1$. We delete this row.

Choose a second row $r = (c_1, \ldots, c_{ms})$ of $OCA_\lambda(N; t, m, s, v + 1)$. In every row except r, where there exist an entry $v + 1$ in column i, replace $v + 1$ by c_i if $c_i \neq v + 1$, otherwise, replace $v + 1$ by any element of $\{1, \ldots, v\}$. Delete row r.

The array A obtained by deleting these two rows of $OCA_\lambda(N; t, m, s, v + 1)$ is an $OCA_\lambda(N - 2; t, m, s, v)$, since each t-tuple that was covered by row r is now covered by one or more of the modified rows. $\qquad \square$

By [6, Theorem 3] and Theorem 1, we derive an upper bound on the ordered covering array number.

Corollary 3. *Let q be a prime power. Then, $OCAN(t, q+1, t, q-1) \leq q^t - 2$.*

5   Constructions of Covering Codes Using Covering Arrays

In this section, ordered covering arrays are used to construct covering codes in RT spaces yielding upper bounds on their size. Theorems 2 and 3 are a generalization of results already discovered for covering codes in Hamming spaces connected with surjective matrices [7].

Let $I = \{i_1, \ldots, i_k\}$ be a subset of $[n] = \{1, \ldots, n\}$. Given an element $x = (x_1, \ldots, x_n) \in \mathbb{Z}_q^n$, the projection of x with respect to I is the element $\pi_I(x) = (x_{i_1}, \ldots, x_{i_k}) \in \mathbb{Z}_q^k$. More generally, for a non-empty subset C of $\mathbb{Z}_q^n$, the projection of C with respect to I is the set $\pi_I(C) = \{\pi_I(c) : c \in C\}$.

In [4, Theorem 13], it was proved that $K_q^{RT}(m, s, ms - t) = q$ if $m \geq (t - 1)q + 1$. What can we say about $K_q^{RT}(m, s, ms - t)$ when $m = (t - 1)q$? In this direction, we have the following result. A reference for item (1) is [7]; the other two items are original results of the present paper.

Theorem 2. *For $t \geq 2$,*

(1) $K_q((t-1)q, (t-1)q - t) \leq q - 2 + CAN(t, (t-1)q, 2)$.

(2) $K_q^{RT}((t-1)q, s, (t-1)qs - t) \leq K_q((t-1)q, (t-1)q - t)$.

(3) $K_q^{RT}((t-1)q, s, (t-1)qs - t) \leq q - 2 + CAN(t, (t-1)q, 2)$.

Proof. (Sketch.) We only prove part (2) here; part (3) is straightforward from (1) and (2). Let M be the set of all maximal elements of the RT poset $[(t-1)q \times s]$ and C' be a $((t-1)q - t)$-covering code of the Hamming space $\mathbb{Z}_q^{(t-1)q}$. Let C be the subset of $\mathbb{Z}_q^{(t-1)qs}$ such that $c \in C$ if and only if $\pi_M(c) \in C'$ and all the other coordinates of c are equal to zero. Given $x \in \mathbb{Z}_q^{(t-1)qs}$, let $\pi_M(x) \in \mathbb{Z}_q^{(t-1)q}$. Since C' is a $((t-1)q - t)$-covering of the Hamming space $\mathbb{Z}_q^{(t-1)q}$, there exists $c' \in C'$ such that $\pi_M(x)$ and c' coincide in at least t coordinates. Let $c \in C$ such that $\pi_M(c) = c'$. Therefore $d_{RT}(x, c) \leq (t-1)qs - t$, and C is a $((t-1)qs - t)$-covering of the RT space $\mathbb{Z}_q^{(t-1)qs-t}$. $\square$

Applying the trivial bounds we have that $3 \leq K_3^{RT}(3, s, 3s - 2) \leq 9$. The upper bound can be improved by Theorem 2.

Corollary 4. $K_3^{RT}(3, s, 3s - 2) \leq 5$.

Proof. Theorem 2 yields $K_3^{RT}(3, s, 3s - 2) \leq 1 + CAN(2, 3, 2)$. On the other hand, $CAN(2, 3, 2) = 4$, according to [12], and the upper bound follows. $\square$

MDS codes have been used to improve upper bounds on $K_q(n, R)$ [1,3,7]. In [4, Theorem 30], MDS codes in RT spaces are used to improve upper bounds for $K_q^{RT}(m, s, R)$. We generalize these results using ordered covering arrays.

Theorem 3. $K_{vq}^{RT}(m, s, R) \leq OCAN(ms - R, m, s, v)K_q^{RT}(m, s, R)$.

Proof. Throughout this proof, the set $\mathbb{Z}_{vq}$ is regarded as the set $\mathbb{Z}_{vq} = \mathbb{Z}_v \times \mathbb{Z}_q$ by setting the bijection $xq + y \rightarrow (x, y)$. This strategy allows us to analyze the information on the coordinates x and y separately.

Let H be an R-covering of the RT space $\mathbb{Z}_q^{ms}$, and let C be the set of the rows of an $OCA(N; ms - R, m, s, v)$. We show that

$$G = \{((c_1, h_1), \ldots, (c_{ms}, h_{ms})) \in \mathbb{Z}_{vq}^{ms} : (c_1, \ldots, c_{ms}) \in C, (h_1, \ldots, h_{ms}) \in H\}$$

is a R-covering of the RT space $\mathbb{Z}_{vq}^{ms}$.

Indeed, for $z = ((x_1, y_1), \ldots, (x_{ms}, y_{ms})) \in \mathbb{Z}_{vq}^{ms}$, let $x = (x_1, \ldots, x_{ms})$ in $\mathbb{Z}_v^{ms}$ and $y = (y_1, \ldots, y_{ms})$ in $\mathbb{Z}_q^{ms}$. Since H is an R-covering of the RT space $\mathbb{Z}_q^{ms}$, for $y \in \mathbb{Z}_q^{ms}$ there exists $h = (h_1, \ldots, h_{ms}) \in H$ such that $d_{RT}(y, h) \leq R$. Let I be the ideal generated by $supp(y - h)$ and I' be an ideal of the RT poset $[m \times s]$ of size R such that $I \subseteq I'$. Then there exists a codeword $c = (c_1, \ldots, c_{ms})$ in C such that x and c coincide in all coordinates of the complementary set of I' (which is an anti-ideal of size $ms - R$). Thus $supp(x - c) \subseteq I'$.

Let $g = ((c_1, h_1), \ldots, (c_{ms}, h_{ms}))$ in G. By construction, z and g coincide in all coordinates of the complementary set of I'. Thus, $d_{RT}(z, g) = |\langle supp(z - g) \rangle| \leq |I'| = R$ and the proof is complete. $\square$

Together with [4, Theorem 13] we get the following consequences of Theorem 3.

Corollary 5. (1) *For* $q < m \leq 2q$ *and* $2 \leq s \leq 3$, $K_{2q}^{RT}(2m, s, 2ms - 3) \leq$
$q(OCAN(3, m, s, 2) + CAN(2, m, 2))$.
(2) *If* $(t-1)q+1 \leq m \leq (t-1)qv$, *then* $K_{qv}^{RT}(m, s, ms-t) \leq qOCAN(t, m, s, v)$.

If $OCAN(t, m, s, v) = v^t$, then there exists an ordered orthogonal array $OOA(t, m, s, v)$. In the following results, ordered orthogonal arrays are used to obtain upper bounds for covering codes in RT spaces. Item (1) has appeared in [4, Theorem 30]. Item (2) is a consequence of Theorem 3 and [6, Theorem 3].

Corollary 6. (1) *If there is an ordered orthogonal array* $OOA(ms - R, m, s, v)$, *then* $K_{vq}^{RT}(m, s, R) \leq v^{ms-R} K_q^{RT}(m, s, R)$.
(2) *Let* q *be a prime power,* $m \leq q+1$ *and* $s \leq t$. *Then, we have* $K_{qv}^{RT}(m, s, ms-$
$t) \leq q^t K_v^{RT}(m, s, ms - t)$.
(3) *Let* q *be a prime power. For* $t \geq 2$, *we have* $K_{(q-1)v}^{RT}(q + 1, t, qt) \leq (q^t -$
$2)K_v^{RT}(q + 1, t, qt)$.

In order to get better upper bounds on $K_{vq}^{RT}(m, s, R)$, we improve the upper bound on $K_v^{RT}(m, s, R)$ for suitable values of m and R. For this purpose, we look at a covering code that gives the trivial upper bound for $K_v^{RT}(m, s, R)$ and modify some of its codewords to reduce the size of the covering code.

Theorem 4. *For* $s \geq 2$,

(1) $K_v^{RT}(2, s, s) \leq v^{s-2}(v^2 - 1)$,
(2) $K_v^{RT}(3, s, 2s - 1) \leq v(v^s - 1)$.

Proof. (1) Let $I = \{1, \ldots, s\}$ ideal of the RT poset $[2 \times s]$. The trivial upper bound for $K_v^{RT}(2, s, s)$ is v^s, and a s-covering of the RT space $\mathbb{Z}_v^{2s}$ of size v^s is

$$C = \{c \in \mathbb{Z}_v^{2s} : \pi_I(c) = 0 \in \mathbb{Z}_v^s\}.$$

For each $z \in \mathbb{Z}_v^{s-2}$, let

$$C_z = \{c \in C : \pi_J(c) = z \in \mathbb{Z}_v^{s-2}\},$$

where $J = \{s + 3, \ldots, 2s\}$ is an anti-ideal of the RT poset $[2 \times s]$. The following properties hold:

(a) $C_z \cap C_{z'} = \emptyset$ if and only if $z \neq z'$;
(b) $|C_z| = v^2$ for all $z \in \mathbb{Z}_v^{s-2}$;
(c) C_z is a s-covering of the RT space $\mathbb{Z}_v^{s+2} \times \{z\}$ over the RT poset $[2 \times m]$;
(d) $C = \bigcup_{z \in \mathbb{Z}_v^{s-2}} C_z$.

For each $z \in \mathbb{Z}_v^{s-2}$, we construct a new set A_z from C_z of size $v^2 - 1$ such that A_z is a s-covering of the RT space $\mathbb{Z}_v^{s+2} \times \{z\}$.

Let $C_z' = C_z \backslash \{(0; 0, 0; z)\}$. For each $c = (0; c_{s+1}, c_{s+2}; z) \in C_z'$ define

$$\phi_z(c) = \begin{cases} (0; c_{s+1}, c_{s+2}; z) & \text{if } c_{s+2} = 0, \\ (0; c_{s+1}, c_{s+2}; c_{s+1}, c_{s+2}; z) & \text{if } c_{s+2} \neq 0. \end{cases}$$

We show that $A_z = \{\phi_z(c) \in \mathbb{Z}_v^{2s} : c \in C_z'\}$ is a s-covering of the RT space $\mathbb{Z}_v^{s+2} \times \{z\}$. Given $x = (x_1, \ldots, x_s; x_{s+1}, x_{s+2}; z) \in \mathbb{Z}_v^{s+2} \times \{z\}$, we divide the proof into three cases.

(a) If $x_{s+2} \neq 0$, then x is covered by $(0; x_{s+1}, x_{s+2}; x_{s+1}, x_{s+2}; z)$;
(b) If $x_{s+1} \neq 0$ and $x_{s+2} = 0$, then x is covered by $(0; x_{s+1}, 0; z)$;
(c) If $x_{s+1} = x_{s+2} = 0$, then we have two subcases:
    (i) If $x_s = 0$, then x is covered by $(0; a, 0; z)$, where $a \neq 0$;
    (ii) If $x_s \neq 0$, then x is covered by $(0; x_{s-1}, x_s; x_{s-1}, x_s; z)$.

Therefore, the set $A = \bigcup_{z \in \mathbb{Z}_v^{s-2}} A_z$ is a s-covering of the RT space $\mathbb{Z}_v^{2s}$ of size $v^{s-2}(v^2 - 1)$.

(2) Let $I = \{1, \ldots, 2s - 1\}$ ideal of the RT poset $[3 \times s]$. The trivial upper bound for $K_v^{RT}(3, s, 2s - 1)$ is v^{s+1}, and a $(2s - 1)$-covering of the RT space $\mathbb{Z}_v^{3s}$ is

$$C = \{c \in \mathbb{Z}_v^{3s} : \pi_I(c) = 0 \in \mathbb{Z}_v^{2s-1}\}.$$

For each $z \in \mathbb{Z}_v$ let $C_z = \{c \in C : \pi_{2s}(c) = z\}$. The following properties hold:

(a) $C_z \cap C_{z'} = \emptyset$ if and only if $z \neq z'$;
(b) $|C_z| = v^s$ for all $z \in \mathbb{Z}_v$;
(c) C_z is a $(2s-1)$-covering of the RT space $\mathbb{Z}_v^{2s-1} \times \{z\} \times \mathbb{Z}_v^s$ over the RT poset $[3 \times s]$;
(d) $C = \bigcup_{z \in \mathbb{Z}_v} C_z$.

For each $z \in \mathbb{Z}_v$, we construct a new set A_z from C_z of size $v^s - 1$ such that A_z is a $(2s - 1)$-covering of the RT space $\mathbb{Z}_v^{2s-1} \times \{z\} \times \mathbb{Z}_v^s$.

Let $C_z' = C_z \backslash \{(0 \ldots 0; 0 \ldots 0z; 0 \ldots 0)\}$. For each $c \in C_z'$ define

$$\phi(c) = \begin{cases} c & \text{if } c_{3s} = 0 \\ (c_{2s+1}, \ldots, c_{3s}; c_{s+1}, \ldots, c_{2s}; c_{2s+1}, \ldots, c_{3s}) & \text{if } c_{3s} \neq 0. \end{cases}$$

We claim that $A_z = \{\phi_z(c) \in \mathbb{Z}_v^{3s} : c \in C_z'\}$ is a $(2s - 1)$-covering of the RT space $\mathbb{Z}_v^{2s-1} \times \{z\} \times \mathbb{Z}_v^s$. Indeed, given $x = (x_1, \ldots, x_{2s-1}, z; x_{2s+1}, \ldots, x_{3s}) \in \mathbb{Z}_v^{2s-1} \times \{z\} \times \mathbb{Z}_v^s$, we divide the proof into two cases.

(a) If $(x_{2s+1}, \ldots, x_{3s}) \neq (0 \ldots 0)$, then $c \in A_z$ such that $\pi_J(c) = (x_{2s+1}, \ldots, x_{3s})$ covers x, where $J = \{2s + 1, \ldots, 3s\}$;
(b) If $(x_{2s+1}, \ldots, x_{3s}) = (0 \ldots 0)$, then we have two subcases:
    (i) If $x_s = 0$, then x is covered by $(0, \ldots, 0; 0, \ldots, 0, z; z', 0, \ldots, 0)$, where $z' \neq 0$;
    (ii) If $x_s \neq 0$, then x is covered by $(x_1, \ldots, x_s; 0, \ldots, 0, z; x_1, \ldots, x_s)$.

Therefore, the set $A = \bigcup_{z \in \mathbb{Z}_v} A_z$ is a $(2s-1)$-covering of the RT space $\mathbb{Z}_v^{3s}$ of size $v(v^s - 1)$. $\qquad\qquad\qquad\qquad\qquad\qquad\qquad\qquad\qquad\qquad\qquad\qquad\quad\square$

Example 3. A 3-covering code of the RT space $\mathbb{Z}_2^6$ (RT poset $[2 \times 3]$) with 2^3 is $C = C_0 \cup C_1$, where

$$C_0 = \{000000, 000010, 000100, 000110\},$$
$$C_1 = \{000001, 000101, 000011, 000111\}.$$

Theorem 4 item (1) improves the upper bound $K_2^{RT}(2,3,3) \leq 8$ by using the 3-covering code $A = A_0 \cup A_1$, where $A_0 = \{000100, 001100, 010110\}$, and $A_1 = \{001001, 011101, 000011\}$. Therefore, $K_2^{RT}(2,3,3) \leq 6$. $\qquad\qquad\quad\square$

Corollary 7. *Let q be a prime power.*

(1) *For $q + 1 \leq (t-1)v$, $K_{qv}^{RT}(q+1, t, qt) \leq q^t v^{t-2}(v^2 - 1)$.*
(2) *For $m \leq q + 1$, $K_{qv}^{RT}(m, s, ms - (s+1)) \leq q^{s+1} v(v^s - 1)$.*
(3) *$K_{(q-1)v}^{RT}(q+1, t, qt) \leq (q^t - 2) v^{t-2}(v^2 - 1)$.*

Proof. (1) Applying [4, Proposition 17] with $n = q - 1$,

$$K_v^{RT}(q+1, t, qt) \leq K_v^{RT}(2, t, t).$$

Theorem 4 item (1) yields $K_v^{RT}(q+1, t, qt) \leq v^{t-2}(v^2 - 1)$. The result follows by Corollary 6 item (2).

(2) Theorem 3 shows that

$$K_{qv}^{RT}(m, s, ms - (s+1)) \leq OCAN(s+1, m, s, q) K_v^{RT}(m, s, ms - (s+1)).$$

Since there exists an $OOA(s+1, q+1, s+1, q)$ then there exists an $OOA(s+1, m, s, q)$ for $m \leq q + 1$. Applying Theorem 4 item (2) and [4, Proposition 17] with $n = m - 2$,

$$K_v^{RT}(m, s, ms - (s+1)) \leq K_v^{RT}(3, s, 2s - 1) \leq v(v^s - 1).$$

Therefore, the upper bound desired is attained.

(3) Applying [4, Proposition 17] with $n = q - 1$,

$$K_v^{RT}(q+1, t, qt) \leq K_v^{RT}(2, t, t).$$

Theorem 4 item (1) implies that $K_v^{RT}(q+1, t, qt) \leq v^{t-2}(v^2 - 1)$. Corollary 6 item (3) completes the proof. $\qquad\qquad\qquad\qquad\qquad\qquad\qquad\qquad\qquad\quad\square$

We compare the upper bounds for $K_{qv}^{RT}(q+1, t, qt)$. By Theorem 4 item (1), $(qv)^{t-2}((qv)^2 - 1)$ is an upper bound for $K_{qv}^{RT}(q+1, t, qt)$. However, Corollary 7 item (1) yields the upper bound $q^t v^{t-2}(v^2 - 1)$ which improves the one given by Theorem 4 item (1).

Acknowledgements. The authors would like to thank the anonymous referees for their suggestions that greatly improved this paper.

A. G. Castoldi was supported by the CAPES of Brazil, Science without Borders Program, under Grant 99999.003758/2014-01. E. L. Monte Carmelo is partially supported by CNPq/MCT grants: 311703/2016-0. The last three authors are supported by discovery grants from NSERC of Canada.

References

1. Blokhuis, A., Lam, C.W.H.: More coverings by rook domains. J. Comb. Theory Ser. A **36**(2), 240–244 (1984)
2. Brualdi, R.A., Graves, J.S., Lawrence, K.M.: Codes with a poset metric. Discret. Math. **147**(1–3), 57–72 (1995)
3. Carnielli, W.A.: On covering and coloring problems for rook domains. Discret. Math. **57**(1–2), 9–16 (1985)
4. Castoldi, A.C., Monte Carmelo, E.L.: The covering problem in Rosenbloom-Tsfasman spaces. Electron. J. Comb. **22**(3), 3–30 (2015)
5. Castoldi, A.C., Monte Carmelo, E.L., da Silva, R.: Partial sums of binomials, intersecting numbers, and the excess bound in Rosenbloom-Tsfasman space. Comput. Appl. Math. **38**, Article: 55 (2019)
6. Castoldi, A.C., Moura, L., Panario, D., Stevens, B.: Ordered orthogonal array construction using LFSR sequences. IEEE Trans. Inform. Theory **63**(2), 1336–1347 (2017)
7. Cohen, G., Honkala, I., Litsyn, S., Lobstein, A.: Covering Codes. North Holland Mathematical Library, vol. 54. Elsevier, Amsterdam (1997)
8. Colbourn, C.J.: Combinatorial aspects of covering arrays. Le Matematiche **59**, 125–172 (2004)
9. Colbourn, C.J.: Strength two covering arrays: existence tables and projection. Discret. Math. **308**(5–6), 772–786 (2008)
10. Colbourn, C.J., Kéri, G., Soriano, P.P.R., Schlage-Puchta, J.-C.: Covering and radius-covering arrays: constructions and classification. Discret. Appl. Math. **158**(11), 1158–1180 (2010)
11. Hedayat, S., Sloane, N.J.A., Stufken, J.: Orthogonal Arrays: Theory and Applications. SSS. Springer, New York (2012). https://doi.org/10.1007/978-1-4612-1478-6
12. Kleitman, D.J., Spencer, J.: Families of k-independent sets. Discret. Math. **6**(3), 255–262 (1973)
13. Krikorian, T.: Combinatorial constructions of ordered orthogonal arrays and ordered covering arrays. MSC thesis, Dept. Math., Ryerson Univ., Toronto, ON, Canada (2011)
14. Lawrence, K.M.: A combinatorial characterization of (t, m, s)-nets in base b. J. Comb. Des. **4**(4), 275–293 (1996)
15. Mullen, G., Schmid, W.: An equivalence between (t, m, s)-nets and strongly orthogonal hypercubes. J. Comb. Theory Ser. A **76**(1), 164–174 (1996)
16. Raaphorst, S.: Variable strength covering arrays. Ph.D. thesis, School Elect. Eng. Comput. Sci., Univ. Ottawa, Ottawa, ON, Canada (2013)
17. Raaphorst, S., Moura, L., Stevens, B.: Variable strength covering arrays. J. Comb. Des. **26**(9), 417–438 (2018)
18. Rosenbloom, M.Y., Tsfasman, M.A.: Codes for the m-metric. Problems Inform. Transmission **33**(1), 45–52 (1997)
19. Yildiz, B., Siap, I., Bilgin, T., Yesilot, G.: The covering problem for finite rings with respect to the RT-metric. Appl. Math. Lett. **23**(9), 988–992 (2010)

Detecting Arrays for Main Effects

Charles J. Colbourn$^{(\boxtimes)}$ and Violet R. Syrotiuk

School of Computing, Informatics, and Decision Systems Engineering,
Arizona State University, Tempe, AZ, USA
{colbourn,syrotiuk}@asu.edu

Abstract. Determining correctness and performance for complex engineered systems necessitates testing the system to determine how its behaviour is impacted by many factors and interactions among them. Of particular concern is to determine which settings of the factors (main effects) impact the behaviour significantly. Detecting arrays for main effects are test suites that ensure that the impact of each main effect is witnessed even in the presence of d or fewer other significant main effects. Separation in detecting arrays dictates the presence of at least a specified number of such witnesses. A new parameter, corroboration, enables the fusion of levels while maintaining the presence of witnesses. Detecting arrays for main effects, having various values for the separation and corroboration, are constructed using error-correcting codes and separating hash families. The techniques are shown to yield explicit constructions with few tests for large numbers of factors.

1 Introduction

Combinatorial testing [21,31] addresses the design and analysis of test suites in order to evaluate correctness (and, more generally, performance) of complex engineered systems. To set the stage, we introduce some basic definitions. There are k factors $F_1, \ldots, F_k$. Each factor F_i has a set $S_i = \{v_{i1}, \ldots, v_{is_i}\}$ of s_i possible *levels* (or *values* or *options*). A *test* is an assignment of a level from $v_{i1}, \ldots, v_{is_i}$ to F_i, for each $1 \le i \le k$. The execution of a test yields a measurement of a *response*. When $\{i_1, \ldots, i_t\} \subseteq \{1, \ldots, k\}$ and $\sigma_{i_j} \in S_{i_j}$, the set $\{(i_j, \sigma_{i_j}) : 1 \le j \le t\}$ is a *t-way interaction*. The value of t is the *strength* of the interaction. A *main effect* is a 1-way interaction. A test on k factors *covers* $\binom{k}{t}$ t-way interactions. A *test suite* is a collection of tests. A test suite is typically represented as an $N \times k$ array $A = (\sigma_{i,j})$ in which $\sigma_{i,j} \in S_j$ when $1 \le i \le N$ and $1 \le j \le k$. The *size* of the test suite is N and its *type* is $(s_1, \ldots, s_k)$. Tests correspond to rows of A, and factors correspond to its columns.

When the response of interest can depend on one or more interactions, each having strength at most t, a test suite must cover each interaction in at least one row (test). To make this precise, let $A = (\sigma_{i,j})$ be a test suite of size N and type $(s_1, \ldots, s_k)$. Let $T = \{(i_j, \sigma_{i_j}) : 1 \le j \le t\}$ be a t-way interaction. Then $\rho_A(T)$ denotes the set $\{r : a_{ri_j} = \sigma_{i_j}, 1 \le j \le t\}$ of rows of A in which the interaction

© Springer Nature Switzerland AG 2019
M. Ćirić et al. (Eds.): CAI 2019, LNCS 11545, pp. 112–123, 2019.
https://doi.org/10.1007/978-3-030-21363-3_10

is covered. A t-way interaction T must have $|\rho_A(T)| \geq 1$ in order to impact the response. For a set $\mathcal{T}$ of interactions, $\rho_A(\mathcal{T}) = \bigcup_{T \in \mathcal{T}} \rho_A(T)$.

When used in practical testing applications, as in [1,18,33], further requirements arise. First, if we suppose that some set $\mathcal{T}$ of interactions are those that significantly impact the response, yet there is another interaction $T \notin \mathcal{T}$ for which $\rho_A(T) \subseteq \bigcup_{S \in \mathcal{T}} \rho_A(S)$, the responses are inadequate to determine whether or not T impacts the response significantly. This requirement was explored in [14], and later in [15,16,27]. Secondly, one or more tests may fail to execute correctly, and yield no response or yield outlier responses. To mitigate this, Seidel et al. [34] impose stronger 'separation' requirements on the test suite.

Extending definitions in [14,16,34], we formally define the test suites with which we are concerned. Let A be a test suite of size N and type $(s_1, \ldots, s_k)$. Let $\mathcal{I}_t$ be the set of all t-way interactions for A. Our objective is to identify the set $\mathcal{T} \subseteq \mathcal{I}_t$ of interactions that have significant impact on the response. In so doing, we assume that at most d interactions impact the response. Without limiting d, it can happen that no test suite of type $(s_1, \ldots, s_k)$ exists for any value of N [27].

An $N \times k$ array A of type $(s_1, \ldots, s_k)$ is (d, t, δ)-detecting if $|\rho_A(T) \backslash \rho_A(\mathcal{T})| < \delta \Leftrightarrow T \in \mathcal{T}$ whenever $\mathcal{T} \subseteq \mathcal{I}_t$, and $|\mathcal{T}| = d$. To record all of the parameters, we use the notation $\mathsf{DA}_\delta(N; d, t, k, (s_1, \ldots, s_k))$. To emphasize that different factors may have different numbers of levels, this is a *mixed* detecting array. When all factors have the same number, v, of levels, the array is *uniform* and the notation is simplified to $\mathsf{DA}_\delta(N; d, t, k, v)$. The parameter δ is the *separation* of the detecting array [34], and the definition in [14] is recovered by setting $\delta = 1$. Rows in $\rho_A(T) \backslash \rho_A(\mathcal{T})$ are *witnesses* for T that are not masked by interactions in $\mathcal{T}$. A separation of δ necessitates δ witnesses, ensuring that fewer than δ missed or incorrect measurements cannot result in an interaction's impact being lost.

Setting $d = 0$ in the definition, $\mathcal{T} = \emptyset$ and $\rho_A(\emptyset) = \emptyset$. Then a $(0, t, \delta)$-detecting array is an array in which each t-way interaction is covered in at least δ rows. This leads to a standard class of testing arrays for testing: A *covering array* $\mathsf{CA}_\delta(N; t, k, (s_1, \ldots, s_k))$ is equivalent to a $\mathsf{DA}_\delta(N; 0, t, k, (s_1, \ldots, s_k))$. Again the simpler notation $\mathsf{CA}_\delta(N; t, k, v)$ is employed when it is uniform.

In this paper we focus on detecting arrays for main effects. In Sect. 2, we develop a further parameter, corroboration, for detecting arrays to facilitate the construction of mixed detecting arrays from uniform ones. In Sect. 3 we briefly summarize what is known about the construction of detecting arrays. In Sect. 4 we develop constructions of $(1, 1)$-detecting arrays with specified corroboration and separation using results on perfect hash families of strength two and higher index, or (equivalently) using certain error-correcting codes. In Sect. 5 we extend these constructions to $(d, 1)$-detecting arrays for $d > 1$ using a generalization of perfect hash families, the separating hash families.

2   Fusion and Corroboration

Covering arrays have been much more extensively studied [10,21,31] than have detecting arrays and their variants; they are usually defined only in the case when

$\delta = 1$, and in a more direct manner than by exploiting the equivalence with certain detecting arrays. Often constructions of covering arrays focus on the uniform cases. In part this is because a $\mathsf{CA}_\delta(N; t, k, (s_1, \ldots, s_{i-1}, s_{i-1}, s_{i+1}, \ldots, s_k))$ can be obtained from a $\mathsf{CA}_\delta(N; t, k, (s_1, \ldots, s_{i-1}, s_i, s_{i+1}, \ldots, s_k))$ by making any two levels of the ith factor identical. This operation is *fusion* (see, e.g., [11]).

When applied to detecting arrays with $\delta \geq 1$, however, fusion may reduce the number of witnesses. Increasing the separation cannot overcome this problem. Because techniques for uniform covering arrays are better developed than for mixed ones, generalizations to detecting arrays can be expected to be again more tractable for uniform cases. As with covering arrays, fusion for detecting arrays promises to extend uniform constructions to mixed cases.

In order to facilitate this, we propose an additional parameter for detecting arrays. We begin with a useful characterization. Let A be an $N \times k$ array. Let $T = \{(i_j, \sigma_{i_j}) : 1 \leq j \leq t\}$ be a t-way interaction for A. Let $C = \{c_i : 1 \leq i \leq d\}$ be a set of d column indices of A with $\{i_1, \ldots, i_t\} \cap \{c_1, \ldots, c_d\} = \emptyset$. A set system $\mathcal{S}_{A,T,C}$ is defined on the ground set $\{(c, f) : c \in C, f \in S_c\}$ containing the collection of sets $\{\{(c_1, v_1), \ldots, (c_d, v_d)\} : T \cup \{(c_1, v_1), \ldots, (c_d, v_d)\}$ is covered in $A\}$.

Lemma 1. *An array A is (d, t, δ)-detecting if and only if for every t-way interaction T and every set C of d disjoint columns, every subset X of elements of the set system $\mathcal{S}_{A,T,C}$, whose removal (along with all sets containing an element of X) leaves fewer than δ sets in $\mathcal{S}_{A,T,C}$, satisfies $|X| > d$.*

Proof. First suppose that for some t-way interaction $T = \{(i_j, \sigma_{i_j}) : 1 \leq j \leq t\}$ and some set $C = \{c_i : 1 \leq i \leq d\}$ of d disjoint columns, in the set system $\mathcal{S}_{A,T,C}$ there is a set of elements $X = \{(c_1, v_1), \ldots, (c_d, v_d)\}$ for which fewer than δ sets in the set system contain no element of X. Define $T_i = \{(i_j, \sigma_{i_j}) : 1 \leq j \leq t - 1\} \cup \{(c_i, v_i)\}$. Set $\mathcal{T} = \{T_1, \ldots, T_d\}$. Then $T \notin \mathcal{T}$ but $|\rho_A(T) \setminus \rho_A(\mathcal{T})| < \delta$, so A is not (d, t, δ)-detecting.

In the other direction, suppose that A is not (d, t, δ)-detecting, and consider a set $\mathcal{T} = \{T_1, \ldots, T_d\}$ of d t-way interactions and a t-way interaction T for which $T \notin \mathcal{T}$ but $|\rho_A(T) \setminus \rho_A(\mathcal{T})| < \delta$. Without loss of generality, there is no interaction $T' \in \mathcal{T}$ for which T and T' share a factor set to different levels in each (and so, because $T \neq T'$, T' contains a factor not appearing in T). For each $T_i \in \mathcal{T}$, let c_i be a factor in T_i that is not in T', and suppose that $(c_i, v_i) \in T_i$ for $1 \leq i \leq d$. Then the set $X = \{(c_i, v_i) : 1 \leq i \leq d\}$, when removed from $\mathcal{S}_{A,T,C}$, leaves fewer than δ sets. $\square$

Lemma 1 implies that a (d, t, δ)-detecting array must cover each t-way interaction at least $d + \delta$ times; indeed when $d \geq 1$, for each t-way interaction T and every column c not appearing in T, interaction T must be covered in at least $d + 1$ rows containing distinct levels in column c. In particular, a necessary condition for a $\mathsf{DA}_\delta(N; d, t, k, (s_1, \ldots, s_{i-1}, s_{i-1}, s_{i_1}, \ldots, s_k))$ to exist is that $d < \min(s_i : 1 \leq i \leq k)$ (see also [14]).

These considerations lead to the parameter of interest. For array A, with t-way interaction T and set C of d disjoint columns, suppose that in $\mathcal{S}_{A,T,C}$, for each column in C one performs fewer than s fusions of elements within those arising from that column. Further suppose that, no matter how these fusions are done, the resulting set system has the property that every subset X of elements of the set system, whose removal (along with all sets containing an element of X) leaves fewer than δ sets, satisfies $|X| \geq d + 1$. Then (T, C) has *corroboration s* in A. When every choice of (T, C) has corroboration (at least) s in a $\mathsf{DA}_\delta(N; d, t, k, (s_1, \ldots, s_k))$, it has *corroboration s*. We extend the notation as $\mathsf{DA}_\delta(N; d, t, k, (s_1, \ldots, s_k), s)$ to include corroboration s as a parameter.

3 Covering Arrays and Sperner Partition Systems

As observed in [14], one method to construct detecting arrays is to use covering arrays of higher strength. The following records consequences for separation and corroboration.

Lemma 2. *A* $\mathsf{CA}_\lambda(N; t, k, v)$ *is*

1. *a* $\mathsf{DA}_\delta(N; d, t - d, k, v, 1)$ *with* $\delta = \lambda(v - d)v^{d-1}$, *and*
2. *a* $\mathsf{DA}_\delta(N; d, t - d, k, v, v - d)$ *with* $\delta = \lambda(d + 1)^{d-1}$

whenever $1 \leq d < \min(t, v)$.

Proof. Let A be a $\mathsf{CA}_\lambda(N; t, k, v)$. Let d satisfy $1 \leq d < \min(t, v)$. Let T be a $(t - d)$-way interaction, and let C be a set of d columns not appearing in T. Using the parameters of the covering array, $\mathcal{S}_{A,T,C}$ contains at least λv^d sets, and each element appears in at least λv^{d-1} of them. Suppose that d elements of $\mathcal{S}_{A,T,C}$ are removed, and further suppose that the numbers of elements deleted for the d factors are $e_1, \ldots, e_d$ (so that $d = \sum_{i=1}^d e_i$). Then the number of remaining sets is $\lambda \prod_{i=1}^d (v - e_i)$, which is minimized at $\delta = \lambda(v - d)v^{d-1}$. This establishes the first statement. For the second, performing at most $v - d - 1$ fusions within each factor of $\mathcal{S}_{A,T,C}$ and then deleting at most d elements leaves at least $\delta = \lambda(d + 1)^{d-1}$ sets by a similar argument. □

The effective construction of detecting arrays is well motivated by practical testing applications, in which the need for higher separation to mitigate the effects of outlier responses, and higher corroboration to support fusion of levels, arise. Despite this, other than the construction from covering arrays of higher strength, few constructions are available. In [43] uniform $(1, t)$-detecting arrays with separation 1, corroboration 1, and few factors are studied. This was extended in [36, 38] to (d, t)-detecting arrays, and further to mixed detecting arrays in [37]. Each of these focuses on the determination of a lower bound on the number of rows in terms of d, t, and v, and the determination of cases in which this bound can be met. For $d + t \geq 2$, however, the number of rows must grow at least logarithmically in k, because every two columns must be distinct. Hence the study of arrays meeting bounds that are independent of k necessarily considers only small values of k. In addition, none of these addresses separation or corroboration.

For larger values of k, algorithmic methods are developed in [34]. The algorithms include randomized methods based on the Stein-Lovász-Johnson framework [20,25,40], and derandomized algorithms using conditional expectations (as in [7,8]); randomized methods based on the Lovász Local Lemma [3,19] and derandomizations using Moser-Tardos resampling [30] (as in [12]). Although these methods produce $(1,t)$-mixed detecting arrays for a variety of separation values, they have not been applied for $d > 1$ or to increase the corroboration. Extensions to larger d for locating arrays are considered in [23].

When $t = 1$, one is considering detecting arrays for main effects. A *Sperner family* is a family of subsets of some ground set such that no set in the family is a subset of any other. Meagher et al. [28] introduced Sperner partition systems as a natural variant of Sperner families. An (n,v)-*Sperner partition system* is a collection of partitions of some n-set, each into v nonempty classes, such that no class of any partition is a subset of a class of any other. In [24,28], the largest number of classes in an (n,v)-Sperner partition system is determined exactly for infinitely many values of n for each v. In [9], lower and upper bounds that match asymptotically are established for all n and each v. As noted there, given an (n,v)-Sperner partition system with k partitions, if we number the elements using $\{1,\ldots,n\}$ and number the sets in each partition with $\{1,\ldots,v\}$, we can form an $n \times k$ array in which cell (r,c) contains the set number to which element r belongs in partition c. This array is a $\mathsf{DA}_1(n;1,1,k,v,1)$, and indeed every such DA arises in this way. Even when $d = t = s = \delta = 1$, the largest value of k as a function of n is not known precisely. Therefore it is natural to seek useful bounds and effective algorithms for larger values of the parameters.

4 $(1, 1, \delta)$-Detecting Arrays

In this section, we consider the case when $d = t = 1$. As noted, Sperner partition systems address the existence of such detecting arrays when the separation $\delta = 1$. A naive way to increase the separation simply forms δ copies of each row in a $\mathsf{DA}_1(N;1,1,k,v,1)$ to form a $\mathsf{DA}_\delta(\delta N;1,1,k,v,1)$. This leaves the corroboration unchanged; in addition, it employs more rows than are needed to obtain the increase in separation. In order to treat larger values of separation and corroboration, we employ further combinatorial arrays.

An $(N;k,v)$-*hash family* is an $N \times k$ array on v symbols. A *perfect hash family* $\mathsf{PHF}_\lambda(N;k,v,t)$ is an $(N;k,v)$-hash family, in which in every $N \times t$ subarray, at least λ rows each consist of distinct symbols. Mehlhorn [29] introduced perfect hash families, and they have subsequently found many applications in combinatorial constructions [41].

Colbourn and Torres-Jiménez [17] relax the requirement that each row have the same number of symbols. An $N \times k$ array is a *heterogeneous hash family*, or $\mathsf{HHF}(N;k,(v_1,\ldots,v_N))$, when the ith row contains (at most) v_i symbols for $1 \le i \le N$. The definition for PHF extends naturally to perfect *heterogeneous* hash families; we use the notation $\mathsf{PHHF}_\lambda(N;k,(v_1,\ldots,v_N),t)$.

Returning to detecting arrays, we first consider larger separation.

Lemma 3. *Whenever a* $\mathsf{PHF}_\delta(N; k, v, 2)$ *exists, a* $\mathsf{DA}_\delta(v(N + \delta); 1, 1, k, v, 1)$ *exists.*

Proof. Let A be a $\mathsf{PHF}_\delta(N; k, v, 2)$ on symbols $\{0, \ldots, v-1\}$. Let A_i be the array obtained from A by adding i modulo v to each entry of A. Let B be the $\delta v \times k$ array consisting of δ rows containing only symbol i, for each $i \in \{0, \ldots, v-1\}$. Vertically juxtapose $A_0, \ldots, A_{v-1}$, and B to form a $v(N + \delta) \times k$ array D. To verify that D is a $\mathsf{DA}_\delta(v(N + \delta); 1, 1, k, v, 1)$, consider a main effect (c, σ) and let $c' \neq c$ be a column. Among the rows of D covering (c, σ), we find σ exactly δ times in the rows of B (and perhaps among rows of one or more of the $\{A_i\}$). Further, each of the δ rows in the PHF having different symbols in columns c and c' yield a row in one of the $\{A_i\}$ in which (c, σ) appears but c' contains a symbol different from σ. Hence no symbol in c' can cover all but $\delta - 1$ rows containing (c, σ). $\qquad\square$

When does a $\mathsf{PHF}_\delta(N; k, v, 2)$ exist? Treating columns as codewords of length N on a v-ary alphabet, two different codewords are at Hamming distance at least δ. Hence such a $\mathsf{PHF}_\delta(N; k, v, 2)$ is exactly a v-ary code of length N and minimum distance δ, having k codewords. (See [26] for definitions in coding theory.) When $\delta = 1$, the set of all v^N codewords provides the largest number of codewords, while for $\delta = 2$, the set of v^{N-1} codewords having sum 0 modulo v provides the largest code. For $\delta \geq 3$, however, the existence question for such codes is far from settled, particularly when $v > 2$ (see [22], for example). As applied here, this fruitful connection with codes permits increase in the separation but not the corroboration. We address this next.

Construction 1 *(h-inflation).* *Let v be a prime power and let $1 \leq h \leq v$. Let $\{e_0, \ldots, e_{v-1}\}$ be the elements of $\mathbb{F}_v$. Let A be an $(N; k, v + 1)$-hash family on $\{e_0, \ldots, e_{v-1}\} \cup \{\infty\}$. Define 2×1 column vectors $\mathcal{C}_h$ containing $\mathbf{c}_\infty = \begin{pmatrix} 1 \\ 0 \end{pmatrix}$ and $\mathbf{c}_x = \begin{pmatrix} x \\ 1 \end{pmatrix}$ for $x \in \mathbb{F}_v$. Form a set of r_h row vectors $\mathcal{R}_h = (\mathbf{r}_1, \ldots, \mathbf{r}_{r_h})$ so that for every $\mathbf{c}_a \in \mathcal{C}_h$, each $\mathbf{d}_a = (\mathbf{r}_i \mathbf{c}_a : 1 \leq i \leq r_h)$ contains each entry of $\mathbb{F}_v$ at least h times. Form B by replacing each element a in array A by the column vector $\mathbf{d}_a^T$. Then B is a $(r_h N; k, v)$-hash family, an h-inflation of A.*

In Construction 1, each column vector $\mathbf{d}_a$ contains each element of the field at least h times. Moreover, if $a \neq b$, the h coordinates in which $\mathbf{d}_a$ contains a specific element of the field contain h different elements in these coordinates in $\mathbf{d}_a$. Both facts can be easily checked.

Lemma 4. *Whenever v is a prime power, a $\mathsf{PHF}_\delta(N; k, v + 1, 2)$ exists, and $1 \leq s \leq v - 1$, a $\mathsf{DA}_{\delta s}((r_{s+1} N; 1, 1, k, v, s)$ exists.*

Proof. Using Construction 1 and the subsequent facts, any $(s + 1)$-inflation, B, of a $\mathsf{PHF}_\delta(N; k, v + 1, 2)$, A, is a $\mathsf{DA}_{\delta s}((r_{s+1} N; 1, 1, k, v, s)$. $\qquad\square$

Given a $\mathsf{PHF}_\delta(N; k, v+1, 2)$, Lemma 4 produces a $\mathsf{DA}_{\delta(v-1)}(v^2 N; 1, 1, k, v, v-1)$ that is, in fact, a covering array $\mathsf{CA}_\delta(Nv; 2, k, v)$. Although this does not lead to the largest number of columns in a covering array with these parameters when $\delta = 1$ (compare with [13]), it is competitive and applies for all δ. More importantly, one can make detecting arrays for a variety of separation and corroboration values.

To illustrate this, we adapted the 'replace-one-column–random extension' randomized algorithm from [12] in order to construct PHFs of index δ. In the interests of space, we do not describe the method here, noting only that it is an heuristic technique that is not expected to produce optimal sizes. In Table 1 we report the largest number of columns found for a $\mathsf{PHF}_\delta(N; k, 6, 2)$ for various values of N and $1 \leq \delta \leq 4$. Recall that each is equivalent to a 6-ary code of length N and minimum distance δ with k codewords.

Table 1. Number k of columns found for a $\mathsf{PHF}_\delta(N; k, 6, 2)$

$\delta \downarrow N \rightarrow$	1	2	3	4	5	6	7	8	9	10
1		6	36	216	1296	7776	46656			
2			6	36	216	1296	7776	46656		
3				6	33	156	704	3156	14007	
4					6	30	116	429	1776	7406	26374

Suppose that we are concerned with a large (but fixed) number of factors, such as 10000. Together with the Lemma 4, the results in Table 1 imply, for example, the existence of the following:

$\mathrm{DA}_1(84; 1, 1, 10000, 5, 1)$ $\mathrm{DA}_2(114; 1, 1, 10000, 5, 2)$ $\mathrm{CA}_1(150; 2, 10000, 5)$
$\mathrm{DA}_2(98; 1, 1, 10000, 5, 1)$ $\mathrm{DA}_4(133; 1, 1, 10000, 5, 2)$ $\mathrm{CA}_2(175; 2, 10000, 5)$
$\mathrm{DA}_3(112; 1, 1, 10000, 5, 1)$ $\mathrm{DA}_6(152; 1, 1, 10000, 5, 2)$ $\mathrm{CA}_3(200; 2, 10000, 5)$
$\mathrm{DA}_4(140; 1, 1, 10000, 5, 1)$ $\mathrm{DA}_8(190; 1, 1, 10000, 5, 2)$ $\mathrm{CA}_4(250; 2, 10000, 5)$

These examples demonstrate not only that increases in both separation and corroboration can be accommodated with a reasonable increase in the number of rows, but also that detecting arrays for main effects can be constructed for very large numbers of factors.

5 $(d, 1, \delta)$-Detecting Arrays

Next we extend these methods to treat higher values of d. To do so, we employ a generalization of PHFs. An $(N; k, v, \{w_1, w_2, \ldots, w_t\})$-*separating hash family* of *index* λ is an $(N; k, v)$-hash family A that satisfies the property: For any $C_1, C_2, \ldots, C_t \subseteq \{1, 2, \ldots, k\}$ such that $|C_1| = w_1$, $|C_2| = w_2$, $\ldots$, $|C_t| = w_t$, and $C_i \cap C_j = \emptyset$ for every $i \neq j$, whenever $c \in C_i$, $c' \in C_j$, and $i \neq j$, different symbols appear in columns c and c' in each of at least λ rows. The notation $\mathsf{SHF}_\lambda(N; k, v, \{w_1, w_2, \ldots, w_t\})$ is used. See, for example, [2,32,39]; and see [4] for the similar notion of 'partially hashing'. When heterogeneous, we use the notation $\mathsf{SHHF}_\lambda(N; k, (v_1, \ldots, v_N), \{w_1, w_2, \ldots, w_t\})$. In the particular case of $\mathsf{SHF}_1(N; k, v, \{1, d\})$, these are *frameproof codes* (see, for example, [39,42]).

Theorem 1. *Let v be a prime power. When an $\mathsf{SHF}_\delta(N; k, v + 1, \{1, d\})$ exists, and $1 \leq s \leq v - d$, a $\mathsf{DA}_{\delta s}(r_{s+d} N; d, 1, k, v, 1)$ and a $\mathsf{DA}_\delta(r_{s+d} N; d, 1, k, v, \lfloor(s + d - 1)/d\rfloor)$ exist.*

Proof. Using Construction 1, let B be an $(s + d)$-inflation of an $\mathsf{SHF}_\delta(N; k, v + 1, \{1, d\})$, A. Then B is a $r_{s+d} N \times k$ array with entries from $\mathbb{F}_v$. Now consider a set of distinct columns $\{c, c_1, \ldots, c_d\}$ of A. Let R be the set of (at least δ) rows of A in which the entry in column c does not appear in any of columns $\{c_1, \ldots, c_d\}$. For each $\sigma \in \mathbb{F}_v$, the inflation of a row in R yields at least $s + d$ rows in which column c contains σ and each of $\{c_1, \ldots, c_d\}$ contains $d + s$ distinct symbols. Indeed, setting $T = \{(c, \sigma)\}$ and $C = \{c_1, \ldots, c_d\}$, the inflation of each row in R places $d + s$ *mutually disjoint* sets in $\mathcal{S}_{B,T,C}$. Consequently, any removal of d elements from $\mathcal{S}_{B,T,C}$ can remove at most d of the $s + d$ sets arising from a row in R. Hence at least δs must remain, and B is a detecting array with separation (at least) δs. Identification of fewer than $\lfloor(s + d - 1)/d\rfloor$ levels for each factor of $\mathcal{S}_{B,T,C}$ leaves at least δ sets, giving the second DA.   □

In order to apply Theorem 1, we require $\{1, d\}$-separating hash families. Their existence is well studied for $\delta = 1$ (see [35] and references therein), but they appear not to have been studied when $\delta > 1$. When $\delta = 1$, Blackburn [6] establishes that an $\mathsf{SHF}_1(N; k, v, \{1, d\})$ can exist only when $k \leq dv^{\lceil \frac{N}{d} \rceil} - d$. Stinson et al. [42] use an expurgation technique to establish lower bounds on k for which an $\mathsf{SHF}_1(N; k, v, \{1, d\})$ exists. One consequence of their results is that an $\mathsf{SHF}_1(N; k, v, \{1, 2\})$ exists for $k = \left\lceil \frac{1}{2} \left(\frac{v^2}{2v-1} \right)^{\frac{N}{2}} \right\rceil$.

Let us consider a concrete set of parameters. Suppose that we are to construct an $\mathsf{SHF}_1(13; k, 6, \{1, 2\})$. The bounds ensure that the largest value of k for which one exists satisfies $1112 \leq k \leq 559870$. A straightforward computation yields such an SHF with $k = 8014$. Naturally one hopes to improve on both the lower and upper bounds, and to generalize them to cases with separation more than $\delta = 1$. Error-correcting codes are not equivalent to the SHF s required when $d > 1$, but they again provide constructions; we leave this discussion for later work. Nevertheless, there appears to be a need to resort to computation as well.

Table 2 gives the largest values of k that we found for an $\mathsf{SHF}_\delta(N; k, 6, \{1, 2\})$ for $1 \leq \delta \leq 4$ and various values of N. Each yields a $\mathsf{DA}_\delta(15N; 2, 1, k, 5, 1)$ (and other detecting arrays, from Theorem 1).

The entries in Table 2 have again been determined using a variant of the 'replace-one-column–random-extension' algorithm developed in [12]. This heuristic method is not expected in general to yield the largest possible number of columns (and the lower and upper bounds on such largest numbers are currently far apart). When the number of rows is small, however, we can make some comparisons, and we do this next. First we establish an upper bound on k when $N \leq d + \delta - 1$.

Table 2. Number k of columns found for an $\mathsf{SHF}_\delta(N; k, 6, \{1,2\})$

$\delta \downarrow N \rightarrow$	1	2	3	4	5	6	7	8	9	10	11	12
1	6	10	36	51	154	201	373	634	1003	1751	2825	4578
2		6	7	34	39	142	152	262	342	529	805	1257
3			6	6	30	32	72	80	168	195	328	486
4				6	6	27	27	56	58	125	134	231

$\delta \downarrow N \rightarrow$	13	14	15	16	17	18	19	20	21	22	23	24
1	8068	10000										
2	2041	3163	4920	8431	10000							
3	716	1086	1695	2543	3891	6290	9878	10000				
4	311	466	696	1005	1540	2310	3387	5181	8242	10000		

Lemma 5. *Let $d \geq 2$, $\delta \geq 1$, and $\alpha \geq 1$. Then*

$$k \leq \max\left(v_1, \ldots, v_{d+\delta-\alpha}, \left\lfloor \frac{\sum_{i=1}^{d+\delta-\alpha}(v_i - 1)}{\delta} \right\rfloor\right)$$

whenever an $\mathsf{SHHF}_\delta(d + \delta - \alpha; k, (v_1, \ldots, v_{d+\delta-\alpha}), \{1, d\})$ exists.

Proof. Let A be an $\mathsf{SHHF}_\delta(d + \delta - 1; k, (v_1, \ldots, v_{d+\delta-\alpha}), \{1, d\})$. An entry in A is a *private* entry if it contains the only occurrence of a symbol in its row. If some row contains only private entries, then $k \leq \max(v_1, \ldots, v_{d+\delta-\alpha})$. If some column c were to contain $d + 1 - \alpha$ entries that are not private, for each of $d+1-\alpha$ such rows choose a column that contains the same symbol as in column c. Let X be the set of at most $d + 1 - \alpha$ columns so chosen. There could be at most $\delta - 1$ rows separating $\{c\}$ from X, which cannot arise. Consequently every column of A contains at least δ private entries, and at most $d - \alpha$ that are not private. Row i employs v_i symbols and hence contains at least $k - v_i + 1$ entries that are not private. It follows that $(d-\alpha)k \geq \sum_{i=1}^{d+\delta-\alpha}(k - v_i + 1)$. Hence $\sum_{i=1}^{d+\delta-\alpha}(v_i - 1) \geq \delta k$ and the bound follows. □

When $\delta = 1$ and N is larger, Blackburn [6] partitions the N rows into d classes; then when the largest class has r rows in it, he amalgamates all rows in the class into a single row on v^r symbols. He employs a version of Lemma 5, using $\delta = 1$ and not exploiting heterogeneity, to obtain the upper bound on k already mentioned. Our heterogeneous bound underlies an improvement in the upper bound in some situations. In particular, in the example given before, an $\mathsf{SHF}_1(13; k, 6, \{1,2\})$ must have $k \leq 326590$. Unfortunately, although the amalgamation strategy cannot reduce a separation $\delta \geq 2$ to zero, it can nonetheless reduce it to 1. Hence Lemma 5 does not lead to an effective upper bound on k as a function of N when $\delta > 1$. Despite this, Lemma 5 implies that the upper bounds on k match the lower bounds found computationally for the entries in Table 2 when $N = 2 + \delta - 1$, showing their optimality.

Proceeding to the next diagonal, when $N = d + \delta$, we employ a general observation: Whenever there exists an $\mathsf{SHF}_{\delta+1}(N; k, v, \{1, d\})$, one can delete any of the N rows to produce an $\mathsf{SHF}_{\delta}(N - 1; k, v, \{1, d\})$. An elementary argument shows that $k \leq v^2$ in an $\mathsf{SHF}_1(d + 1; k, v, \{1, d\})$ when $d \leq v$, and hence this upper bound on k extends to $\mathsf{SHF}_{\delta}(d + \delta; k, v, \{1, d\})$. Equality is met if and only if there exist $d + \delta - 2$ mutually orthogonal latin squares of side v (via their equivalence with "$(d + \delta)$-nets", see [5]); we omit the proof here. The non-existence of two orthogonal latin squares of side 6 explains in part the entries on this diagonal in Table 2.

For few rows, these observations indicate that the SHF s found in Table 2 are optimal, or nearly so. We do not anticipate that the numbers of columns given are optimal for larger numbers of rows, but they provide explicit solutions that are better than known general lower bounds, and often substantially better.

6   Concluding Remarks

Certain separating hash families, the frameproof codes, can be used to produce detecting arrays for main effects supporting larger separation (to cope with outlier and missing test results) and corroboration (to permit fusion of some levels). Although such SHFs have been extensively researched for index one, the generalization to larger indices is not well studied. Because we require explicit presentations of detecting arrays for testing applications, we examine constructions for SHFs for small indices, and demonstrate that a randomized algorithm can be used to provide useful detecting arrays.

Acknowledgements. This work is supported in part by the U.S. National Science Foundation grants #1421058 and #1813729, and in part by the Software Test & Analysis Techniques for Automated Software Test program by OPNAV N-84, U.S. Navy. Thanks to Randy Compton, Ryan Dougherty, Erin Lanus, and Stephen Seidel for helpful discussions. Thanks also to three very helpful reviewers.

References

1. Aldaco, A.N., Colbourn, C.J., Syrotiuk, V.R.: Locating arrays: a new experimental design for screening complex engineered systems. SIGOPS Oper. Syst. Rev. **49**(1), 31–40 (2015)
2. Alon, N., Cohen, G., Krivelevich, M., Litsyn, S.: Generalized hashing and parent-identifying codes. J. Comb. Theory Ser. A **104**, 207–215 (2003)
3. Alon, N., Spencer, J.H.: The Probabilistic Method. Wiley, Hoboken (2008)
4. Barg, A., Cohen, G., Encheva, S., Kabatiansky, G., Zémor, G.: A hypergraph approach to the identifying parent property: the case of multiple parents. SIAM J. Discret. Math. **14**, 423–431 (2001)
5. Beth, T., Jungnickel, D., Lenz, H.: Design Theory. Encyclopedia of Mathematics and its Applications, vol. 69, 2nd edn. Cambridge University Press, Cambridge (1999)
6. Blackburn, S.R.: Frameproof codes. SIAM J. Discret. Math. **16**(3), 499–510 (2003)

7. Bryce, R.C., Colbourn, C.J.: The density algorithm for pairwise interaction testing. Softw. Test. Verif. Reliab. **17**, 159–182 (2007)
8. Bryce, R.C., Colbourn, C.J.: A density-based greedy algorithm for higher strength covering arrays. Softw. Test. Verif. Reliab. **19**, 37–53 (2009)
9. Chang, Y., Colbourn, C.J., Gowty, A., Horsley, D., Zhou, J.: New bounds on the maximum size of Sperner partition systems (2018)
10. Colbourn, C.J.: Covering arrays and hash families. In: Information Security and Related Combinatorics, pp. 99–136. NATO Peace and Information Security, IOS Press (2011)
11. Colbourn, C.J., Kéri, G., Rivas Soriano, P.P., Schlage-Puchta, J.C.: Covering and radius-covering arrays: constructions and classification. Discret. Appl. Math. **158**, 1158–1190 (2010)
12. Colbourn, C.J., Lanus, E., Sarkar, K.: Asymptotic and constructive methods for covering perfect hash families and covering arrays. Des. Codes Cryptogr. **86**, 907–937 (2018)
13. Colbourn, C.J., Martirosyan, S.S., Mullen, G.L., Shasha, D.E., Sherwood, G.B., Yucas, J.L.: Products of mixed covering arrays of strength two. J. Comb. Des. **14**, 124–138 (2006)
14. Colbourn, C.J., McClary, D.W.: Locating and detecting arrays for interaction faults. J. Comb. Optim. **15**, 17–48 (2008)
15. Colbourn, C.J., Syrotiuk, V.R.: Coverage, location, detection, and measurement. In: 2016 IEEE Ninth International Conference on Software Testing, Verification and Validation Workshops (ICSTW), pp. 19–25. IEEE Press (2016)
16. Colbourn, C.J., Syrotiuk, V.R.: On a combinatorial framework for fault characterization. Math. Comput. Sci. **12**(4), 429–451 (2018)
17. Colbourn, C.J., Torres-Jiménez, J.: Heterogeneous hash families and covering arrays. Contemp. Math. **523**, 3–15 (2010)
18. Compton, R., Mehari, M.T., Colbourn, C.J., De Poorter, E., Syrotiuk, V.R.: Screening interacting factors in a wireless network testbed using locating arrays. In: IEEE INFOCOM International Workshop on Computer and Networking Experimental Research Using Testbeds (CNERT) (2016)
19. Erdős, P., Lovász, L.: Problems and results on 3-chromatic hypergraphs and some related questions. In: Infinite and Finite Sets, pp. 609–627. North-Holland, Amsterdam (1975)
20. Johnson, D.S.: Approximation algorithms for combinatorial problems. J. Comput. Syst. Sci. **9**, 256–278 (1974)
21. Kuhn, D.R., Kacker, R., Lei, Y.: Introduction to Combinatorial Testing. CRC Press, Boca Raton (2013)
22. Laaksonen, A., Östergård, P.R.J.: New lower bounds on error-correcting ternary, quaternary and quinary codes. In: Barbero, Á.I., Skachek, V., Ytrehus, Ø. (eds.) ICMCTA 2017. LNCS, vol. 10495, pp. 228–237. Springer, Cham (2017). https://doi.org/10.1007/978-3-319-66278-7_19
23. Lanus, E., Colbourn, C.J., Montgomery, D.C.: Partitioned search with column resampling for locating array construction. In: 2019 IEEE Ninth International Conference on Software Testing, Verification and Validation Workshops (ICSTW). IEEE Press (2019, to appear)
24. Li, P.C., Meagher, K.: Sperner partition systems. J. Comb. Des. **21**(7), 267–279 (2013)
25. Lovász, L.: On the ratio of optimal integral and fractional covers. Discret. Math. **13**(4), 383–390 (1975)

26. MacWilliams, F.J., Sloane, N.J.A.: The Theory of Error-Correcting Codes. North-Holland Publishing Co., Amsterdam (1977)
27. Martínez, C., Moura, L., Panario, D., Stevens, B.: Locating errors using ELAs, covering arrays, and adaptive testing algorithms. SIAM J. Discret. Math. **23**, 1776–1799 (2009)
28. Meagher, K., Moura, L., Stevens, B.: A Sperner-type theorem for set-partition systems. Electron. J. Comb. **12**, 20 (2005)
29. Mehlhorn, K.: Data Structures and Algorithms 1: Sorting and Searching. Springer, Berlin (1984). https://doi.org/10.1007/978-3-642-69672-5
30. Moser, R.A., Tardos, G.: A constructive proof of the general Lovász local lemma. J. ACM **57**(2), 11 (2010)
31. Nie, C., Leung, H.: A survey of combinatorial testing. ACM Comput. Surv. **43**(2), 11 (2011)
32. Sarkar, P., Stinson, D.R.: Frameproof and IPP codes. In: Rangan, C.P., Ding, C. (eds.) INDOCRYPT 2001. LNCS, vol. 2247, pp. 117–126. Springer, Heidelberg (2001). https://doi.org/10.1007/3-540-45311-3_12
33. Seidel, S.A., Mehari, M.T., Colbourn, C.J., De Poorter, E., Moerman, I., Syrotiuk, V.R.: Analysis of large-scale experimental data from wireless networks. In: IEEE INFOCOM International Workshop on Computer and Networking Experimental Research Using Testbeds (CNERT), pp. 535–540 (2018)
34. Seidel, S.A., Sarkar, K., Colbourn, C.J., Syrotiuk, V.R.: Separating interaction effects using locating and detecting arrays. In: Iliopoulos, C., Leong, H.W., Sung, W.-K. (eds.) IWOCA 2018. LNCS, vol. 10979, pp. 349–360. Springer, Cham (2018). https://doi.org/10.1007/978-3-319-94667-2_29
35. Shangguan, C., Wang, X., Ge, G., Miao, Y.: New bounds for frameproof codes. IEEE Trans. Inf. Theory **63**(11), 7247–7252 (2017)
36. Shi, C., Tang, Y., Yin, J.: The equivalence between optimal detecting arrays and super-simple OAs. Des. Codes Cryptogr. **62**(2), 131–142 (2012)
37. Shi, C., Tang, Y., Yin, J.: Optimum mixed level detecting arrays. Ann. Stat. **42**(4), 1546–1563 (2014)
38. Shi, C., Wang, C.M.: Optimum detecting arrays for independent interaction faults. Acta Math. Sin. (Engl. Ser.) **32**(2), 199–212 (2016)
39. Staddon, J.N., Stinson, D.R., Wei, R.: Combinatorial properties of frameproof and traceability codes. IEEE Trans. Inf. Theory **47**, 1042–1049 (2001)
40. Stein, S.K.: Two combinatorial covering theorems. J. Comb. Theory Ser. A **16**, 391–397 (1974)
41. Stinson, D.R., Van Trung, T., Wei, R.: Secure frameproof codes, key distribution patterns, group testing algorithms and related structures. J. Stat. Plan. Inference **86**, 595–617 (2000)
42. Stinson, D.R., Wei, R., Chen, K.: On generalized separating hash families. J. Comb. Theory (A) **115**, 105–120 (2008)
43. Tang, Y., Yin, J.X.: Detecting arrays and their optimality. Acta Math. Sin. (Engl. Ser.) **27**(12), 2309–2318 (2011)
44. Wigmore, J.H.: Required numbers of witnesses: a brief history of the numerical system in England. Harv. Law Rev. **15**(2), 83–108 (1901)

Regular Languages as Local Functions
with Small Alphabets

Stefano Crespi Reghizzi and Pierluigi San Pietro[✉]

Dipartimento di Elettronica, Informazione e Bioingegneria (DEIB),
Politecnico di Milano, Piazza Leonardo da Vinci 32, 20133 Milan, Italy
{stefano.crespireghizzi,pierluigi.sanpietro}@polimi.it

Abstract. We extend the classical characterization (a.k.a. Medvedev theorem) of any regular language as the homomorphic image of a local language over an alphabet of cardinality depending on the size of the language recognizer. We allow strictly locally testable (slt) languages of degree greater than two, and instead of a homomorphism, we use a rational function of the local type. By encoding the automaton computations using comma-free codes, we prove that any regular language is the image computed by a length-preserving local function, which is defined on an alphabet that extends the terminal alphabet by just one additional letter. A binary alphabet suffices if the local function is not required to preserve the input length, or if the regular language has polynomial density. If, instead of a local function, a local relation is allowed, a binary input alphabet suffices for any regular language. From this, a new simpler proof is obtained of the already known extension of Medvedev theorem stating that any regular language is the homomorphic image of an slt language over an alphabet of double size.

1 Introduction

The family of regular languages has different characterizations using regular expressions, logical formulas or finite automata (FA). In the latter approach the more abstract formulation, often named Medvedev theorem [6,8], uses a local language (i.e., a *strictly locally testable* (slt) [5] language of testability degree $k = 2$) and a letter-to-letter homomorphism: every regular language $R \subseteq \Sigma^*$ is the homomorphic image of a local language, called the *source*, over another alphabet Λ; the *alphabetic ratio* $\frac{|\Lambda|}{|\Sigma|}$ is in the order of the square of the number of FA states. Continuing a previous investigation [3] motivated by the attractive properties of slt encoding, we address the following question: how small can the source alphabet, or, better, the alphabetic ratio, be? We recall the answer provided by the generalized Medvedev theorem [3]: any regular language is the homomorphic image of a k-slt language over an alphabet Λ of cardinality $2|\Sigma| -$ but in general not less – where k is in the order of the logarithm of the FA size. Thus the minimal alphabetic ratio is independent from the FA size.

Work partially supported by CNR - IEIIT.

M. Ćirić et al. (Eds.): CAI 2019, LNCS 11545, pp. 124–137, 2019.
https://doi.org/10.1007/978-3-030-21363-3_11

The present study concerns new possibilities of reducing the source alphabet size, while generalizing Medvedev theorem in a different direction: the homomorphism is replaced by a rational function [1] (also known as transduction) of the local type [9]. Loosely speaking, a *local function* defines a mapping from a source language $L \subseteq \Lambda^*$ to a target language $R \subseteq \Sigma^*$ by means of a partial local mapping from words of fixed length $k \geq 1$, over Λ, to letters of Σ: parameter k is called the degree of locality of the function. To the best of our knowledge, the approach to regular language characterization using local rational functions instead of homomorphisms, has never been considered before, in this context.

Since a homomorphism is a function of locality degree one, the main question we address is whether every regular language is the image of a local function, defined on a source alphabet of cardinality smaller that $2|\Sigma|$; the latter, as said, is the minimum needed for a characterization using homomorphism.

Exploiting the properties of comma-free codes [2] to encode the computations of an FA, we obtain a series of results. First, the main question above bifurcates depending on the local function being length-preserving or not. If the local function is allowed to be length-decreasing, we show that every regular language is the target of a local function defined on a binary alphabet. Second, assuming that the local function preserves the input length up to a fixed constant value, we prove that the source alphabet of size $|\Sigma| + 1$ suffices to characterize any regular language using a local function. Moreover, for the subfamily of regular languages having polynomial density, we show that a binary source alphabet permits to define every language using a local length-preserving function.

In a further generalization, the second part of the paper moves from a local function to a *local relations*, i.e., a set of pairs of source and target words. Again, we assume the relation to be length-preserving, and we prove that the source alphabet can be taken to be binary, independently of the complexity of the target regular language. At last, the latter results permits to obtain a new, simpler proof of the already mentioned homomorphic characterization theorem in [3].

It is noteworthy that although the theorems differ with respect to their use of local functions/relations and on the length-preserving feature, all the proofs have a common structure and rely on a formal property of comma-free codes when they are mapped by a morphism and a local function/relation. Stating such property as a preliminary lemma permitted considerable saving in the following proofs.

Altogether, a rather complete picture results about the minimum alphabet size needed to characterize regular languages by means of local functions (including homomorphism as special case) and relations.

Paper Organization. Section 2 lists the basic definitions for slt languages and rational local functions/relations; it also includes the definition and the number of comma-free codes, and states and proves the preliminary lemma mentioned above. Section 3 defines the local function that encode the labelled paths of an FA, proves the results for length-decreasing and then for length-preserving functions, and finishes with the case of languages having polynomial density. Section 4 presents the characterization of regular languages based on local relations, and the new proof of the homomorphic characterization result in [3]. Section 5 summarizes the main results.

2   Preliminaries

For brevity, we omit the basic classical definitions for language and automata theory and just list our notations. The empty word is denoted ε. The Greek upper-case letters $\Gamma, \Delta, \Theta, \Lambda$ and Σ denote finite terminal alphabets. For clarity, when the alphabet elements are more complex than single letters, e.g., when a finite set of words is used as alphabet, we may also embrace the alphabet name and its elements between "$\langle$" and "$\rangle$". For a word x, $|x|$ denotes the length of x. The i-th letter of x is $x(i)$, $1 \leq i \leq |x|$, i.e., $x = x(1)x(2)\ldots x(|x|)$. For any alphabet, $\Sigma^{\leq k}$ stands for $\bigcup_{1 \leq i \leq k} \Sigma^i$. Let $\#$ be a new character not present in the alphabets, to be used as word *delimiter* to shorten some definitions, but not to be counted as true input symbol.

A *homomorphism* $\xi : \Lambda^* \to \Sigma^*$ is called *letter-to-letter* if for every $b \in \Lambda$, $\xi(b)$ is in Σ.

A *finite automaton* (FA) A is defined by a 5-tuple $(\Sigma, Q, \to, I, F)$ where Q is the set of states, $\to$ the state-transition relation (or graph) $\to \subseteq Q \times \Sigma \times Q$; I and F are resp. the subsets of Q comprising the initial and final states. If $(q, a, q') \in \to$, we write $q \xrightarrow{a} q'$. The transitive closure of $\to$ is defined as usual, e.g., we also write $q \xrightarrow{x} q'$ with $x \in \Sigma^+$ with obvious meaning, and call it a *path*, with an abuse of language (for a nondeterministic FA, $q \xrightarrow{x} q'$ may actually correspond to more than one path in the transition graph). We denote the *label* x of the path $\alpha = q \xrightarrow{x} q'$ by $lab(\alpha)$. The starting and ending states are resp. denoted by $in(\alpha) = q$ and $out(\alpha) = q'$. If $q \in I$ and $q' \in F$, the path is called *accepting*.

Strictly Locally Testable Language Family. There are different equivalent definitions of the family of strictly locally testable (*slt*) languages [4,5]; without loss of generality, the following definition is based on bordered words and disregards for simplicity a finite number of short words that may be present in the language.

The following short notation is useful: given an alphabet Λ and for all $k \geq 2$, let

$$\Lambda_\#^k = \#\Lambda^{k-1} \cup \Lambda^k \cup \Lambda^{k-1}\#.$$

For all words x, $|x| \geq k$, let $F_k(x) \subseteq \Lambda_\#^k$ be the *set of factors of length k* present in $\#x\#$. The definition of F_k is extended to languages as usual.

Definition 1 (Strict local testability). A language $L \subseteq \Lambda^*$ is *k-strictly locally testable* (*k-slt*), if there exist a set $M_k \subseteq \Lambda_\#^k$ such that, for every word $x \in \Lambda^*$, x is in L if, and only if, $F_k(x) \subseteq M_k$. Then, we write $L = L(M_k)$. A language is *slt* if it is k-slt for some value k, which is called the testability *degree*. A *forbidden factor* of M_k is a word in $\Lambda_\#^k - M_k$.

The degree $k = 2$ yields the family of *local* languages. The k-slt languages form an infinite hierarchy under inclusion, ordered by k.

Local Relations and Functions. Let Λ and Σ be finite alphabets, called the source and target alphabet, respectively. A *rational relation* (also called a transduction) [1,8,9] over Λ and Σ is a rational (i.e., regular) subset $r \subseteq \Lambda^+ \times \Sigma^*$. The image of a word $x \in \Lambda^+$ is the set of words $y \in \Sigma^*$ such that $(x, y) \in r$. The *source* and *target* languages of a rational relation are respectively defined as $\{x \in \Lambda^+ \mid \exists y \in \Sigma^* : (x, y) \in r\}$ and as $\{y \in \Sigma^+ \mid \exists x \in \Lambda^+ : (x, y) \in r\}$.

A rational relation r is *length-preserving* if, for all pair of related words, the length of the words differ by at most a constant value, i.e., there exists $m \geq 0$ such that for all $(x, y) \in r$, $abs(|x| - |y|) \leq m$.

Let r be a rational relation such that, for all $x \in \Lambda^*$, $|\{y \in \Sigma^+ \mid (x, y) \in r\}| \leq 1$. Then the mapping $f : \Lambda^* \to \Sigma^*$ defined by $f(x) = y$ is a (partial) *function*.

Next, we focus on the rational relations/functions called *local*[1] [9], where there exists $k > 0$ such that the image of each word $x \in \Lambda^+$ only depends on its factors of length k; such factors may be visualized as the contents of window of width k that slides from left to right on the source word. More precisely, for every word $w \in \Lambda^* \cup \#\Lambda^* \cup \Lambda^*\# \cup \#\Lambda^*\#$, with $|w| \geq k$, we define the *scan* [9], denoted by $\Phi_k(w)$, as the sequence:

$$\Phi_k(w) = \langle w(1) \dots w(k) \rangle, \langle w(2) \dots w(k+1) \rangle, \dots, \langle w(|w| - k + 1) \dots w(|w|) \rangle.$$

Clearly, a scan $\Phi_k(w)$ can be viewed as a word over the "alphabet" $\Lambda_\#^k$, that we denote $\langle \Lambda_\#^k \rangle$ to prevent confusion. Such alphabet comprises all k-tuples in $\#\Lambda^{k-1} \cup \Lambda^k \cup \Lambda^{k-1}\#$. For instance, $\Phi_3(\#abbab\#)$ is the word $\langle \#ab \rangle \langle abb \rangle \langle bba \rangle \langle bab \rangle \langle ab\# \rangle$.

Definition 2 (local function/relation). A (partial) *function* $f : \Lambda^* \to \Sigma^*$ is *local of degree k*, $k \geq 1$, if there exist a finite set $T \subseteq \langle \Lambda_\#^k \rangle$, and a homomorphism $\nu : T^* \to \Sigma^*$, called *associated*, such that $f(x) = \nu(\Phi_k(\#x\#))$.

A *local relation* $r \subseteq \Lambda^* \times \Sigma^*$ of *degree k* is similarly defined, using a *finite substitution* $\sigma : T^* \to 2^{\Sigma^*}$ instead of a homomorphism, as: $r = \{(x, \sigma(\Phi_k(\#x\#)))\}$.

A function (a relation) is called *local* if it is local of degree k for some $k \geq 1$.

It is obvious that the source language of a local function/relation is a k-slt language, defined by the finite set T of factors.

Comma-Free Codes. A finite set $X \subset \Lambda^+$ is a *code* [2] if every word in Λ^+ has at most one factorization in words (also known as *codewords*) of X, more precisely: for any $u_1 u_2 \dots u_m$ and $v_1 v_2 \dots v_n$ in X, where the u and v are codewords, the identity $u_1 u_2 \dots u_m = v_1 v_2 \dots v_n$ holds only if $m = n$ and $u_i = v_i$ for $1 \leq i \leq n$. We use a code X to represent a finite alphabet Γ by means of a one-to-one homomorphism, denoted by $[\![\ ]\!]_X : \Gamma^+ \to \Lambda^+$, called *encoding*, such that $[\![\alpha]\!]_X \in X$ for every $\alpha \in \Gamma$.

Let $n \geq 1$. A set $X \subset \Lambda^n$ is a *comma-free* code, if, intuitively, no codeword overlaps the concatenation of two codewords: more precisely, for any $t, u, v, w \in \Lambda^*$, if tu, uv, vw are in X, then $u = w = \varepsilon$, or $t = v = \varepsilon$.

---

[1] Unfortunately, the adjective "local", for slt languages means of testability degree two, whereas for the locality degree of functions, it means any integer value.

Number of Words of Comma-Free Code. We need the following result (see [7] and its references) on the number of codewords in a comma-free code of length k over an alphabet with cardinality $|\Lambda| = n$. Let $\ell_k(n) = \frac{1}{k} \sum \mu(d) n^{k/d}$, where the summation is extended over all divisors d of k, and μ is the Möbius function defined by

$$\mu(d) = \begin{cases} 1 \text{ if } d = 1 \\ 0 \text{ if } d \text{ has any square factor} \\ (-1)^r \text{ if } d = p_1 p_2 \ldots p_r \text{ where } p_1 p_2 \ldots p_r \text{ are distinct primes.} \end{cases}$$

Proposition 1. *For every alphabet with n letters and for every odd integer $k \geq 1$ there is a comma-free code of length k with $\ell_k(n)$ words.*

The definition of the Möbius function μ is such that if k is a prime number the summation in the formula is just equal to $n^k - n$, i.e., for k prime:

$$\ell_k(n) = \frac{n^k - n}{k}. \tag{1}$$

Comma-Free Codes and Local Functions/Relations. The next lemma will be repeatedly invoked in later proofs.

Lemma 1. *Let Λ, Γ and Σ be finite alphabets and $X \subset \Lambda^k$ be a comma-free code of length k, for some $k > 1$, such that $|X| = |\Gamma|$. Let $L \subseteq \Gamma^+$ be the 2-slt language $L(M_2)$ defined by a set $M_2 \subset \Gamma_\#^2$.*

1. *The encoding of L by means of code X, i.e., the language $[\![L]\!]_X$, is a 2k-slt language included in $(\Lambda^k)^*$.*
2. *Given a homomorphism $\pi : \Gamma^* \to \Sigma^*$, the language $\pi(L)$ is the target language of a local function $f : \Lambda^* \to \Sigma^*$ of degree 2k, having $[\![L]\!]_X$ as source language.*
3. *Given a finite substitution $\sigma : \Gamma^* \to 2^{\Sigma^*}$, the language $\sigma(L)$ is the target of a local relation $r \subseteq \Lambda^* \times \Sigma^*$ of degree 2k, having $[\![L]\!]_X$ as source language.*

Proof. We first claim that XX^+ is a $(2k)$-slt language. Let $F_{2k}(XX^+)$ be the set of the factors of length $2k$ of $\#XX^+\#$, hence it is obvious that $XX^+ \subseteq L(F_{2k}(XX^+))$. We prove the converse inclusion by contradiction. Let $z \in \Lambda^+$ be such that $F_2(z) \subseteq F_{2k}(XX^+)$ but $z \notin XX^+$. Since every word in $F_{2k}(XX^+)$ must have a code $x \in X$ has a factor, then for z not to be in $L(F_{2k}(XX^+))$, the set $F_{2k}(z)$ must include a word of the form $xy \in \Lambda^{2k}$, with $x \in X, y \notin X, y \in \Lambda^k$, or a word of the form $xy\#$, with $x \in X, y \in \Lambda^{<n}$. We only consider the former case, since the latter is analogous. Since $xy \in F_{2k}(XX^+)$, there is a word $p \in XX^+$ including xy as a factor of length $2k$.

Since $p \in XX^+$, p must be of the form $X^*txy\Lambda^+$, with $t \neq \varepsilon$ (otherwise $y \in X$), $|t| < k$, and there exist $u, v \in \Lambda^+$ such that $uv = x \in X$ and $tu \in X$; therefore, p has the form X^*tuvwX^*, with w being a non empty prefix of y and such that also $vw \in X$. By definition of comma-free code, since the three words tu, uv and vw are in X, either $t = v = \varepsilon$, or $u = w = \varepsilon$, a contradiction with the assumption that all those words are in Λ^+.

We need a few definitions. Let $\Psi_{2k} \subset \Lambda_{\#}^{2k}$ be the set of forbidden factors of M_2 when encoded with X, i.e., the set:

$$\Psi_{2k} = \{\#[\![\alpha]\!]_X \Lambda^{k-1} \mid \alpha \in \Gamma, \#\alpha \notin M_2\} \cup \{[\![\alpha\beta]\!]_X \mid \alpha,\beta \in \Gamma, \alpha\beta \notin M_2\} \cup \\ \{\Lambda^{k-1}[\![\beta]\!]_X\# \mid \beta \in \Gamma, \beta\# \notin M_2\}. \tag{2}$$

To define language $[\![L]\!]_X$ we use the following set which avoids the forbidden factors:

$$M_{2k} = F_{2k}(XX^+) - \Psi_{2k}. \tag{3}$$

Clearly, the inclusion $L(M_{2k}) \subseteq X^+$ holds since $M_{2k} \subseteq F_{2k}(XX^+)$.

Part (1). We claim that $L(M_{2k})$ is exactly the language $[\![L]\!]_X$, i.e., for any $x \in \Lambda^+$, $x \in [\![L]\!]_X$ if, and only if, $F_{2k}([\![x]\!]_X) \subseteq M_{2k}$.

We prove $L(M_{2k}) \subseteq [\![L]\!]_X$. Let $x \in L(M_{2k})$ therefore $F_{2k}(x) \subseteq M_{2k}$ and, by contradiction, let $x \notin [\![L]\!]_X$. Since $x \in XX^+$ and $[\![L]\!]_X \subseteq X^+$, x must contain a factor of one of the forbidden forms (2) in Ψ_{2k}, a contradiction.

We prove $[\![L]\!]_X \subseteq L(M_{2k})$. Let $x \in [\![L]\!]_X$; it is enough to show that $F_{2k}(x) \subseteq M_{2k}$. By contradiction, assume that there is $w \in F_{2k}(x)$, with $w \notin M_{2k}$. Since $w \in F_{2k}(XX^+)$, it must be $w \notin \Psi_{2k}$. Therefore, w can only be of the form $y[\![\alpha]\!]_X z$, with $yz \in \Lambda^k$, for some $\alpha \in \Gamma$, with both $y, z \neq \varepsilon$, otherwise x could not be the comma-free encoding of a word of L while having a factor not in Ψ_{2k}. However, since $x \in X^+$, there exist β, γ such that $w' = [\![\gamma]\!]_X [\![\alpha]\!]_X [\![\beta]\!]_X$ is a factor of x, with y a suffix of γ and z a prefix of β. If at least one of $[\![\gamma]\!]_X [\![\alpha]\!]_X$, $[\![\alpha]\!]_X [\![\beta]\!]_X$ is in Ψ_{2k}, then $x \notin [\![L]\!]_X$, a contradiction. If both $[\![\gamma]\!]_X [\![\alpha]\!]_X$, $[\![\alpha]\!]_X [\![\beta]\!]_X \notin \Psi_{2k}$, then by definition of $F_{2k}(XX^+)$ it is necessary that $y[\![\alpha]\!]_X z \in M_{2k}$, also a contradiction.

Part (2). Define a homomorphism $\nu' : \langle \Lambda_{\#}^{2k} \rangle^* \to \Sigma^*$ for every $z \in \langle \Lambda_{\#}^{2k} \rangle$, by means of the following cases, for all $u \in \Lambda^+$:

> H_1 : *if z has the form $\langle \#u \rangle$, let $\nu'(z) = \varepsilon$*
> H_2 : *if z has the form $\langle [\![\alpha]\!]_X [\![\beta]\!]_X \rangle$, for some $\alpha, \beta \in \Gamma$, let $\nu'(z) = \pi(\alpha)$*
> H_3 : *if z has the form $\langle u \rangle$, with $u \neq \langle [\![\alpha]\!]_X [\![\beta]\!]_X \forall \alpha, \beta \in \Gamma$, let $\nu'(z) = \varepsilon$*
> H_4 : *if z has the form $\langle u[\![\alpha]\!]_X \# \rangle$, let $\nu'(z) = \pi(\alpha)$.*
> $\qquad\qquad\qquad\qquad\qquad\qquad\qquad\qquad\qquad\qquad\qquad\qquad (4)$

Loosely speaking, the image is a non-empty word in two cases: H_2, when the "sliding window" contains two codewords, and H_4, when the window ends with a codeword followed by $\#$.

The local function $f' : \Lambda^* \to \Sigma^*$, defined (as in Definition 2) by applying morphism ν' to the scan Φ_{2k}, is total, since it is defined for every $\gamma \in \Lambda^{2k}\Lambda^*$. This is useful in the following proof.

If, as usual, we consider M_{2k} as an alphabet, denoted as $\langle M_{2k} \rangle$, we can define a homomorphism $\nu : \langle M_{2k} \rangle^* \to \Sigma^*$, as $\nu(z) = \nu'(z)$ for every $z \in \langle M_{2k} \rangle \subseteq \langle \Lambda_{\#}^{2k} \rangle$. Let $f : \Lambda^* \to \Sigma^*$ be the local function defined as $f(x) = \nu(\Phi_{2k}(\#x\#))$, for all $x \in L(M_{2k})$, which is thus defined only over $L(M_{2k})$. We claim that the target language of f is $\pi(L)$.

(i) We first prove that $\pi(L) \subseteq f(\Lambda^*)$. The proof is by induction on the length $n \geq 2$ of words in L (ignoring shorter words as usual). Precisely, the induction hypothesis is:

$$\text{if } z \in \Gamma^+ \text{ has length } n \geq 2, \text{ then } \nu'(\Phi_{2k}(\#[\![z]\!]_X)) = \pi(z).$$

From this the thesis follows immediately: if $y \in \pi(L)$, then $y = \pi(z)$ for some $z \in L \subseteq \Gamma^+$; obviously, $F_{2k}([\![z]\!]_X) \subseteq M_{2k}$ so f' is defined. Since f, f' have the same value where they are both defined, it follows that $f([\![z]\!]_X) = f'([\![z]\!]_X) = \pi(z)$.

Base Case: if $|z| = 2$, then $z = \alpha\beta$, for $\alpha, \beta \in \Gamma$. By definition, the set M_2 contains $\#\alpha, \alpha\beta, \beta\#$. Thus, the set $F_{2k}([\![\alpha\beta]\!]_X) \subseteq M_{2k}$ comprises three words: $t_1 = \#[\![\alpha]\!]_X v$, $t_2 = [\![\alpha\beta]\!]_X$, $t_3 = u[\![\beta]\!]_X\#$, for suitable $u, v \in \Lambda^{k-1}$. By Eq. (4), $\nu'(t_1) = \varepsilon$, $\nu'(t_2) = \pi(\alpha)$, $\nu'(t_3) = \pi(\beta)$. Since $\Phi_{2k}(x) = t_1 t_2 t_3$, we have $\nu'(t_1 t_2 t_3) = \pi(\alpha)\pi(\beta)$.

Inductive Step: assume now $|z| > 2$ and that the induction hypothesis holds for every word $z' \in \Gamma^+$ with $|z'| < |z|$. Word z can be factored into $\delta\alpha\beta\gamma$, where $\delta \in \Gamma^*$, $\alpha, \beta, \gamma \in \Gamma$. Let $z' = \delta\alpha\beta$: by induction hypothesis, $\nu'(\Phi_{2k}(\#[\![z']\!]_X)) = \pi(z') = \pi(\delta\alpha\beta)$. Let u be the suffix of length $k - 1$ of β: Then, $\nu'(\Phi_{2k}(\#[\![\delta\alpha\beta\gamma]\!]_X\#)) = \nu'(\Phi_{2k}(\#[\![\delta\alpha\beta]\!]_X)) \cdot \nu'(\Phi_{2k}(u[\![\gamma]\!]_X\#)) = \pi(\delta\alpha\beta) \cdot \nu'(\Phi_{2k}(u[\![\gamma]\!]_X\#))$.

By definition of ν' (case H_4), $\nu'(\Phi_{2k}(u[\![\gamma]\!]_X\#)) = \pi(\gamma)$, hence the thesis follows.

(ii) We now show that $f(\Lambda^*) \subseteq \pi(L)$. It is enough to prove by induction on $n \geq 2k$ that

$$\text{for every } x \in XX^+ \subseteq \Lambda^+ \text{ of length } n, \text{ there exists } z \in \Gamma^+ \text{ s.t. } f'(x) = \pi(z). \quad (5)$$

In fact, to prove (ii), it suffices to notice that if $y \in f(\Lambda^*)$, then there is $x \in X^+$ such that $y = f(x)$ is defined (i.e., $F_{2k}(x) \subseteq M_{2k}$): since by (5) $f'(x) = \pi(z)$, we have $f(x) = f'(x) = \pi(z)$ (functions f and f' are the same where f is defined).

We prove the base case $n = 2k$ of (5). Let $x = [\![\alpha]\!]_X[\![\beta]\!]_X$ for $\alpha, \beta \in \Gamma$. As in the proof of Part (1), $F_{2k}(x) \subseteq M_{2k}$ is composed of three words: $t_1 = \#[\![\alpha]\!]_X u$, $t_2 = [\![\alpha\beta]\!]_X$, $t_3 = v[\![\beta]\!]_X\#$, for suitable $u, v \in \Lambda^{k-1}$. Since ν' is total, $f'(x) = \nu'(\Phi_{2k}(\#x\#))$ is by definition $\nu'(t_1)\nu'(t_2)\nu'(t_3) = \pi(\alpha)\pi(\beta) = \pi(\alpha\beta)$.

The inductive case is also trivial. Let $x \in XX^+$, $|x| = n$, with the induction hypothesis holding for words of length less than n. Word x can be factored into $x'x''$, with $x' \in X^+$, $x'' \in X$. By induction hypothesis, there exists $z' \in \Gamma^+$ such that $f'(x') = z'$. The proof is then analogous to the base case.

Part (3). (Sketch) We notice that for every $z \in \Gamma$, the substitution $\sigma(z)$ is a finite set of words over Σ, and we let m be the length of the longest word in $\sigma(\Gamma)$. We can thus define a finite alphabet $\langle \Theta \rangle$, whose elements are the subsets in $2^{(\Sigma^{\leq m})}$, and a new finite substitution $\tau : \langle \Theta \rangle^* \to 2^{(\Sigma^{\leq m})}$, associating every symbol in $\langle \Theta \rangle$ with its corresponding set of words. We define the homomorphism $\pi : \Gamma^* \to \langle \Theta \rangle^*$, as $\forall z \in \Gamma, \pi(z) = \langle \sigma(z) \rangle$.

Then, the substitution σ can be defined as the composition of substitution τ with homomorphism π, i.e., $\sigma(L) = \tau(\pi(L))$.

By Part (2), there is a local function $f : \Lambda^* \to \Theta^*$ such that its target language is equal to $\pi(L)$. It is then clear that $\tau(f(L))$ is a local relation.   □

Example 1. We first illustrate Definition 2. Let $\Lambda = \{a, b\}$ and $\Sigma = \{0, 1\}$. We define a local function $f : \{a, b\}^* \to \{0, 1\}^*$ of degree 4. Let the set $T \subseteq \Lambda_\#^4$ be $T = F_4(\{aab, bab\}^+)$. To finish, let the associated homomorphism $\nu : T^* \to \Sigma^*$ be:

$$\begin{cases} \nu(\#aab) = 0, \ \nu(baab) = 0, \ \nu(bbab) = 1, \\ \text{for all other } z \in F_4(\{aab, bab\}^+) : \ \nu(z) = \varepsilon. \end{cases}$$

Notice that ν is undefined for all other words in $\Lambda_\#^4$, such as $\#a^3$ and $abab$. The target language of f is $0\{0, 1\}^*$; we show how to compute a value of f:

$f(aab\,bab) = \nu\left(\Phi_4(\#aab\,bab\#)\right)$

$\qquad\qquad = \nu(\langle\#aab\rangle)\,\nu(\langle aabb\rangle)\,\nu(\langle abba\rangle)\,\nu(\langle bbab\rangle)\,\nu(\langle bab\#\rangle)$

$\qquad\qquad = 0\varepsilon\varepsilon 1\varepsilon = 01.$

Observe that $X = \{aab, bab\}$ is a comma-free code of length 3, therefore $\{aab, bab\}^+$ is a 6-slt language, although in this particular case is also 4-slt. If we encode 0 and 1 resp. with the codewords aab and bab, then the function f can be defined as follows: $f([\![z]\!]_X) = \begin{cases} z, & \textit{if } z \in 0(0 \cup 1)^* \\ \bot, & \textit{otherwise} \end{cases}$. Clearly function f is not length-preserving, because of the definition of ν.

To illustrate Part 1 of Lemma 1, observe that $L = 0(0 \cup 1)^*$ is 2-slt, with $L = L(M_2)$ and $M_2 = \{\#0, 00, 01, 10, 11, 0\#, 1\#\}$. Since the code length is 3, the language $[\![L]\!]_X$ is 6-slt; its defining set M_6 has the form of Eq. (3); we just list some factors: $M_6 = \{\#\mathbf{aab}aa, \#\mathbf{aab}ba, \mathbf{aabaab}, \dots, ab\mathbf{aaba}, \dots ab\mathbf{aab}\#\}$ where codewords are evidenced in bold.

3   Characterization of Regular Languages by Local Functions

By the extended Medvedev theorem [3] (reproduced below in Theorem 5), every regular language over Σ is the homomorphic image of an slt source language over an alphabet Λ, where $|\Lambda| = 2|\Sigma|$, and a smaller alphabet does not suffice in general. Instead of a homomorphism, we study the use of a local function (of degree greater than one) such that its target language is exactly the regular language to be defined. Then, the main question is how small the source alphabet can be. The first answer (Theorem 1) is that a binary source alphabet suffices if the local function is not required to be length-preserving. Second, Theorem 2 says that for a local length-preserving function, a source alphabet containing just one more letter than the target alphabet suffices. Then, a specialized result (Theorem 3) for regular languages of polynomial density, says that a length-preserving local function over a binary source alphabet suffices, irrespectively of the size of Σ.

Theorem 1. *For every regular language $R \subseteq \Sigma^*$, there exist a binary alphabet Δ and a local function $f : \Delta^* \to \Sigma^*$, such that the target language of f is R.*

Proof. Let $A = (\Sigma, Q, \to, I, F)$ be an FA recognizing R and let $\Gamma = \to$ be the set comprising the edges of A; let $m = |\Gamma|$. Choose a prime k such that in Eq. (1) $\ell_k(2) \geq m$: this is always possible since $\ell_k(2) = \frac{2^k - 2}{k}$. Therefore, there exists a comma-free code $Z \subset \Delta^k$ such that $|Z| = m$, and $[\![q \overset{a}{\to} q']\!]_Z$ is the codeword for $\langle q \overset{a}{\to} q' \rangle$.

Define (as in the classical proof of Medvedev theorem) the 2-slt language $L = L(M_2) \subseteq \Gamma^+$, where $M_2 \subseteq \langle \Gamma_\#^2 \rangle$ is the set:

$$M_2 = \left\{ \#\langle q \overset{a}{\to} q' \rangle \mid q \in I, a \in \Sigma, q' \in Q \right\} \cup$$
$$\left\{ \langle q \overset{a}{\to} q' \rangle \langle q' \overset{b}{\to} q'' \rangle \mid a, b \in \Sigma, q, q', q'' \in Q \right\} \cup$$
$$\left\{ \langle q \overset{a}{\to} q' \rangle \# \mid q \in Q, a \in \Sigma, q' \in F \right\}.$$

Define the homomorphism $\pi : \Gamma^* \to \Sigma^*$ by means of $\pi(\langle q \overset{a}{\to} q' \rangle) = a$. It is obvious that $\pi(L) = R$. From Lemma 1, Part (2), we have that $\pi(L)$ is the target language of a local function of degree $2k$. $\qquad \square$

In general, the local function of Theorem 1 is not length-preserving. A length-preserving function may require a source alphabet size depending on the target alphabet size. We prove that a source alphabet barely larger than the target one is sufficient, also improving on the alphabetic ratio of the generalized Medvedev theorem [3].

Theorem 2. *For every regular language $R \subseteq \Sigma^*$, there exist an alphabet Λ of size $|\Sigma| + 1$ and a length-preserving local function $f : \Lambda^* \to \Sigma^*$ such that the target language of f is R.*

We need some definitions and intermediate properties to prove the thesis. First, we define certain sets of paths of bounded length in the graph of the FA A that recognizes the language $R \subseteq \Sigma^*$.

Definition 3 (Bounded paths). *Let $A = (\Sigma, Q, \delta, I, F)$ and let $k \geq 1$. For $\sim \in \{<, \leq, =\}$, let $\Sigma^{\sim k}$ be the set of words in Σ^+ of length, respectively, less than, less or equal to, or equal to k. We define the following sets:*

$$P_{\sim k} = \{ q \overset{y}{\to} q' \mid q, q' \in Q, y \in \Sigma^{\sim k} \}, \quad P_{\sim k, F} = \{ q \overset{y}{\to} q_F \mid q \in Q, q' \in F, y \in \Sigma^{\sim k} \}.$$

We view the sets $P_{\sim k}, P_{\sim k, F}$ as finite alphabets, to be respectively written as $\langle P_{\sim k} \rangle$ and $\langle P_{\sim k, F} \rangle$. The language of the accepting paths of automaton A, of length $\geq k$, is denoted by $\mathcal{P}_k \subseteq \langle P_{=k} \rangle^+ \langle P_{\leq k, F} \rangle$.

Of course, $P_{<k} \subseteq P_{\leq k}$, $P_{=k} \subseteq P_{\leq k}$ and $P_{\sim k, F} \subseteq P_{\sim k}$.
The following statement is obvious.

Lemma 2. *The language of the accepting paths of an FA A, $\mathcal{P}_k \subseteq$ $\langle P_{=k} \rangle^+ \langle P_{\leq k,F} \rangle$, is the 2-slt language $\mathcal{P}_k = L(M_2)$ defined by the following set:*

$$M_2 = \{\#\alpha \mid \alpha \in \langle P_{=k} \rangle, in(\alpha) \in I\} \cup$$
$$\{\alpha\alpha' \mid \alpha \in \langle P_{=k} \rangle, \alpha' \in \langle P_{=k} \rangle \cup \langle P_{\leq k,F} \rangle, out(\alpha) = in(\alpha')\} \cup$$
$$\{\alpha\# \mid \alpha \in \langle P_{\leq k,F} \rangle\}. \qquad (6)$$

Next, we define the homomorphism

$$\pi : (\langle P_{>k} \rangle \cup \langle P_{\leq k,F} \rangle)^* \to \Sigma^* \text{ as: } \pi(\alpha) = lab(\alpha). \qquad (7)$$

It is obvious that $\pi(\mathcal{P}_k) = L(A) \cap \Sigma^{\geq k}$.

Now, we encode every path in $P_{\leq k}$ with a comma-free code X of the same length k.

Proposition 2. *There exist $k > 0$, an alphabet Λ of cardinality $|\Sigma| + 1$ and a comma-free code $X \subset \Lambda^k$ such that $|P_{\leq k}| = |X|$.*

Proof. The set $P_{\leq k}$ can be viewed as a subset of $Q \times (\cup_{1 \leq i \leq k} \Sigma^i) \times Q$. By posing $n = |\Sigma|$, it follows that $|P_{\leq k}| \leq |Q|^2 \sum_{1 \leq i \leq k} n^i \leq |Q|^2 n^{k+1}$. By Eq. (1), if k is prime then $\ell_k(n+1) = \frac{(n+1)^k - n - 1}{k}$. To have $\ell_k(n+1) \geq |P_{\leq k}|$, we need to choose k so that $|Q|^2 n^{k+1} \leq \frac{(n+1)^k - n - 1}{k}$, i.e., $|Q|^2 k n^{k+1} + n + 1 \leq (n+1)^k$. For fixed n and fixed Q, the inequality holds for all sufficiently large k. $\qquad \square$

Thus, each path $\alpha \in P_{\leq k}$ is encoded by a word $[\![\alpha]\!]_X$ of X and the following inequality holds, to be used to prove the length-preserving property of the local function:

$$\forall \beta \in P_{\leq k} : |lab(\beta)| \leq |[\![\beta]\!]_X| \leq |lab(\beta)| + k - 1. \qquad (8)$$

Proof of Theorem 2. To finish the proof, we apply Lemma 1, Part (2) with the following correspondence between mathematical entities:

- The alphabet Γ is the set of paths $P_{\leq k}$ of Definition 3
- The code X is the one defined in Proposition 2
- The language $L \subseteq \Gamma^+$ is $\mathcal{P}_k$ of Lemma 2
- The homomorphism π is defined in Eq. (7).

Hence a local function f of degree $2k$ exists, length-preserving by inequality (8). $\qquad \square$

The Case of Polynomial Density Languages. Here we focus on the family of regular languages that have polynomial density. The *density function* [10] of a language $R \subseteq \Sigma^*$ counts the number of words of length n in R and is defined as $\rho_R(n) = |R \cap \Sigma^n|$. Language R has *polynomial density* if $\rho_R(n) = \mathcal{O}(n^k)$ for some integer $k \geq 0$. Clearly, a language R has polynomial density if, and only if, a deterministic trim FA that recognizes R is such that, for any states $q, q' \in Q$, the number of distinct paths of length n from q to q' is polynomial. We prove that if a regular language has polynomial density, then in Theorem 2 a binary source alphabet suffices.

Theorem 3. *Let $R \subseteq \Sigma^*$ be regular language of polynomial density. There is a binary alphabet Δ and a length-preserving local function $f : \Delta^* \to \Sigma^*$ such that $f(\Delta^*) = R$.*

Proof. The number of words of length h is $\mathcal{O}(h^{m-1})$, where m is the number of states of a deterministic FA recognizing R. By letting in Eq. (1) (Proposition 1) $n = |\Lambda| = 2$ and choosing a prime value for k, we have that $\ell_k(2) = \frac{2^k - 2}{k}$, which is $\mathcal{O}(2^k)$, i.e., there is a comma-free code X with $|X|$ being $\mathcal{O}(2^k)$. If the FA is trim, the number of different k-paths is at most polynomial in k, hence for suitably large k it will be smaller than $|X|$. Therefore, the proof of Theorem 2 still holds with a binary comma-free code. $\square$

4   Other Results

Theorem 2 above says that any regular language is the result of a local length-preserving function applied to words over an alphabet containing one more letter. The next theorem positively answers the question whether any improvement over the previous result is possible if the image is defined by means of a local relation instead of a function.

Theorem 4. *For every regular language $R \subseteq \Sigma^*$, there exist a binary alphabet Δ and a length-preserving local relation $r \subseteq \Delta^+ \times \Sigma^+$ such that the target language of r is R.*

Proof. Let $A = (\Sigma, Q, \to, I, F)$ be an FA. Refer to Lemma 1, and assume that $\Lambda = \Delta$, $X \subseteq \Delta^k$ is a comma-free code of length k, and $\Gamma = \{(q, q') \mid q, q' \in Q, \exists \alpha \in P_{\le k}, q = in(\alpha), q' = out(\alpha)\}$. We can safely assume that k is large enough so that $|X| = |Q|^2$, hence we can define a codeword $[\![(q, q')]\!]_X$ for every pair (q, q') of states of Q. The proof resembles the proof of Theorem 2 but, instead of encoding every labelled accepting path, we just encode the two end states of the same path, omitting the path label. Let $\xi : \langle P_{\le k} \rangle^* \to \Delta^*$ be the homomorphism that erases the label of a path $\alpha \in \langle P_{\le k} \rangle$, and returns its encoding by X, more precisely: $\xi(\alpha) = [\![\langle in(\alpha), out(\alpha) \rangle]\!]_X$. Define the 2-slt language $L = L(M_2)$ specified by the following set M_2 over the alphabet $\langle \Gamma_\#^2 \rangle$:

$$M_2 = \{\#\langle q, q' \rangle \mid q \in I, \exists \alpha \in P_{=k}, \langle q, q' \rangle = \xi(\alpha)\} \cup$$
$$\{\langle q, q' \rangle \langle q', q'' \rangle \mid \exists \alpha \in P_{=k}, \beta \in P_{\le k}, \xi(\alpha) = \langle q, q' \rangle, \xi(\beta) = \langle q', q'' \rangle\} \cup$$
$$\{\langle q, q' \rangle \# \mid q' \in F, \exists \alpha \in P_{=k}, \langle q, q' \rangle = \xi(\alpha)\} .$$

We define a finite substitution $\sigma : \langle \Gamma \rangle^* \to 2^{\Sigma^*}$ as follows: $\forall z \in \langle \Gamma \rangle^*$, $\sigma(z) = lab\left(\xi^{-1}([\![z]\!]_X)\right)$. From Lemma 1, Part (3), we have that $\sigma(L)$ is the target language of a local relation of degree $2k$. $\square$

Characterization of Regular Languages as Homomorphic Images of slt Languages. Our last contribution is a new simpler proof, based on Theorem 4, of the known result (Theorem 8 of [3]) that every regular language over an alphabet Σ is the homomorphic image of an slt language over an alphabet of size $2|\Sigma|$. The new proof sets a connection between the old result and the preceding theorems. Overall, we obtain a fairly complete picture of the alphabetic ratio needed for computing regular language by means of local functions, local relations, and homomorphic images of slt languages.

It is convenient to introduce a binary operation that merges two strings into one. Given two alphabets Δ, Σ, define the operator $\otimes : \Delta^+ \times \Sigma^+ \to (\Delta \times (\Sigma \cup \varepsilon))^+$ as follows.

For every $u \in \Delta^+, v \in \Sigma^+$ such that $j = |u| \geq |v| = k$, let

$$u \otimes v = \langle u(1), v(1) \rangle \ldots \langle u(k), v(k) \rangle \langle u(k+1), \varepsilon \rangle \ldots \langle u(j), \varepsilon \rangle.$$

E.g., if $u = 010001$ and $v = abbab$, then $u \otimes v = \langle 0, a \rangle \langle 1, b \rangle \langle 0, b \rangle \langle 0, a \rangle \langle 0, b \rangle \langle 1, \varepsilon \rangle$. The operator can be extended to languages over the two alphabets as usual. We also need the projections, resp. denoted by $[ ]_\Delta$ and $[ ]_\Sigma$ onto the alphabets Δ and Σ, defined as: $[u \otimes v]_\Delta = u$, $[u \otimes v]_\Sigma = v$.

Proposition 3. *If $X \subset \Delta^k$ is a comma-free code of length $k > 1$, then every subset Z of $X \otimes \Sigma^{\leq k}$ is also a comma-free code of length k.*

Proof. By contradiction, assume that a word $w \in Z^+$ can be factored as $w = uzv$ and as $w = uu'z'v'$, where $|u'| < k$ and both $z, z' \in Z$, i.e., z, z' do overlap in w. By definition, $z = x \otimes y$ and $z' = x' \otimes y'$, for $x, x' \in X, y, y' \in \Sigma^{\leq k}$; therefore $[z]_\Delta$ and $[z']_\Delta$ are codewords of X, but they also overlap in $[w]_\Delta$, the projection of w to Δ, a contradiction of the definition of comma-free code. □

Theorem 5 (part of Theorem 8 of [3]). *For any language $R \subseteq \Sigma^*$, there exists an slt language $L \subseteq \Lambda^*$, where Λ is a finite alphabet of size $|\Lambda| = 2|\Sigma|$, and a letter-to letter homomorphism $\vartheta : \Lambda^* \to \Sigma^*$, such that $R = \vartheta(L)$.*

Proof. For the sake of simplicity, we prove a looser bound, namely $|\Lambda| = 2(|\Sigma| + 1)$. The tighter bound is proved in [3]. Let $\Delta = \{0, 1\}$, the homomorphism $\xi : \langle P_{\leq k} \rangle^* \to \Delta^*$ and the comma-free code $X \subset \Delta^+$ be defined as in the proof of Theorem 4. Let $\Lambda = \Delta \times (\Sigma \cup \varepsilon)$. Let $Z \subset \Lambda^k$ be a comma-free code of length k, such that the encoding of each $\alpha \in \langle P_{\leq k} \rangle$ is defined as $[\![\alpha]\!]_Z = \xi(\alpha) \otimes lab(\alpha)$.

Referring to Lemma 1, we consider Γ to be the alphabet $\langle P_{\leq k} \rangle$ and the homomorphism $\pi : \Gamma^* \to \Sigma^*$ to be the projection $\pi(\alpha) = lab(\alpha)$ for every $\alpha \in \langle P_{\leq k} \rangle$. Therefore, there exists a local function $f : \Lambda^* \to \Sigma^*$ whose source language is a $2k$-slt language $L \subseteq \Lambda^*$ and whose target language is R.

Define a letter-to-letter homomorphism $\vartheta : \Lambda^* \to \Sigma^*$ as the projection to the alphabet Σ, i.e., $\vartheta(z) = [z]_\Sigma$ for every $z \in \Lambda$. Let $z \in L, \alpha \in \langle P_{\leq k} \rangle^+$ be such that $z = [\![\alpha]\!]_Z$. It is clear that $\vartheta(z) = lab(\alpha)$, and $f(z) = lab(\alpha)$ as well. Therefore, $R = \vartheta(L)$. □

In comparison, the proof in [3] used an *ad hoc* encoding paying the price of computing its size; moreover, it did not take advantage of the properties in Lemma 1 about comma-free codes, slt languages and local relations, that have permitted to shorten and simplify all the proofs in this paper.

5   Conclusion

We sum up the known results about characterizations of regular languages through local mappings (local function, local relation, homomorphic image of strictly locally testable language) in the following diagram:

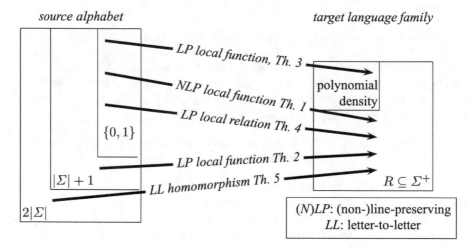

We add that the lower limit $2|\Sigma|$ for the case of homomorphism is tight [3]. On the other hand, it is likely but not proved that the $|\Sigma| + 1$ limit for length-preserving local functions is tight.

Acknowledgements. D. Perrin directed us to comma-free codes. We thank the anonymous referees for their helpful suggestions.

References

1. Berstel, J.: Transductions and Context-Free Languages. Teubner, Stuttgart (1979)
2. Berstel, J., Perrin, D., Reutenauer, C.: Codes and Automata. Cambridge University Press, Cambridge (2015)
3. Crespi Reghizzi, S., San Pietro, P.: From regular to strictly locally testable languages. Int. J. Found. Comput. Sci. **23**(8), 1711–1728 (2012)
4. De Luca, A., Restivo, A.: A characterization of strictly locally testable languages and its applications to subsemigroups of a free semigroup. Inf. Control **44**(3), 300–319 (1980)
5. McNaughton, R., Papert, S.: Counter-Free Automata. MIT Press, Cambridge (1971)

6. Medvedev, Y.T.: On the class of events representable in a finite automaton. In: Moore, E.F. (ed.) Sequential Machines – Selected Papers, pp. 215–227. Addison-Wesley, Reading (1964)
7. Perrin, D., Reutenauer, C.: Hall sets, Lazard sets and comma-free codes. Discrete Math. **341**(1), 232–243 (2018)
8. Eilenberg, S.: Automata, Languages, and Machines, vol. A. Academic Press, New York (1974)
9. Sakarovitch, J.: Elements of Automata Theory. Cambridge University Press, Cambridge (2009)
10. Szilard, A., Yu, S., Zhang, K., Shallit, J.: Characterizing regular languages with polynomial densities. In: Havel, I.M., Koubek, V. (eds.) MFCS 1992. LNCS, vol. 629, pp. 494–503. Springer, Heidelberg (1992). https://doi.org/10.1007/3-540-55808-X_48

Rational Weighted Tree Languages with Storage and the Kleene-Goldstine Theorem

Zoltán Fülöp[1] and Heiko Vogler[2(✉)]

[1] University of Szeged, Szeged, Hungary
fulop@inf.u-szeged.hu
[2] Technische Universität Dresden, Dresden, Germany
heiko.vogler@tu-dresden.de

Abstract. We introduce rational weighted tree languages with storage over commutative, complete semirings and show a Kleene-Goldstine theorem.

1 Introduction

In [39], Kleene published his famous result on the coincidence of rational languages of words and regular languages of words, i.e., languages recognized by one-way nondeterministic finite-state automata.

Soon there was a need for more powerful automata which could better model, e.g., programming languages or natural languages. Thus a number of models of automata with auxiliary data store were introduced; examples of such data stores are pushdown [5], stack [30], checking-stack [18], checking-stack pushdown [48], nested stack [1], iterated pushdown [8,19,36,42], queue, and monoid [12, Ex. 3] (cf. [38]). Each of these automata models follows the recipe "finite-state automaton + data store" [46] where (i) the finite-state automaton uses in its transitions predicates and instructions and (ii) the data store [19,33,46] is a set D of configurations on which the predicates and instructions are interpreted. A string $u \in \Sigma^*$ is accepted if there is a sequence d of transitions such that (i) d leads from some initial to some final state, (ii) the projection of d to Σ^* is u, and (iii) the projection of d to the set I^* of sequences of instructions yields a sequence v which is *executable* on D (e.g., pop pop is not executable on the pushdown, because initially it contains only one symbol).

In [33–35] Goldstine advocated a new approach to the recipe "finite-state automaton + data store" by applying the implication "regular ⇒ rational" of Kleene's theorem to the automaton part of the recipe. He illustrated the benefit by contrasting the specification of the context-free language $L = \{a^n b^n \mid n \in \mathbb{N}\}$ by means of a usual pushdown automaton with his new approach, cf. Fig. 1.

Z. Fülöp—Research was supported by the "Integrated program for training new generation of scientists in the fields of computer science", no EFOP-3.6.3-VEKOP-16-2017-0002.

© Springer Nature Switzerland AG 2019
M. Ćirić et al. (Eds.): CAI 2019, LNCS 11545, pp. 138–150, 2019.
https://doi.org/10.1007/978-3-030-21363-3_12

In general, he defined an *automaton with data store* [34, p. 276] as a rational subset A of the monoid $(\Sigma^* \times I^*)^*$. The *language* defined by A is the set

$$L(A) = \{u \in \Sigma^* \mid \text{ there is a sequence } v \in \hat{A}(u) \text{ which is executable}\}$$

finite-state automaton + pushdown [46]	rational expression + pushdown [33]
$A = (\{q_0, q_1, q_2\}, \{a, b\}, \{Z_0, Z\}, \delta, q_0, Z_0, \{q_2\})$	
$\delta(q_0, \varepsilon, Z_0) = \{(q_2, \varepsilon)\}$	
$\delta(q_0, a, Z_0) = \{(q_0, Z_0 Z), (q_1, Z_0 Z)\}$	$A = (a, Z)^*(b, Z^{-1})^*(\varepsilon, Z_0^{-1})$
$\delta(q_0, a, Z) = \{(q_0, ZZ), (q_1, ZZ)\}$	
$\delta(q_1, b, Z) = \{(q_1, \varepsilon)\}$	
$\delta(q_1, \varepsilon, Z_0) = \{(q_2, \varepsilon)\}$	

Fig. 1. Comparison of the specification of $L = \{a^n b^n \mid n \in \mathbb{N}\}$. In the right-hand side, Z, Z^{-1}, Z_0^{-1} mean "push Z", "pop Z", and "pop Z_0", respectively.

where $(\hat{.})$ is the natural morphism from the free monoid $(\Sigma^* \times I^*)^*$ to the product monoid $\Sigma^* \times I^*$ (with string concatenation in both components), i.e., $\hat{A} \subseteq \Sigma^* \times I^*$. For instance $\hat{A}(abb) = \{ZZ^{-1}Z^{-1}Z_0^{-1}\}$, thus $abb \notin L(A)$ because $ZZ^{-1}Z^{-1}Z_0^{-1}$ is not executable; and $\hat{A}(ab) = \{ZZ^{-1}Z_0^{-1}\}$, thus $ab \in L(A)$ because $ZZ^{-1}Z_0^{-1}$ is executable. The following is very important to notice: although each sequence $v \in \hat{A}(u)$ is built up according to the rational set A, its executability is checked *outside* of $\hat{A}$. This is understandable because, in general, executability is not decomposable under rational operations.

Motivated by the wish to model not only qualitative properties of formal string languages but also quantitative ones, Schützenberger introduced weighted automata [45]. Also, there was a need to analyse tree-structured objects, which gave rise to finite-state tree automata [10,11,47]. By now, the theories of weighted string automata [4,13,15,41,43–45], finite-state tree automata [6,17,27,28], and weighted tree automata [2,3,23,25,40] are very well established. In particular, Kleene's theorem was proved for finite-state tree automata [47] and for weighted tree automata [2,14]. The combination of weighted string automata and data store was studied in [26,37,49].

In this paper we apply Goldstine's approach to the recipe "weighted regular tree grammar + storage" [24] (cf. [19] for the unweighted case). More precisely, we apply the implication "regular ⇒ rational" of Kleene's theorem [14] to the weighted regular tree grammar part and check executability of instruction sequences outside of the rational weighted tree languages. Actually, sequences of instructions turn now into trees, which we call behaviours (cf. [21] where they are called approximations). This leads to the new concept of *rational weighted tree language with storage*.

Due to Kleene's theorem for weighted tree automata [14] and a decomposition theorem for "weighted regular tree grammar + storage" (cf. Theorem 1), we can conclude that, for each weighted tree language $s : \mathrm{T}_\Sigma \to K$ the following equivalence holds: s can be defined by a weighted regular tree grammar with storage iff s is a rational weighted tree language with storage (cf. Theorem 3). This characterization might be called a *Kleene-Goldstine theorem*.

2   Preliminaries

We let $\mathbb{N} = \{0, 1, \ldots\}$ (set of all natural numbers) and $\mathbb{N}_+ = \mathbb{N} \setminus \{0\}$. The powerset of a set A is denoted by $\mathcal{P}(A)$.

A *ranked alphabet* is a finite, non-empty set Σ in which each element σ has a rank in $\mathbb{N}$, denoted by $\mathrm{rk}_\Sigma(\sigma)$. As usual, we set $\Sigma^{(k)} = \{\sigma \in \Sigma \mid \mathrm{rk}_\Sigma(\sigma) = k\}$ and write $\sigma^{(k)}$ to denote that $\sigma \in \Sigma^{(k)}$ for each $k \in \mathbb{N}$ and $\sigma \in \Sigma$. We abbreviate $\max\{k \mid \Sigma^{(k)} \neq \emptyset\}$ by $\mathrm{maxrk}(\Sigma)$. For any set H, we denote by $\mathrm{T}_\Sigma(H)$ the set of trees over Σ indexed by H [24, Sec. 2.3]. Let $\xi \in \mathrm{T}_\Sigma$. We denote the set of *positions of* ξ (or: Gorn-addresses of ξ) by $\mathrm{pos}(\xi)$; we note that $\mathrm{pos}(\xi) \subseteq \mathbb{N}_+^*$ with ε denoting the root of ξ. For every $\xi \in \mathrm{T}_\Sigma$ and $w \in \mathrm{pos}(\xi)$, we denote by $\xi(w)$ the *label of* ξ at w. A *0-extension of* Σ is a ranked alphabet Θ such that $\Sigma \subseteq \Theta$, $\mathrm{rk}_\Theta(\sigma) = \mathrm{rk}_\Sigma(\sigma)$ for each $\sigma \in \Sigma$, and $\mathrm{rk}_\Theta(\sigma) = 0$ for each $\sigma \in \Theta \setminus \Sigma$.

A *semiring* (cf. [32] and [13, Ch. 1]) is an algebra $(K, +, \cdot, 0, 1)$ such that $(K, +, 0)$ is a commutative monoid and $(K, \cdot, 1)$ is a monoid, multiplication $\cdot$ distributes over addition $+$, and $a \cdot 0 = 0 \cdot a = 0$ for every $a \in K$. The semiring K is *commutative* if $\cdot$ is commutative. A semiring is *complete* if, for each countable index set I, it has a sum operation $\sum_I : K^I \to K$ which coincides with $+$ when I is finite and for which the axioms of [15, p. 124] hold for infinite index sets (guaranteeing commutativity, associativity, and distributivity).

In the sequel, if K or Σ are left unspecified, then they stand for an arbitrary commutative complete semiring and an arbitrary ranked alphabet, respectively.

A (Σ, K)-*weighted tree language* is a mapping $s : \mathrm{T}_\Sigma \to K$. We call s a *polynomial (monome)* if the set $\mathrm{supp}(s) = \{\xi \in \mathrm{T}_\Sigma \mid s(\xi) \neq 0\}$ is finite (has at most one element). A monome s with $\mathrm{supp}(s) \subseteq \{\xi\}$ for some $\xi \in \mathrm{T}_\Sigma$ is also denoted by $s(\xi).\xi$. A *0-extension of* s is a weighted tree language $s' : \mathrm{T}_\Theta \to K$ such that Θ is a 0-extension of Σ, $s'|_{\mathrm{T}_\Sigma} = s$ and $s'(\xi) = 0$ for every $\xi \in \mathrm{T}_\Theta \setminus \mathrm{T}_\Sigma$. We denote the set of all (Σ, K)-weighted tree languages by $K\langle\langle \mathrm{T}_\Sigma \rangle\rangle$.

Let Δ be a ranked alphabet, $\tau : \mathrm{T}_\Delta \to \mathcal{P}(\mathrm{T}_\Sigma)$ a mapping, and $s : \mathrm{T}_\Sigma \to K$ a (Σ, K)-weighted tree language. Then we define the (Δ, K)-weighted tree language $(\tau; s) : \mathrm{T}_\Delta \to K$ for each $\zeta \in \mathrm{T}_\Delta$ by $(\tau; s)(\zeta) = \sum_{\xi \in \tau(\zeta)} s(\xi)$.

3   Storage Types and Behaviour

We recall the (slightly modified) concept of storage type from [19]. Storage types are a reformulation of the concept of machines [46] and data stores [34,35].

A *storage type* is a tuple $S = (C, P, F, c_0)$, where C is a set (*configurations*), $c_0 \in C$ (*initial configuration*), P is a set of total functions each having the type $p : C \to \{0, 1\}$ (*predicates*), and F is a nonempty set of partial functions $f : C \to C$ (*instructions*). Moreover, we assume that F contains the *identity instruction (on C)*, i.e., the total function $\mathrm{id}_C : C \to C$ such that $\mathrm{id}_C(c) = c$ for each $c \in C$.

The *trivial storage type* is the storage type $\mathrm{TRIV} = (\{c\}, \emptyset, \{\mathrm{id}_{\{c\}}\}, c)$, where c is some arbitrary but fixed symbol. Another example of a storage type is the *counter* $\mathrm{COUNT} = (\mathbb{N}, \{\mathrm{zero}?\}, \{\mathrm{id}_\mathbb{N}, \mathrm{inc}, \mathrm{dec}\}, 0)$, where for each $n \in \mathbb{N}$, we let $\mathrm{zero}?(n) = 1$ iff $n = 0$, $\mathrm{inc}(n) = n+1$, and $\mathrm{dec}(n) = n-1$ if $n \geq 1$ and undefined otherwise.

We define the predicates true and false by $\text{true}(c) = 1$ and $\text{false}(c) = 0$ for every $c \in C$. We denote by $\text{BC}(P)$ the Boolean closure of P, i.e., the set which consists of true, false, and all predicates which can be obtained by finitely many applications of negation $\neg$, disjunction $\vee$, and conjunction $\wedge$ on elements P. For each $p \in \text{BC}(P)$ and $c \in C$, the truth value $p(c)$ is defined in the usual way. If P is finite, then $\text{BC}(P)$ is also finite because it is a finitely generated subalgebra of the Boolean algebra of all predicates over C (cf. [31, Cor. 2]).

In the sequel, if S is left unspecified, then it stands for an arbitrary storage type $S = (C, P, F, c_0)$. Also, if P and F are left unspecified, then they stand for the sets of predicates and instructions, respectively, of some storage type S.

Let $P' \subseteq P$ a finite and $F' \subseteq F$ be a finite and nonempty subset. Moreover, let $n \in \mathbb{N}$. We define the ranked alphabet

$$\Delta = \bigcup_{0 \le k \le n} \Delta^{(k)} \text{ with } \Delta^{(k)} = \text{BC}(P') \times (F')^k.$$

We call Δ the *ranked alphabet n-corresponding to P' and F'*. We write elements $(p, f_1, \ldots, f_k)$ of Δ in the slightly shorter form $(p, f_1 \ldots f_k)$. Obviously, the parameter n is used to put an upper bound on the rank of symbols in Δ, i.e., $\text{maxrk}(\Delta) = n$. The *ranked alphabet corresponding to Σ, P', and F'* is the ranked alphabet n-corresponding to P' and F' where $n = \max\{\text{maxrk}(\Sigma), 1\}$.

The concept of behaviour is inspired by the set L_D of all executable sequences of instructions (defined on [29, p. 148]; also cf. the notion of storage tracks in [34]). In [21, Def. 3.23] a family of behaviours is put together into a tree by sharing common prefixes; such a tree is called approximation. Here we recall the concept of approximation from [24], but we keep the original name "behaviour".

Formally, let $c \in C$, $n \in \mathbb{N}$, and Δ be the ranked alphabet n-corresponding to $P' \subseteq P$ and $F' \subseteq F$. Then a tree $b \in T_\Delta$ is a *(Δ, c)-behaviour* if there is a family $(c_w \in C \mid w \in \text{pos}(b))$ of configurations such that $c_\varepsilon = c$ and for every $w \in \text{pos}(b)$: if $b(w) = (p, f_1 \ldots f_k)$, then $p(c_w) = 1$ and for every $1 \le i \le k$, the configuration $f_i(c_w)$ is defined and $c_{wi} = f_i(c_w)$. If b is a (Δ, c)-behaviour, then we call $(c_w \in C \mid w \in \text{pos}(b))$ the *family of configurations determined by b and c*. A *Δ-behaviour* is a (Δ, c_0)-behaviour. (Figure 2 right shows an example b of a Δ-behaviour; the grey-shaded tree is the family of configurations determined by b and 0.) We denote the set of all Δ-behaviours by $\mathcal{B}(\Delta)$. We refer the reader to [24, Fig. 2] for an example of a behaviour of the pushdown storage.

4   Rational Weighted Tree Languages with Storage

In this section we generalize the approach of [34, 35] from the unweighted case to the weighted case and from strings to trees. For the definition of rational weighted tree languages, (disregarding storage for the time being) we first recall the usual rational operations [14, 25]. Let Θ be a ranked alphabet.

Let $k \in \mathbb{N}$, $\theta \in \Theta^{(k)}$, and $s_1, \ldots, s_k \in K\langle\!\langle T_\Theta \rangle\!\rangle$. The *top concatenation of $s_1, \ldots, s_k$ with θ* is the (Θ, K)-weighted tree language $\text{top}_\theta(s_1, \ldots, s_k)$ defined for each $\xi \in T_\Theta$ by $\text{top}_\theta(s_1, \ldots, s_k)(\xi) = s_1(\xi_1) \cdot \ldots \cdot s_k(\xi_k)$ if $\xi = \theta(\xi_1, \ldots, \xi_k)$ for some $\xi_1, \ldots, \xi_k \in T_\Theta$, and 0 otherwise. For $k = 0$ we have $\text{top}_\theta() = 1.\theta$.

Let $s \in K\langle\!\langle T_\Theta \rangle\!\rangle$ and $a \in K$. The *scalar multiplication of s with a* is the (Θ, K)-weighted tree language $a \cdot s$ defined for each $\xi \in T_\Theta$ by $(a \cdot s)(\xi) = a \cdot s(\xi)$.

Let $s_1, s_2 \in K\langle\!\langle T_\Theta \rangle\!\rangle$. The *sum of s_1 and s_2* is the (Θ, K)-weighted tree language $(s_1 + s_2)$ defined for each $\xi \in T_\Theta$ by $(s_1 + s_2)(\xi) = s_1(\xi) + s_2(\xi)$.

For the definition of concatenation of weighted tree languages, we need the following concept. Let $\alpha \in \Theta^{(0)}$ and $\zeta \in T_\Theta$ with exactly $r \geq 0$ occurrences of α. Moreover, let $\xi_1, \ldots, \xi_r \in T_\Theta$. We define $\zeta[\alpha \leftarrow (\xi_1, \ldots, \xi_r)]$ to be the tree obtained from ζ by replacing the ith occurrence of α by ξ_i for every $1 \leq i \leq r$ (cf. [14, p. 7]). Let $s_1, s_2 \in K\langle\!\langle T_\Theta \rangle\!\rangle$ and $\alpha \in \Theta^{(0)}$. The *α-concatenation of s_1 and s_2* [14] is the (Θ, K)-weighted tree language $(s_1 \circ_\alpha s_2)$ defined for every $\xi \in T_\Theta$ by

$$(s_1 \circ_\alpha s_2)(\xi) = \sum_{\substack{\zeta, \xi_1, \ldots, \xi_r \in T_\Theta \\ \xi = \zeta[\alpha \leftarrow (\xi_1, \ldots, \xi_r)]}} s_1(\zeta) \cdot s_2(\xi_1) \cdot \ldots \cdot s_2(\xi_r).$$

Let $s \in K\langle\!\langle T_\Theta \rangle\!\rangle$ and $\alpha \in \Theta^{(0)}$. We define the (Θ, K)-weighted tree language s_α^n (called the *n-th iteration of s at α*) for every $n \in \mathbb{N}$ inductively as follows [20] (cf. [14, Def. 3.9]): $s_\alpha^0 = \tilde{0}$ and $s_\alpha^{n+1} = (s \circ_\alpha s_\alpha^n) + 1.\alpha$ for every $n \geq 0$, where $\tilde{0}$ is the weighted tree language which associates 0 to each tree. Let us assume that s is α-proper, i.e., $s(\alpha) = 0$. Then, for every $\xi \in T_\Theta$ and $n \in \mathbb{N}$, if $n \geq \text{height}(\xi)+1$, then $s_\alpha^{n+1}(\xi) = s_\alpha^n(\xi)$ [14, Lm. 3.10]. This justifies to define the operation *α-Kleene star of s* as follows: it is the (Θ, K)-weighted tree language s_α^* defined for every $\xi \in T_\Theta$ by $s_\alpha^*(\xi) = s_\alpha^{\text{height}(\xi)+1}(\xi)$ [20] (cf. [14, Def. 3.11]).

We denote by $\mathcal{L}(\Theta, K)$ the smallest class of (Θ, K)-weighted tree languages which is closed under top-concatenation with Θ, scalar multiplication with K, sum, Θ-tree concatenation, and Θ-Kleene star. By iterating top-concatenations and using scalar multiplication and sum we can build up each polynomial, hence each (Θ, K)-polynomial is in $\mathcal{L}(\Theta, K)$.

Definition 1. The *set of (Θ, K)-rational weighted tree languages*, denoted by $\text{Rat}(\Theta, K)$, contains each (Θ, K)-weighted tree language s such that there is a 0-extension $s' \in \mathcal{L}(\Theta', K)$ of s for some 0-extension Θ' of Θ.

In particular, $\mathcal{L}(\Theta, K) \subseteq \text{Rat}(\Theta, K)$. We use the concept of 0-extension for the following purpose: for the analysis of chain-free (Θ, K)-regular tree grammars (cf. Sect. 5), extra nullary symbols for tree concatenation are needed (see [14, Thm. 5.2], cf. also [47, Thm. 9]), and we provide them by the 0-extensions.

In our definition of rational weighted tree languages over Σ with storage S, we will use the rational operations to build up trees and each such tree ζ combines a tree $\xi \in T_\Sigma$ and a tree b over the ranked alphabet Δ corresponding to Σ and some finite sets $P' \subseteq P$ and $F' \subseteq F$. Then, according to Goldstine's idea, we check outside of the building process whether b is a behaviour. In order to allow manipulation of the storage via P' and F' also independently from the generation of a Σ-symbol, we use $*$ as a padding symbol of rank 1 such that $* \notin \Sigma$. Formally, we define the *Σ-extension of Δ*, denoted by $\langle \Delta, \Sigma \rangle$, to be the ranked alphabet where $\langle \Delta, \Sigma \rangle^{(1)} = \Delta^{(1)} \times (\Sigma^{(1)} \cup \{*\})$ and $\langle \Delta, \Sigma \rangle^{(k)} = \Delta^{(k)} \times \Sigma^{(k)}$ for $k \neq 1$.

Additionally, let $\mathcal{R}$ be the term rewriting system having for every $k \in \mathbb{N}$ and $\sigma \in \Sigma^{(k)}$ the rules:

$$\sigma(x_1, \ldots, x_k) \to \langle (p, f_1 \ldots f_k), \sigma \rangle (x_1, \ldots, x_k) \quad \text{for every } (p, f_1 \ldots f_k) \in \Delta^{(k)}$$

$$\sigma(x_1, \ldots, x_k) \to \langle (p, f), * \rangle (\sigma(x_1, \ldots, x_k)) \quad \text{for every } (p, f) \in \Delta^{(1)}.$$

Then we define the mapping $\mathcal{B}_\Delta : T_\Sigma \to \mathcal{P}(T_{\langle \Delta, \Sigma \rangle})$ for each $\xi \in T_\Sigma$ by

$$\mathcal{B}_\Delta(\xi) = \{\zeta \in T_{\langle \Delta, \Sigma \rangle} \mid \xi \Rightarrow_\mathcal{R}^* \zeta \text{ and } \mathrm{pr}_1(\zeta) \in \mathcal{B}(\Delta)\} \ ,$$

where $\mathrm{pr}_1 : T_{\langle \Delta, \Sigma \rangle} \to T_\Delta$ is the relabeling [16] defined by $\mathrm{pr}_1((\delta, \_)) = \delta$ for every $(\delta, \_) \in \langle \Delta, \Sigma \rangle$. We call $\mathcal{B}_\Delta(\xi)$ the set of Δ-behaviours on ξ.

Definition 2. Let $s : T_\Sigma \to K$ be a weighted tree language. Then s is (S, Σ, K)-rational if there are finite sets $P' \subseteq P$ and $F' \subseteq F$ and a weighted tree language $t \in \mathrm{Rat}(\langle \Delta, \Sigma \rangle, K)$ where Δ is the ranked alphabet corresponding to Σ, P', and F' such that

$$s = \mathcal{B}_\Delta; t, \quad \text{i.e., for each } \xi \in T_\Sigma : \quad s(\xi) = \sum_{\zeta \in \mathcal{B}_\Delta(\xi)} t(\zeta) \ .$$

We denote by $\mathrm{Rat}(S, \Sigma, K)$ the class of all (S, Σ, K)-rational weighted tree languages or for short: the class of rational weighted tree languages with storage.

Now we compare Definition 2 with the automata with data store [34]. For every set B and each $B' \subseteq B$, the characteristic function of B' in B is the mapping $\chi(B, B') : B \to K$ defined for every $a \in B$ by $\chi(B, B')(a) = 1$ if $a \in B'$ and $\chi(B, B')(a) = 0$ otherwise. If we transcribe the definition of the language $L(A) \subseteq \Sigma^*$ defined by a Goldstine-automaton A (as given in [34, p. 276] for some alphabet Σ) by replacing the membership test $u \in L(A)$ (for some string $u \in \Sigma^*$) by the equation $\chi_{(\Sigma^*, L(A))}(u) = 1$, then the definition of $L(A)$ reads:

$$\chi_{(\Sigma^*, L(A))}(u) = \bigvee_{\substack{v \in I^*: \\ \iota(v): D_0 \mapsto D_1}} \chi_{(\Sigma^* \times I^*, \hat{A})}(u, v)$$

where $\iota(v) : D_0 \mapsto D_1$ says that there is an initial configuration $c_0 \in D_0$ and a final configuration $c_1 \in D_1$ such that the sequence v of instructions can transform c_0 into c_1. This can be easily compared to our definition

$$s(\xi) = \sum_{\zeta \in \mathcal{B}_\Delta(\xi)} t(\zeta) \ .$$

Thus our concept of (S, Σ, K)-rational weighted tree languages generalizes the concept of automata over data store [34] from the unweighted to the weighted case and from strings to trees.

In the following two examples we drop the parentheses corresponding to unary symbols occurring in trees.

Example 1. Let $(\mathbb{N} \cup \{\infty\}, +, \cdot, 0, 1)$ be the commutative, complete semiring of natural numbers and $\Sigma = \{\sigma^{(2)}, \delta^{(1)}, \alpha^{(0)}\}$. We consider the weighted tree language $s \colon T_\Sigma \to \mathbb{N}_\infty$ (derived from [24, Ex. 3.1]) where, for each $\xi \in T_\Sigma$, we let

$$s(\xi) = 8^n \text{ if } \xi = \sigma(\delta^{2n}(\alpha), \delta^{2n}(\alpha)) \text{ for some } n \in \mathbb{N}, \text{ and } 0 \text{ otherwise .}$$

We show that s is $(\mathrm{COUNT}, \Sigma, \mathbb{N})$-rational. For this we show that there is a weighted tree language $t \in \mathrm{Rat}(\langle \Delta, \Sigma \rangle, K)$ such that $s = \mathcal{B}_\Delta; t$, where Δ is the ranked alphabet corresponding to Σ, P, and F. Thus,

$$
\begin{aligned}
\langle \Delta, \Sigma \rangle = {} & \{ \langle (p, \varepsilon), \alpha \rangle^{(0)} \mid p \in \mathrm{BC}(P) \} \cup \{ \langle (p, f), * \rangle^{(1)} \mid (p, f) \in \mathrm{BC}(P) \times F \} \\
& \cup \{ \langle (p, f), \delta \rangle^{(1)} \mid (p, f) \in \mathrm{BC}(P) \times F \} \\
& \cup \{ \langle (p, f_1 f_2), \sigma \rangle^{(2)} \mid (p, f_1 f_2) \in \mathrm{BC}(P) \times F^2 \} .
\end{aligned}
$$

We define the weighted tree language $t \in \mathrm{Rat}(\langle \Delta, \Sigma \rangle, \mathbb{N})$ as follows (where we abbreviate $\langle (\mathrm{true}, \varepsilon), \alpha \rangle$ by $\boldsymbol{\alpha}$):

$$t = (t_1)^*_\alpha \circ_\alpha \mathrm{top}_{\langle (\mathrm{true}, \mathrm{id}_\mathbb{N} \mathrm{id}_\mathbb{N}), \sigma \rangle}(t_2, t_2),$$
where $t_1 = 2.\langle (\mathrm{true}, \mathrm{inc}), * \rangle(\boldsymbol{\alpha})$, $t_2 = (t_3)^*_\alpha \circ_\alpha 1.\langle (\mathrm{zero}?, \varepsilon), \alpha \rangle$, and
$$t_3 = 2.\langle (\mathrm{true}, \mathrm{id}), \delta \rangle \big(\langle (\neg \mathrm{zero}?, \mathrm{dec}), \delta \rangle (\boldsymbol{\alpha}) \big) .$$

Note that true and $\neg \mathrm{zero}?$ are in $\mathrm{BC}(\{\mathrm{zero}?\})$ but they are not predicates of COUNT.

Thus, for each $\zeta \in T_{\langle \Delta, \Sigma \rangle}$, we have $(t_1)^*_\alpha(\zeta) = 2^n$ if $\zeta = \langle (\mathrm{true}, \mathrm{inc}), * \rangle^n(\boldsymbol{\alpha})$ for some $n \in \mathbb{N}$, and $(t_1)^*_\alpha(\zeta) = 0$ otherwise. Moreover, $t_2(\zeta) = 2^n$ if $\zeta = \big(\langle (\mathrm{true}, \mathrm{id}), \delta \rangle \langle (\neg \mathrm{zero}?, \mathrm{dec}), \delta \rangle \big)^n \big(\langle (\mathrm{zero}?, \varepsilon), \alpha \rangle \big)$ for some $n \in \mathbb{N}$, and $t_2(\zeta) = 0$ otherwise.

Now let $\xi \in T_\Sigma$ and $\zeta \in \mathcal{B}_\Delta(\xi)$ such that $t(\zeta) \neq 0$. By the definition of t, there are $n, n_1, n_2 \in \mathbb{N}$ and $\zeta_i = \big(\langle (\mathrm{true}, \mathrm{id}), \delta \rangle \langle (\neg \mathrm{zero}?, \mathrm{dec}), \delta \rangle \big)^{n_i} \big(\langle (\mathrm{zero}?, \varepsilon), \alpha \rangle \big)$ for each $i \in \{1, 2\}$ such that

$$\zeta = \langle (\mathrm{true}, \mathrm{inc}), * \rangle^n \big(\langle (\mathrm{true}, \mathrm{id}_\mathbb{N} \mathrm{id}_\mathbb{N}), \sigma \rangle (\zeta_1, \zeta_2) \big) .$$

Moreover, $t(\zeta) = 2^n \cdot 2^{n_1} \cdot 2^{n_2}$. Since $\xi \Rightarrow^*_\mathcal{R} \zeta$, we have $\xi = \sigma(\delta^{2n_1}(\alpha), \delta^{2n_2}(\alpha))$. Since $\mathrm{pr}_1(\zeta) \in \mathcal{B}(\Delta)$, we have $n = n_1 = n_2$. Hence $\sum_{\zeta \in \mathcal{B}_\Delta(\xi)} t(\zeta) = s(\xi)$ for every $\xi \in T_\Sigma$, i.e., $s \in \mathrm{Rat}(\mathrm{COUNT}, \Sigma, \mathbb{N})$. $\square$

5   Weighted Regular Tree Grammars with Storage and the Main Result

In this section we define weighted regular tree grammars with storage and prove our Kleene-Goldstine theorem. Our grammar model is the weighted version of regular tree grammar with storage [21], where we take the weights from a commutative, complete semiring. Our concept slightly extends the form of rules of (S, Σ, K)-regular tree grammar as defined in [24, Sec. 3.1]; on the other hand, the weight algebras of our concept are commutative, complete semirings and not the more general complete M-monoids as in [24].

A *weighted regular tree grammar over Σ with storage S and weights in K* (for short: (S, Σ, K)-rtg) is a tuple $\mathcal{G} = (N, Z, R, \mathrm{wt})$, where N is a finite set (*nonterminals*) such that $N \cap \Sigma = \emptyset$, $Z \subseteq N$ (set of *initial nonterminals*), R is a finite and nonempty set of *rules*; each rule has the form $A(p) \to \xi$, where $A \in N$, $p \in \mathrm{BC}(P)$, and $\xi \in T_\Sigma(N(F))$ with $N(F) = \{A(f) \mid A \in N, f \in F\}$, and $\mathrm{wt} : R \to K$ is the *weight function*.

If r is a rule of the form $A(p) \to B(f)$, then it called a *chain rule*. If $\mathcal{G}$ does not have chain rules, then we call it *chain-free*. We call $\mathcal{G}$ *start-separated* if it has exactly one initial nonterminal and this nonterminal does not occur in the right-hand side of a rule. We say that $\mathcal{G}$ is *in normal form* if each rule contains at most one symbol from Σ.

Let $P_\mathcal{G} \subseteq P$ be the finite set of all predicates which occur in Boolean combinations in rules in R. (We assume that each of such Boolean combinations is given by a Boolean expression, hence $P_\mathcal{G}$ can be determined by checking the rules in R.) Moreover, let $F_\mathcal{G} \subseteq F$ be the finite set of all instructions which occur in rules in R, and let $\Delta_\mathcal{G}$ be the ranked alphabet corresponding to Σ, $P_\mathcal{G}$, and $F_\mathcal{G}$.

For the definition of the semantics of $\mathcal{G}$, we view R as a ranked alphabet by associating with each rule r the number of nonterminals occurring in the right-hand side of r (*rank of r*). We define the mappings $\pi : T_R \to T_\Sigma$, $\beta : T_R \to T_{\Delta_\mathcal{G}}$, and $\mathrm{wt}_\mathcal{G} : T_R \to K$ inductively. Let $r(d_1, \ldots, d_k) \in T_R$:

- The tree $\pi(r(d_1, \ldots, d_k))$ is obtained from the right-hand side of r by replacing the ith occurrence of an element in $N(F)$ (in the order left-to-right) by $\pi(d_i)$.
- The tree $\beta(r(d_1, \ldots, d_k))$ is $(p, f_1 \ldots f_k)(\beta(d_1), \ldots, \beta(d_k))$ if r has the form $A(p) \to \xi$ and the ith occurrence of a nonterminal in ξ is associated with f_i.
- The value $\mathrm{wt}_\mathcal{G}(r(d_1, \ldots, r_k))$ is $\mathrm{wt}_\mathcal{G}(d_1) \cdot \ldots \cdot \mathrm{wt}_\mathcal{G}(d_k) \cdot \mathrm{wt}(r)$.

We refer to Fig. 2 for examples of π and β.

A *derivation tree of $\mathcal{G}$* is a tree $d \in T_R$ which satisfies the conditions: (i) the left-hand side of $d(\varepsilon)$ must be an initial nonterminal, (ii) for each position $w \in \mathrm{pos}(d)$, if A is the ith occurrence of a nonterminal in the right-hand side of $d(w)$, then A is the left-hand side of $d(wi)$, and (iii) $\beta(d) \in \mathcal{B}(\Delta_\mathcal{G})$. Then we say that d is a *derivation tree of $\mathcal{G}$ for $\pi(d)$*. For each $\xi \in T_\Sigma$, we denote the set of all derivation trees of $\mathcal{G}$ for ξ by $D_\mathcal{G}(\xi)$.

The *weighted tree language generated by $\mathcal{G}$* is the mapping $[\![\mathcal{G}]\!] : T_\Sigma \to K$ defined for each $\xi \in T_\Sigma$ by

$$[\![\mathcal{G}]\!](\xi) = \sum\nolimits_{d \in D_\mathcal{G}(\xi)} \mathrm{wt}_\mathcal{G}(d).$$

Two (S, Σ, K)-rtg $\mathcal{G}_1$ and $\mathcal{G}_2$ are *equivalent* if $[\![\mathcal{G}_1]\!] = [\![\mathcal{G}_2]\!]$.

Let s be a (Σ, K)-weighted tree language. It is called (S, Σ, K)-*regular* if there is an (S, Σ, K)-rtg $\mathcal{G}$ such that $s = [\![\mathcal{G}]\!]$. The class of all (S, Σ, K)-regular tree languages is denoted by $\mathrm{Reg}(S, \Sigma, K)$. We denote the class of all $s \in \mathrm{Reg}(S, \Sigma, K)$ such that there is a chain-free (S, Σ, K)-rtg $\mathcal{G}$ with $s = [\![\mathcal{G}]\!]$ by $\mathrm{Reg}_{\mathrm{nc}}(S, \Sigma, K)$ (*no* chain). We call a $(\mathrm{TRIV}, \Sigma, K)$-rtg just (Σ, K)-rtg, and we abbreviate $\mathrm{Reg}(\mathrm{TRIV}, \Sigma, K)$ by $\mathrm{Reg}(\Sigma, K)$.

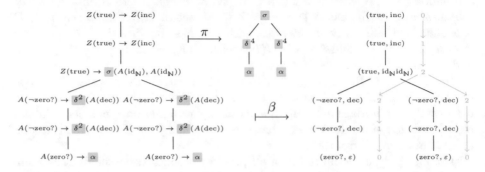

Fig. 2. A derivation tree $d \in D_{\mathcal{G}}(\xi)$ (left), the input tree $\xi = \pi(d) = \sigma(\delta^4(\alpha), \delta^4(\alpha))$ (up middle), the $\Delta_{\mathcal{G}}$-behaviour $b = \beta(d)$ (right), the family $(c_w \mid w \in \mathrm{pos}(b))$ of configurations (in grey) determined by b and 0.

Example 2. We consider the weighted tree language $s : T_\Sigma \to \mathbb{N}_\infty$ from Example 1. It is easy to see that $s = [\![\mathcal{G}]\!]$ for the $(\mathrm{COUNT}, \Sigma, \mathbb{N}_\infty)$-rtg $\mathcal{G} = (N, \{Z\}, R, \mathrm{wt})$ where $N = \{Z, A\}$ and R contains the following rules and weights:

rule r:	$\mathrm{wt}(r)$:	rule r:	$\mathrm{wt}(r)$:
$Z(\text{true}) \to Z(\text{inc})$	2	$A(\neg\text{zero?}) \to \delta^2(A(\text{dec}))$	2
$Z(\text{true}) \to \sigma(A(\text{id}_\mathbb{N}), A(\text{id}_\mathbb{N}))$	1	$A(\text{zero?}) \to \alpha$	1

Figure 2 shows the trees $d \in D_{\mathcal{G}}(\xi)$, $\pi(d) = \xi$, and $\beta(d)$ for $\xi = \sigma(\delta^4(\alpha), \delta^4(\alpha))$. □

Lemma 1. *For each (S, Σ, K)-rtg $\mathcal{G}$ there is an equivalent start-separated (S, Σ, K)-rtg $\mathcal{G}'$ which is in normal form.*

Proof. As in [24, Lm. 3.2], we can first transform $\mathcal{G}$ into an equivalent start-separated (S, Σ, K)-rtg $\mathcal{G}_1$. Then we can apply the usual construction (cf., e.g., [17, Thm. 3.22]) to $\mathcal{G}_1$ in order to decompose rules with more than one terminal symbol into several rules with at most one terminal symbol.   □

In [24, Thm. 5.3] a crucial decomposition theorem was proved for particular (S, Σ, K)-rtg: each such rtg is in normal form and, for each rule $A(p) \to \xi$, the inclusion $p \in P$ holds. Next we will lift this theorem to (S, Σ, K)-rtg of the present paper.

Theorem 1. (cf. [24, Thm. 5.3]) *For every* $s\colon T_\Sigma \to K$ *the following two statements are equivalent:*

(i) $s = [\![\mathcal{G}]\!]$ *for some* (S, Σ, K)-*rtg* $\mathcal{G}$.
(ii) *There are finite sets* $P' \subseteq P$ *and* $F' \subseteq F$, *and a chain-free* $(\langle \Delta, \Sigma \rangle, K)$-*rtg* $\mathcal{G}$ *such that* Δ *is the ranked alphabet corresponding to* Σ, P', *and* F', *and* $s = \mathcal{B}_\Delta; [\![\mathcal{G}]\!]$.

Proof. For the proof of both implications (i) $\Rightarrow$ (ii) and (ii) $\Rightarrow$ (i), we can assume by Lemma 1 that $\mathcal{G}$ is in normal form. Then the proof of [24, Thm. 5.3] can be extended easily to rtg in normal form with an arbitrary Boolean combination in the left-hand side of rules. $\square$

Theorem 2. *For each ranked alphabet* Θ *we have* $\mathrm{Reg}_{\mathrm{nc}}(\Theta, K) = \mathrm{Rat}(\Theta, K)$ [14, Thm. 5.2 and Thm. 6.8(2)].

Now we can prove our Kleene-Goldstine theorem for weighted regular tree grammars with storage. This theorem generalizes [14, Thm. 7.1] from TRIV to an arbitrary storage type.

Theorem 3. $\mathrm{Reg}(S, \Sigma, K) = \mathrm{Rat}(S, \Sigma, K)$.

Proof. Let $s \in \mathrm{Reg}(S, \Sigma, K)$. By Theorem 1, there is a $t \in \mathrm{Reg}_{\mathrm{nc}}(\langle \Delta, \Sigma \rangle, K)$ such that $s = \mathcal{B}_\Delta; t$. By Theorem 2, $t \in \mathrm{Rat}(\langle \Delta, \Sigma \rangle, K)$, hence by definition $s \in \mathrm{Rat}(S, \Sigma, K)$. This argumentation also holds in the reverse order. $\square$

If S has finitely many configurations (e.g. if $S = \mathrm{TRIV}$), then by Theorem 3 $\mathrm{Rat}(S, \Sigma, K)$ is the class of weighted tree languages recognized by K-weighted tree automata over Σ as defined in, e.g., [25] (cf. [24, Cor. 6.5]; note that S contains an identity and each (S, Σ, K)-rtg can have the always true predicate in its rules; each M-monoid associated with a complete semiring is compressible).

As another application, we mention the n-iterated pushdown storage type P^n where $n \in \mathbb{N}$ [19,21,36]. Theorem 3 provides a Kleene-Goldstine theorem for $(\mathrm{P}^n, \Sigma, K)$-weighted tree languages, in particular, for the infinite hierarchy $(\mathrm{Rec}(\mathrm{P}^n, \Sigma, \mathbb{B}) \mid n \in \mathbb{N})$ of classes of tree languages, the OI-hierarchy of n-level tree languages [7] (cf. [9, Thm. 1] and [22, Thm. 6.15]). This starts with the regular tree languages ($n = 0$) and the OI context-free tree languages ($n = 1$).

If we choose Σ to be monadic (i.e., $\Sigma = \Sigma^{(1)} \cup \Sigma^{(0)}$ and $|\Sigma^{(0)}| = 1$), then Theorem 3 provides a Kleene-Goldstine theorem for K-weighted regular string grammars with storage S, which are equivalent to K-weighted automata with storage S.

References

1. Aho, A.V.: Nested stack automata. J. ACM **16**, 383–406 (1969)
2. Alexandrakis, A., Bozapalidis, S.: Weighted grammars and Kleene's theorem. Inf. Process. Lett. **24**(1), 1–4 (1987)

3. Berstel, J., Reutenauer, C.: Recognizable formal power series on trees. Theoret. Comput. Sci. **18**(2), 115–148 (1982)

4. Berstel, J., Reutenauer, Ch.: Rational Series and Their Languages. EATCS Monographs on Theoretical Computer Science, vol. 12. Springer, Heidelberg (1988)

5. Chomsky, N.: Context-free grammars and pushdown storage. Technical report, MIT Research Lab of Electronics, Quaterly Progress Report 65 (1962)

6. Comon, H., et al.: Tree automata techniques and applications (2007). http://www. grappa.univ-lille3.fr/tata

7. Damm, W.: The IO- and OI-hierarchies. Theoret. Comput. Sci. **20**, 95–208 (1982)

8. Damm, W., Goerdt, A.: An automata-theoretical characterization of the OI-hierarchy. Inf. Control **71**, 1–32 (1986)

9. Damm, W., Guessarian, I.: Combining T and level-N. In: Gruska, J., Chytil, M. (eds.) MFCS 1981. LNCS, vol. 118, pp. 262–270. Springer, Heidelberg (1981). https://doi.org/10.1007/3-540-10856-4_92

10. Doner, J.E.: Decidability of the weak second-order theory of two successors, Abstract 65T-468. Not. Am. Math. Soc. **12**, 819 (1965)

11. Doner, J.E.: Tree acceptors and some of their applications. J. Comput. Syst. Sci. **4**, 406–451 (1970)

12. Droste, M., Herrmann, L., Vogler, H.: Weighted automata with storage. Inf. Comput. (accepted for publication)

13. Droste, M., Kuich, W., Vogler, H. (eds.): Handbook of Weighted Automata. EATCS Monographs in Theoretical Computer Science. Springer, Heidelberg (2009). https://doi.org/10.1007/978-3-642-01492-5

14. Droste, M., Pech, Chr., Vogler, H.: A Kleene theorem for weighted tree automata. Theory Comput. Syst. **38**, 1–38 (2005)

15. Eilenberg, S.: Automata, Languages, and Machines. Pure and Applied Mathematics, vol. 59A. Academic Press, New York (1974)

16. Engelfriet, J.: Bottom-up and top-down tree transformations - a comparison. Math. Syst. Theory **9**(3), 198–231 (1975)

17. Engelfriet, J.: Tree automata and tree grammars. Technical report DAIMI FN-10, Institute of Mathematics, University of Aarhus, Department of Computer Science, Ny Munkegade, 8000 Aarhus C, Denmark (1975). See also: arXiv:1510.02036v1 [cs.FL], 7 October 2015

18. Engelfriet, J.: Two-way automata and checking automata. Math. Centre Tracts **108**, 1–69 (1979)

19. Engelfriet, J.: Context-free grammars with storage. Technical report 86-11, University of Leiden (1986). See also: arXiv:1408.0683 [cs.FL] (2014)

20. Engelfriet, J.: Alternative Kleene theorem for weighted automata. Pers. Commun. (2003)

21. Engelfriet, J., Vogler, H.: Pushdown machines for the macro tree transducer. Theoret. Comput. Sci. **42**(3), 251–368 (1986)

22. Engelfriet, J., Vogler, H.: High level tree transducers and iterated pushdown tree transducers. Acta Inform. **26**, 131–192 (1988)

23. Ésik, Z., Kuich, W.: Formal tree series. J. Autom. Lang. Comb. **8**(2), 219–285 (2003)

24. Fülöp, Z., Herrmann, L., Vogler, H.: Weighted regular tree grammars with storage. Discrete Math. Theoret. Comput. Sci. **20**(1), #26 (2018). http://arxiv.org/abs/1705.06681

25. Fülöp, Z., Vogler, H.: Weighted tree automata and tree transducers. In: Droste, M., Kuich, W., Vogler, H. (eds.) Handbook of Weighted Automata. Monographs in Theoretical Computer Science. An EATCS Series, pp. 313–403. Springer, Heidelberg (2009). https://doi.org/10.1007/978-3-642-01492-5_9
26. Fülöp, Z., Vogler, H.: Weighted iterated linear control. Acta Inform. (2018) https://doi.org/10.1007/s00236-018-0325-x
27. Gécseg, F., Steinby, M.: Tree Automata. Akadémiai Kiadó, Budapest (1984). See also: arXiv:1509.06233v1 [cs.FL], 21 September 2015
28. Gécseg, F., Steinby, M.: Tree languages. In: Rozenberg, G., Salomaa, A. (eds.) Handbook of Formal Languages. Beyond Words, vol. 3, pp. 1–68. Springer, Heidelberg (1997). https://doi.org/10.1007/978-3-642-59126-6_1
29. Ginsburg, S.: Algebraic and Automata-theoretic Properties of Formal Languages. North-Holland, Amsterdam (1975)
30. Ginsburg, S., Greibach, S.A., Harrison, M.A.: One-way stack automata. J. ACM 14(2), 389–418 (1967)
31. Givant, S., Halmos, P.: Introduction to Boolean Algebras. UTM. Springer, New York (2009). https://doi.org/10.1007/978-0-387-68436-9
32. Golan, J.S.: Semirings and Their Applications. Kluwer Academic Publishers, Dordrecht (1999)
33. Goldstine, J.: Automata with data storage. In: Proceedings of the Conference on Theoretical Computer Science, University of Waterloo, Ontario, Canada, August 1977, pp. 239–246 (1977)
34. Goldstine, J.: A rational theory of AFLs. In: Maurer, H.A. (ed.) ICALP 1979. LNCS, vol. 71, pp. 271–281. Springer, Heidelberg (1979). https://doi.org/10.1007/3-540-09510-1_21
35. Goldstine, J.: Formal languages and their relation to automata: what Hopcroft & Ullman didn't tell us. In: Book, R.V. (ed.) Formal Language Theory: Perspectives and Open Problems, pp. 109–140. Academic Press, New York (1980)
36. Greibach, S.: Full AFLs and nested iterated substitution. Inf. Control 16, 7–35 (1970)
37. Herrmann, L., Vogler, H.: A Chomsky-Schützenberger theorem for weighted automata with storage. In: Maletti, A. (ed.) CAI 2015. LNCS, vol. 9270, pp. 115–127. Springer, Cham (2015). https://doi.org/10.1007/978-3-319-23021-4_11
38. Kambites, M.: Formal languages and groups as memory. Commun. Algebra 37(1), 193–208 (2009)
39. Kleene, S.C.: Representation of events in nerve nets and finite automata. In: McCarthy, J., Shannon, C.E. (eds.) Automata Studies, pp. 3–42 (1956)
40. Kuich, W.: Formal power series over trees. In: Bozapalidis, S. (ed.) Proceedings of the 3rd International Conference on Developments in Language Theory, DLT 1997, Thessaloniki, Greece, pp. 61–101. Aristotle University of Thessaloniki (1998)
41. Kuich, W., Salomaa, A.: Semirings, Automata, Languages. EATCS Monographs on Theoretical Computer Science, vol. 5. Springer, Heidelberg (1986). https://doi.org/10.1007/978-3-642-69959-7
42. Maslov, A.N.: Multilevel stack automata. Probl. Inf. Transm. 12, 38–43 (1976)
43. Sakarovitch, J.: Elements of Automata Theory. Cambridge University Press, Cambridge (2009)
44. Salomaa, A., Soittola, M.: Automata-Theoretic Aspects of Formal Power Series. Texts and Monographs in Computer Science. Springer, New York (1978). https://doi.org/10.1007/978-1-4612-6264-0
45. Schützenberger, M.P.: On the definition of a family of automata. Inf. Control 4, 245–270 (1961)

46. Scott, D.: Some definitional suggestions for automata theory. J. Comput. Syst. Sci. **1**, 187–212 (1967)
47. Thatcher, J.W., Wright, J.B.: Generalized finite automata theory with an application to a decision problem of second-order logic. Math. Syst. Theory **2**(1), 57–81 (1968)
48. Van Leeuwen, J.: Variations of a new machine model. In: Proceedings of the 17th FOCS, pp. 228–235 (1976)
49. Vogler, H., Droste, M., Herrmann, L.: A weighted MSO logic with storage behaviour and its Büchi-Elgot-Trakhtenbrot theorem. In: Dediu, A.-H., Janoušek, J., Martín-Vide, C., Truthe, B. (eds.) LATA 2016. LNCS, vol. 9618, pp. 127–139. Springer, Cham (2016). https://doi.org/10.1007/978-3-319-30000-9_10

Commutative Regular Languages – Properties and State Complexity

Stefan Hoffmann$^{(\boxtimes)}$

Informatikwissenschaften, FB IV,
Universität Trier, Universitätsring 15, 54296 Trier, Germany
hoffmanns@informatik.uni-trier.de

Abstract. We consider the state complexity of intersection, union and the shuffle operation on commutative regular languages for arbitrary alphabets. Certain invariants will be introduced which generalize known notions from unary languages used for refined state complexity statements and existing notions for commutative languages used for the subclass of periodic languages. Our bound for shuffle is formulated in terms of these invariants and shown to be optimal, from this we derive the bound of $(2nm)^{|\Sigma|}$ for commutative languages of state complexities n and m respectively. This result is a considerable improvement over the general bound $2^{mn-1} + 2^{(m-1)(n-1)}(2^{m-1} - 1)(2^{n-1} - 1)$.

We have no improvement for union and intersection for any alphabet, as was to be expected from the unary case. The general bounds are optimal here. Seeing commutative languages as generalizing unary languages is a guiding theme. For our results we take a closer look at a canonical automaton model for commutative languages.

Keywords: Commutative language · State complexity · Shuffle · Automata theory

1 Introduction

The state complexity of some regular language L is the minimal number of states needed in a complete deterministic automaton accepting L, or equivalently it is the number of Nerode right-equivalence classes. We will denote the state complexity of L by $\mathrm{sc}(L)$. Investigating the state complexity of the result of an operation on languages was first initiated in [9] and systematically started in [14], for a survey of this important and vast field see [5]. Here we consider the state complexity of the shuffle operation, and of union and intersection, on the class of commutative languages, which will be introduced below. Commutative automata, which accept commutative languages, were introduced in [3]. They are precisely the permutation-closed languages. First we recap some notions and fix notations of the theory of formal languages and automata. Then we state some results for unary languages which will be needed in the sequel. The reader

© Springer Nature Switzerland AG 2019
M. Ćirić et al. (Eds.): CAI 2019, LNCS 11545, pp. 151–163, 2019.
https://doi.org/10.1007/978-3-030-21363-3_13

might notice that several results, notions and methods of proofs are generalisations from unary languages. Then we look at commutative regular languages, we define the minimal commutative automaton first introduced in [7] and investigate some properties. This automaton is crucial for many of the following definitions. With its help we also give a new characterization of the class of periodic languages first introduced in [4] and used for a sufficient condition for context-free commutative languages to be regular. In the course of this we introduce the index and period vectors of some commutative regular language, and use these invariant extensively in all the following statements. We then give state complexity results, first by relating our invariants to the number of states, then by proving bounds and showing them to be tight. Lastly we look at aperiodic commutative languages, and commutative group languages, and show that every commutative language could be written as a union of shuffle products of an aperiodic and a group commutative language.

2   Prerequisites

Let $\Sigma = \{a_1, \ldots, a_k\}$ be a finite set of symbols[1], called an alphabet. The set Σ^* denotes the set of all finite sequences, i.e., of all words. The finite sequence of length zero, or the empty word, is denoted by ε. For a given word we denote by $|w|$ its length, and for $a \in \Sigma$ by $|w|_a$ the number of occurrences of the symbol a in w. Subsets of Σ^* are called languages. With $\mathbb{N} = \{0, 1, 2, \ldots\}$ we denote the set of natural numbers, including zero. A finite deterministic and complete automaton will be denoted by $\mathcal{A} = (\Sigma, S, \delta, s_0, F)$ with $\delta : S \times \Sigma \to S$ the state transition function, S a finite set of states, $s_0 \in S$ the start state and $F \subseteq S$ the set of final states. The properties of being deterministic and complete are implied by the definition of δ as a total function. The transition function $\delta : S \times \Sigma \to S$ could be extended to a transition function on words $\delta^* : S \times \Sigma^* \to S$ by setting $\delta^*(s, \varepsilon) := s$ and $\delta^*(s, wa) := \delta(\delta^*(s, w), a)$ for $s \in S$, $a \in \Sigma$ and $w \in \Sigma^*$. In the remainder we drop the distinction between both functions and will also denote this extension by δ. Herein we do not use other automata models. Hence all automata considered in this paper will be complete, deterministic and initially connected, the last notion meaning for every $s \in S$ there exists some $w \in \Sigma^*$ such that $\delta(s_0, w) = s$. The language accepted by some automaton $\mathcal{A} = (\Sigma, S, \delta, s_0, F)$ is $L(\mathcal{A}) = \{w \in \Sigma^* \mid \delta(s_0, w) \in F\}$. A language $L \subseteq \Sigma^*$ is called regular if $L = L(\mathcal{A})$ for some finite automaton. If $u, v \in L$ for some language $L \subseteq \Sigma^*$ we define the Nerode right-congruence with respect to L by $u \equiv_L v$ if and only if $\forall x \in \Sigma : ux \in L \leftrightarrow vx \in L$. The equivalence class for some $w \in \Sigma^*$ is denoted by $[w]_{\equiv_L} := \{x \in \Sigma^* \mid x \equiv_L w\}$. A language is regular if and only if the above right-congruence has finite index, and it could be used to define the minimal deterministic automaton $\mathcal{A}_L = (\Sigma, Q, \delta, [\varepsilon]_{\equiv_L}, F)$ with $Q := \{[w]_{\equiv_L} \mid w \in \Sigma^*\}$, $\delta([w]_{\equiv_L}, a) := [wa]_{\equiv_L}$ for $a \in \Sigma$, $w \in \Sigma^*$ and $F := \{[w]_{\equiv_L} \mid w \in L\}$. It is indeed the smallest automaton accepting L in terms

---

[1] If not otherwise stated we assume that our alphabet has the form $\Sigma = \{a_1, \ldots, a_k\}$ and k denotes the number of symbols.

of states. Given two automata $\mathcal{A} = (\Sigma, S, \delta, s_0, F)$ and $\mathcal{B} = (\Sigma, T, \mu, t_0, E)$, an automaton homomorphism $h : \mathcal{A} \to \mathcal{B}$ is a map between the state sets such that for each $a \in \Sigma$ and state $s \in S$ we have $h(\delta(s, a)) = \delta(h(s), a)$, $h(s_0) = t_0$ and $h^{-1}(E) = F$. If $\mathcal{B}$ is a surjective homomorphic image of $\mathcal{A}$ as above then $L(\mathcal{B}) = L(\mathcal{A})$. The minimal deterministic automaton has the additional property that every accepting automaton could be homomorphically mapped onto it. Let $\Sigma = \{a_1, \ldots, a_k\}$ be an alphabet. The map $\psi : \Sigma^* \to \mathbb{N}^k$ given by $\psi(w) = (|w|_{a_1}, \ldots, |w|_{a_k})$ is called the *Parikh-morphism*. For a given word $w \in \Sigma^*$ we define $\mathrm{perm}(w) := \{u \in \Sigma^* : \psi(u) = \psi(w)\}$ and for languages $L \subseteq \Sigma^*$ we set $\mathrm{perm}(L) := \bigcup_{w \in L} \mathrm{perm}(w)$. We also define the one-letter projection mapping $\pi_j : \Sigma^* \to \{a_j\}^*$ by $\pi_j(w) := a_j^{|w|_{a_j}}$. A language is called *commutative* if $\mathrm{perm}(L) = L$, i.e., with every word each permutation of this word is also in the language.

Definition 1. The *shuffle operation*, denoted by $\sqcup\!\sqcup$, is defined by

$$u \sqcup\!\sqcup v := \left\{ x_1 y_1 x_2 y_2 \cdots x_n y_n \; \middle| \; \begin{array}{l} u = x_1 x_2 \cdots x_n, v = y_1 y_2 \cdots y_n, \\ x_i, y_i \in \Sigma^*, 1 \le i \le n, n \ge 1 \end{array} \right\},$$

for $u, v \in \Sigma^*$ and $L_1 \sqcup\!\sqcup L_2 := \bigcup_{x \in L_1, y \in L_2} (x \sqcup\!\sqcup y)$ for $L_1, L_2 \subseteq \Sigma^*$.

The shuffle operation is commutative, associative and distributive with respect to union. We will use these properties without further mention. In writing formulas without brackets we suppose that the shuffle operation binds stronger than the set operations, and the concatenation operator has the strongest binding.

Theorem 1. *The class of commutative languages is closed under union, intersection, complement and the shuffle operation.*

The state complexity of the above operations will be a major concern of our paper. A regular language is called *aperiodic* if it is accepted by some complete automaton in which no word induces a permutation of some subset of states. More formally if for all $w \in \Sigma^*$, states $s \in S$ and $n \ge 1$ we have $\delta(s, w^n) = s$ implies $\delta(s, w) = s$. Notice that by this condition all minimal cycles in the automaton must be labelled by primitive words, where a word is primitive if it is not a non-trivial power of another word, and conversely if a non-trivial power of a word labels a minimal cycle then the above condition cannot hold. The class of aperiodic languages was introduced in [11] and admits a wealth of other characterizations in terms of logic, regular expressions and other means. We call a language a *(pure-)group language*[2] if it is accepted by a complete automaton where every letter acts as a permutation on the state set. Such automata are also called permutation automata, and the name stems from the fact that the transformation monoid of such an automaton forms a group. Note some ambiguity here in the sense that if $\Sigma = \{a, b\}$ then $(aa)^*$ is not a group language over this

---

[2] These were introduced in [10] under the name of pure-group events.

alphabet, but it is over the unary alphabet $\{a\}$. Hence we mean the existence of a permutation automaton over any alphabet. By definition $\{\varepsilon\}$ is considered to be a group language[3], this will unify the statements of some results.

Lemma 1. *An automaton $\mathcal{A} = (\Sigma, S, \delta, s_0, F)$ is aperiodic if and only if there exists $n \geq 0$ such that for all states $s \in S$ and any word $w \in \Sigma$ we have $\delta(s, w^n) = \delta(s, w^{n+1})$.*

2.1   Unary Languages

Let $\Sigma = \{a\}$ be a unary alphabet. In this section we collect some results about unary languages, and in a sense our results for commutative regular languages are strongly motivated by generalizing from unary languages. Let $L \subseteq \Sigma^*$ be regular with an accepting complete deterministic automaton $\mathcal{A} = (\Sigma, S, \delta, s_0, F)$, then by considering the sequence of states $\delta(s_0, a^1), \delta(s_0, a^2), \delta(s_0, a^3), \ldots$ we find numbers $i \geq 0, p > 0$ with $i + p$ minimal such that $\delta(s_0, a^i) = \delta(s_0, a^{i+p})$. We call these numbers the index i and the period p of the automaton $\mathcal{A}$, note that $i + p = |S|$. For the unique minimal complete automaton we call these numbers the index and period of the language.

Lemma 2. *Let L be unary regular with accepting automata $\mathcal{A} = (\Sigma, S, \delta, s_0, F)$ and $\mathcal{B} = (\Sigma, T, \mu, t_0, R)$ and automaton homomorphism $h : \mathcal{A} \to \mathcal{B}$. If $\mu(t_0, a^{\hat{i}}) = \mu(t_0, a^{\hat{i}+p})$ with $p > 0$, $i + p$ minimal and $\delta(s_0, a^{\hat{i}}) = \delta(s_0, a^{\hat{i}+\hat{p}})$ with $\hat{p} > 0$, then $i \leq \hat{i}$ and p divides $\hat{p}$.*

As the minimal automaton is a homomorphic image of every accepting automaton we get the following corollary.

Corollary 1. *For unary regular languages L with index i and period p and any accepting deterministic automaton $\mathcal{A} = (\Sigma, S, \delta, s_0, F)$, if $\delta(s_0, a^{\hat{i}}) = \delta(s_0, a^{\hat{i}+\hat{p}})$ for $\hat{i} \geq 0, \hat{p} > 0$, then $i \leq \hat{i}$ and p divides $\hat{p}$.*

As done in [13] the index and period could be used for refined state complexity statements in the unary case. Later we will need the following result from [13].

Theorem 2 ([13]). *Let $U, V \subseteq \Sigma^*$ be two unary languages with U accepted by an automaton with index i and period p, and V accepted by an automaton with index j and period q. Then the concatenation $U \cdot V$ could be accepted by an automaton of index $i + j + \mathrm{lcm}(p, q) - 1$ and period $\mathrm{lcm}(p, q)$, and this result is optimal in the case of $\gcd(p, q) > 1$ in the sense that there exists languages that reach these bounds.*

In our discussion several times unary languages which are accepted by automata with a single final state appear.

---

[3] It is not possible to give such an automaton for $|\Sigma| \geq 1$, but allowing $\Sigma = \emptyset$ the single-state automaton will do, or similar as $\Sigma^* = \{\varepsilon\}$ in this case.

Lemma 3. *Let $L \subseteq \{a\}^*$ be a unary language which is accepted by an automaton with a single final state and index i and period p. Then either $L = \{u\}$ with $|u| < i$ (and if the automaton is minimal we would have $p = 1$), or L is infinite with $L = a^{i+m}(a^p)^*$ and $0 \leq m < p$. Hence two words u, v with $\min\{|u|, |v|\} \geq i$ are both in L or not if and only if $|u| \equiv |v| \pmod{p}$.*

2.2   Commutative Regular Languaes

Let $\Sigma = \{a_1, \ldots, a_k\}$ be our finite alphabet. The minimal commutative automaton for a commutative language was introduced in [7].

Definition 2. Let L be a commutative language, we define the minimal commutative automaton $\mathcal{C}_L = (\Sigma, S_1 \times \ldots \times S_k, \delta, s_0, F)$ with

$$S_j := \{[a_j^m]_{\equiv_L} : m \geq 0\}, \quad F := \{([\pi_1(w)]_L, \ldots, [\pi_k(w)]_L) : w \in L\}$$

and $\delta((s_1, \ldots, s_j, \ldots, s_k), a_j) := (s_1, \ldots, \delta_j(s_j, a_j), \ldots, s_k)$ with one-letter transitions $\delta_j([a_j^m]_{\equiv_L}, a_j) := [a_j^{m+1}]_{\equiv_L}$ for $j = 1, \ldots, k$ and $s_0 := ([\varepsilon]_{\equiv_L}, \ldots, [\varepsilon]_{\equiv_L})$.

In [7] it was shown that this notion is well-defined and accepts L. And surely is finite if and only if L is regular. But it was also noted that in general the minimal commutative automaton is not equal to the minimal deterministic and complete automaton for L.

Theorem 3. *For a commutative language its minimal commutative automaton accepts this language. A commutative language is regular if and only if its commutative automaton is finite.*

The following definition on the one side generalizes a well-known notion from unary regular languages [12,13], and a notion of periodic languages as introduced in [4], at which we will also take a closer look later.

Definition 3 (Index and period vector). Let L be a commutative regular language with minimal commutative automaton $\mathcal{C}_L = (\Sigma, S_1 \times \ldots \times S_k, \delta, s_0, F)$. For $1 \leq j \leq k$ consider the sequence of states $\delta(s_0, a_j^m)$ for $m = 0, 1, \ldots$ with respect to the input letter a_j. By finiteness there exists $i_j \geq 0$ and $p_j > 0$ with $i_j + p_j$ minimal such that $\delta(s_0, a_j^{i_j}) = \delta(s_0, a_j^{i_j + p_j})$. As this state sequence could be identified with S_j we have $|S_j| = i_j + p_j$. The vector $(i_1, \ldots, i_k)$ we call the *index vector* and the vector $(p_1, \ldots, p_k)$ the *period vector* of L.

The next Lemma is helpful for deciding if a given word is in a given regular commutative language.

Lemma 4. *Let L be commutative regular with index vector $(i_1, \ldots, i_k)$, periodic vector $(p_1, \ldots, p_k)$ and minimal commutative automaton $\mathcal{C}_L = (\Sigma, S_1 \times \ldots \times S_k, \delta, s_0, F)$. Then $\delta(s_0, u) = \delta(s_0, v)$ if and only if*

$$|u|_{a_j} = |v|_{a_j} \vee (\min\{|u|_{a_j}, |v|_{a_j}\} \geq i_j \wedge |u|_{a_j} \equiv |v|_{a_j} \pmod{p_j})$$

for all $j \in \{1, \ldots, k\}$.

In the following the languages whose minimal commutative automaton has a single final state are of particular importance. Related to those is the following definition of unary languages derived from the minimal commutative automaton of a given commutative regular language.

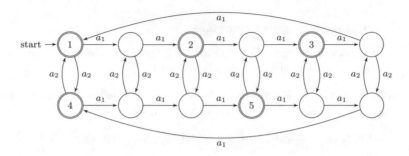

Fig. 1. The minimal commutative automaton for $L = (a_1a_1)^* \sqcup (a_2a_2)^* \cup (a_1a_1a_1)^* \sqcup a_2(a_2a_2)^*$. The final states are enumerated according to the sets $U_j^{(l)}$ for $j \in \{1,2\}$, $l \in \{1,\ldots,5\}$. See Example 1.

Definition 4. Let L be a commutative regular language with minimal commutative automaton $\mathcal{C}_L = (\Sigma, S_1 \times \ldots \times S_k, \delta, s_0, F)$. Suppose $F = \{(s_1^{(l)}, \ldots, s_k^{(l)}) \mid l = 1, \ldots |F|\}$. Then $U_j^{(l)} := \pi_j(\{w \in L \mid \delta(s_0, w) = (s_1^{(l)}, \ldots, s_k^{(l)})\})$. These are unary languages with $U_j^{(l)} \subseteq \{a_j\}^*$.

If we write the sets $U_j^{(l)}$ with $1 \le l \le |F|$ from the minimal commutative automaton and $j \in \{1, \ldots, k\}$ we always implicitly assume a certain order represented by the l-parameter is given on the final state set F of the minimal commutative automaton.

Example 1. Consider $L = (a_1a_1)^* \sqcup (a_2a_2)^* \cup (a_1a_1a_1)^* \sqcup a_2(a_2a_2)^*$. See Fig. 1 for the minimal commutative automaton. The final state sets are numbered in concordance with the sets $U_j^{(l)}$ from Definition 4. We have for example $U_1^{(2)} = a_1a_1(a_1a_1a_1a_1a_1a_1)^*$, $U_2^{(1)} = (a_2a_2)^*$ and $\pi_1(L) = (a_1a_1)^* \cup (a_1a_1a_1)^*$ and $\pi_2(L) = (a_2)^*$. Note that in this case, the minimal commutative automaton equals the minimal automaton, for if two states are in the top row and are both final or non-final, we see that after reading a_2 we could distinguish them by some word from $\{a_1\}^*$ on the bottom row, and a similar reasoning applies to the bottom row.

Corollary 2. *For a commutative regular language L we have $L = \bigcup_{l=1}^{n} U_1^{(l)} \sqcup \ldots \sqcup U_k^{(l)}$ for some $n \ge 0$.*

Lemma 5. *The minimal commutative automaton $\mathcal{C}_L = (\Sigma, S_1 \times \ldots \times S_k, \delta, s_0, F)$ for L accepts the languages $U_j^{(l)}$ for $j = 1, \ldots, k$, and for this purpose only $|S_j|$ states and a single final state are used.*

The minimal commutative automaton could also be used to accept the one-letter projection languages.

Lemma 6. *The minimal commutative automaton $C_L = (\Sigma, S_1 \times \ldots \times S_k, \delta, s_0, F)$ for L accepts the projection languages $\pi_j(L) = \bigcup_{l=1}^{n} U_j^{(l)}$ for $j = 1, \ldots, k$, and for this purpose only $|S_j|$ states are used.*

The index and periods of the projected one-letter languages may differ in general. For example let $L = (a_1 a_1)^* \shuffle (a_2 a_2)^* \cup (a_1 a_1 a_1 a_1)^* \shuffle a_2^*$. We have $(i_1, i_2) = (0, 0)$ and $(p_1, p_2) = (4, 2)$, but $\pi_1(L) = (a_1 a_1)^*$ and $\pi_2(L) = a_2^*$. If our commutative languages have a special form, two unary words over the same letter a_j are Nerode right-equivalent with respect to L exactly if they are Nerode right-equivalent with respect to the one-letter projection language $\pi_j(L)$ over the unary alphabet $\{a_j\}$.

Lemma 7. *Let L be language with $L = \pi_1(L) \shuffle \ldots \shuffle \pi_k(L)$. Then for each $j \in \{1, \ldots k\}$ we have $a_j^m \equiv_L a_j^n \Leftrightarrow a_j^m \equiv_{\pi_j(L)} a_j^n$, where on the right side the equivalence is considered with respect to the unary alphabet $\{a_j\}$.*

The next notion is taken from [4]. Here we give a different definition, but show its equivalence to the one from [4] in Lemma 8.

Definition 5. A commutative regular language L is called *periodic* if its minimal commutative automaton has a single final state.

By our notation, this implies that L is periodic iff $L = U_1^{(1)} \shuffle \ldots \shuffle U_k^{(1)}$, but we can say even more, namely that if we have a language as a shuffle product of unary languages acceptable by an automaton with a single final state, then it is periodic. Note that for languages of the form $L = U_1 \shuffle \ldots \shuffle U_k$ as $U_j = \pi_j(L)$ those languages U_j are uniquely determined. Hence for periodic regular languages we have a normal form theorem.

Theorem 4. *A commutative regular language L is periodic if and only if $L = U_1 \shuffle U_2 \shuffle \ldots \shuffle U_k$ for unique unary regular languages $U_j \subseteq \{a_j\}^*$ that are acceptable by a unary automaton with a single final state.*

Example 2. Note that we could not state something along the lines that L is periodic if we could write $L = U_1 \shuffle \ldots \shuffle U_k$. For example $L = ((a_1 a_1 a_1)^* a_1 \cup (a_1 a_1 a_1)^*) \shuffle (a_2 a_2)^*$ is not periodic, if we write it in terms of the $U_j^{(l)}$-sets we would have $L = (a_1 a_1 a_1)^* a_1 \shuffle (a_2 a_2)^* \cup (a_1 a_1 a_1)^* \shuffle (a_2 a_2)^*$.

Corollary 3. *Let L be a commutative regular language with minimal commutative automaton $C_L = (\Sigma, S_1 \times \ldots \times S_k, \delta, s_0, F)$. We have $L = \bigcup_{l=0}^{|F|} U_1^{(l)} \shuffle \ldots \shuffle U_k^{(l)}$ with the languages $U_j^{(l)}$ from Definition 4. Then the languages of the form $U_1^{(l)} \shuffle \ldots \shuffle U_k^{(l)}$ are periodic.*

In [4] a sequence of vectors $\rho = v_0, v_1, \ldots, v_k$ from $\mathbb{N}^k$ was called a *base* if $v_i(j) = 0$ for $i, j \geq 1$ such that $i \neq j$. The ρ-set was defined as $\Theta(\rho) = \{v \in \mathbb{N}^k : v = v_0 + l_1 v_1 + \ldots + l_k v_k$ for some $l_1, \ldots, l_k \in \mathbb{N}\}$. In some sense the index and period vectors extend these notions to arbitrary regular commutative languages. For periodic languages the entries in the period vector are precisely the non-zero entries in the base vectors $v_1, \ldots, v_k$. But v_0 does not equal the index vector, but its entries are at least the size of the index vector plus some number determined by the position of the final state among the loop in the minimal accepting automaton for infinite unary languages.

Lemma 8. *A commutative regular language L is periodic if and only $\psi(L) = \Theta(\rho)$ for some base ρ.*

In [4] it was further noted that the base is unique, but follows also with the uniqueness from Theorem 4. A language L is called *strictly bounded* if $L \subseteq a_1^* a_2^* \ldots a_k^*$. The maps $L \mapsto \text{perm}(L)$ and $L \mapsto L \cap a_1^* \cdots a_k^*$ are mutually inverse bijections between the class of strictly bounded languages and the class of commutative languages. Hence both language classes are closely related, and the next Theorem 5 was firstly given in [6] formulated for strictly bounded languages, and in [4] formulated for commutative languages. We note that it is easily implied by utilizing the minimal commutative automaton.

Theorem 5. *A commutative language is regular if and only if it is a finite union of periodic languages.*

Example 3. Note that we do not have uniqueness in Theorem 5, for example $\{a_1, a_2\}^* = (a_1)^* \sqcup (a_2)^* = (a_1)^* \sqcup (a_2 a_2)^* \cup (a_1)^* \sqcup a_2 (a_2 a_2)^*$.

The periodic languages fulfill the equation $L = \pi_1(L) \sqcup \ldots \sqcup \pi_2 k(L)$, and the class of all languages for which this equation holds true is a proper subclass of all commutative languages, which occurs in later statements.

Example 4. The language $aa^* \sqcup b$ is periodic as its minimal commutative automaton has a single final state. The language $U = (a(aaa)^* \cup aa(aaa)^*) \sqcup b = a(aaa)^* \sqcup b \cup aa(aaa)^* \sqcup b$ is not periodic, as its minimal commutative automaton has more than a single final state, but we have $U = \pi_1(U) \sqcup \pi_2(U)$ here. For the language $V = aa(aaa)^* \sqcup b(bb)^* \cup a(aaa)^* \sqcup (bb)^*$ we have $(a(aaa)^* \sqcup b(bb)^*) \cap V = \emptyset$ but as $\pi_1(V) = aa(aaa)^* \cup a(aaa)^*$ and $\pi_2(V) = b^*$ the language V is properly contained in $\pi_1(V) \sqcup \pi_2(V)$ as the latter contains $a(aaa)^* \sqcup b(bb)^*$.

3   State Complexity Results

For a general commutative regular language we have the following inequality for its state complexity.

Lemma 9. *Let L be a commutative regular language with index vector $(i_1, \ldots, i_k)$ and period vector $(p_1, \ldots, p_k)$. Then $i_j + p_j \leq \text{sc}(L) \leq \prod_{r=1}^{k}(i_r + p_r)$ for $j \in \{1, \ldots, k\}$.*

By Lemma 6 we can bound the state complexity of the projection languages by the invariants i_j and p_j.

Lemma 10. *If L is commutative regular with index vector $(i_1, \ldots, i_k)$ and period vector $(p_1, \ldots, p_k)$ we have $\mathrm{sc}(\pi_j(L)) \leq i_j + p_j$ for each $j \in \{1, \ldots, k\}$.*

The state complexity of the one-letter projection languages $\pi_j(L)$ could be strictly smaller than $i_j + p_j$, as is shown by L from Example 1, here we have $\pi_2(L) = a_2^*$, but $i_2 = 0, p_2 = 2$. But equality holds for periodic languages L and more general by Lemma 7 we have the next corollary.

Corollary 4. *If $L = \pi_1(L) \sqcup \ldots \sqcup \pi_k(L)$ is regular with index vector $(i_1, \ldots, i_k)$ and period vector $(p_1, \ldots, p_k)$ then $\mathrm{sc}(\pi_j(L)) = i_j + p_j$ and so $\mathrm{sc}(\pi_j(L)) \leq \mathrm{sc}(L) \leq \prod_{j=1}^{k} \mathrm{sc}(\pi_j(L))$. In particular this holds for periodic languages.*

In [8] it was shown that for an arbitrary (not necessarily commutative) regular language L we have $\mathrm{sc}(\pi_j(L)) \leq e^{(1+o(1))\sqrt{\mathrm{sc}(L)\ln \mathrm{sc}(L)}}$. In the case of commutative regular languages combining Lemma 10 and Lemma 9 gives the following improvement.

Corollary 5. *For a regular commutative language L and $j \in \{1, \ldots, k\}$ we have $\mathrm{sc}(\pi_j(L)) \leq \mathrm{sc}(L)$.*

Next our result on the state complexity of the shuffle for commutative regular languages.

Theorem 6. *Let U, V be commutative regular languages with index and period vectors $(i_1, \ldots, i_k), (j_1, \ldots, j_k)$ and $(p_1, \ldots, p_k), (q_1, \ldots, q_k)$. Then the state complexity of the shuffle $U \sqcup V$ is at most $\prod_{l=1}^{k}(i_l + j_l + 2 \cdot \mathrm{lcm}(p_l, q_l) - 1)$.*

The following corollary is a little bit less involved in its statement, and helps in giving a inequalities solely in terms of the state complexities of the original languages, without using the index and period vectors, as in Corollaries 7 and 8.

Corollary 6. *Let U, V be commutative regular languages with index and period vectors $(i_1, \ldots, i_k), (j_1, \ldots, j_k)$ and $(p_1, \ldots, p_k), (q_1, \ldots, q_k)$. Then the state complexity of the shuffle $U \sqcup V$ is at most $\prod_{l=1}^{k} 2(i_l + p_l)(j_l + q_l)$.*

Corollary 7. *For periodic commutative languages U and V, we have $\mathrm{sc}(U \sqcup V) \leq 2 \prod_{j=1}^{k}(\mathrm{sc}(\pi_j(U)) \mathrm{sc}(\pi_j(V)))$.*

Using Lemma 9 and Corollary 6 the state complexity of the shuffle of two regular commutative languages U, V is at most $(2\,\mathrm{sc}(U)\,\mathrm{sc}(V))^{|\Sigma|}$. Note that the size of the alphabet appears here, and for fixed alphabets this is an improvement over the general upper bound $2^{mn-1} + 2^{(m-1)(n-1)}(2^{m-1} - 1)(2^{n-1} - 1)$ with $n = \mathrm{sc}(U), m = \mathrm{sc}(V)$ as given in [1] for complete deterministic automata. But note that the general bound is only tight for $|\Sigma| \geq mn - 1$ as shown in [1], hence if the alphabet is fixed this bound is not reached by automata with sufficiently large state sets. Later we will show that our bound given in Theorem 6 is tight for any alphabet size.

Corollary 8. *The state complexity of the shuffle of two regular commutative languages is at most* $(2\operatorname{sc}(U)\operatorname{sc}(V))^{|\Sigma|}$.

For unary languages with states complexities n, m by results from [13] we find that $n\cdot m$ states are sufficient and necessary for the intersection and union, and so we have no improvement over the general case. Similar for regular commutative languages with more than one alphabet symbol we need that many states in the worst case. But first we need some assertions about the minimal automaton for a special class of commutative regular languages to prove this claim.

Lemma 11. *If $L = \pi_1(L) \sqcup \ldots \sqcup \pi_k(L)$ is regular and the projection languages are not aperiodic, then the minimal automaton equals the minimal commutative automaton.*

With Lemma 11 we can construct examples that reach the bound for intersection and shuffle in the next statements.

Theorem 7. *For regular commutative languages U, V with state complexities n and m respectively and $|\Sigma| \geq 1$ arbitrary, $n\cdot m$ states are sufficient and necessary to accept their union or intersection.*

Theorem 8. *The bound in Theorem 6 is sharp, i.e., there exists languages of arbitrary large state complexities such that the state complexity of their shuffle reaches the bound for every alphabet size.*

4   A Decomposition Result

In this section we take a closer look at the aperiodic commutative languages and at the commutative group languages. We show that for aperiodic commutative languages we can improve our bound on the state complexity of the shuffle, but also on the state complexities of the boolean operations. Also we show that every commutative language could be written as a union of shuffle products of an aperiodic language and a group language. It is easy to see that the standard product automaton construction gives a permutation automaton if the original automata are permutation-automata, and likewise yields an aperiodic automaton if applied to aperiodic automata. Hence the following closure result is implied.

Theorem 9. *The class of aperiodic and of group-languages is closed under the boolean set operations union, intersection and complement.*

For some results we need the following equivalence.

Lemma 12. *A language is aperiodic (respectively a group language) if and only if its minimal automaton is aperiodic (respectively a permutation-automaton).*

Theorem 10. *A commutative language is aperiodic if and only if its period vector equals $(1, \ldots, 1)$. And it is a group-language if and only if the index vector equals $(0, \ldots, 0)$.*

Corollary 9. *If a commutative language L is aperiodic then its projection languages $\pi_j(L)$ for $j = 1, \ldots, k$ are aperiodic. Similarly if L is a group language, then the projection languages are also group languages.*

Example 5. The reverse implication does not hold in the above corollary. Let $L = a_1(a_1a_1)^* \sqcup a_2(a_2a_2)^* \cup a_1a_1(a_1a_1)^* \sqcup (a_2a_2)^*$. This language is not aperiodic, but $\pi_1(L) = a_1a_1^*$ and $\pi_2(L) = a_2^*$ are aperiodic languages. Similar for group languages $L = a_1^* \sqcup \{\varepsilon\} \cup \{\varepsilon\} \sqcup a_2^*$ would be a counter-example for the reverse implication.

But we can prove the reverse direction for languages of the form $L = \pi_1(L) \sqcup \ldots \sqcup \pi_k(L)$, in particular for periodic languages.

Lemma 13. *If $L = \pi_1(L) \sqcup \ldots \sqcup \pi_k(L)$ then L is aperiodic (respectively a group-language) if and only if each one-letter projection language is aperiodic (respectively a group-language).*

As every regular commutative language is a finite union of periodic languages[4] by Theorem 5, and as we can decompose periodic languages in the shuffle product of an aperiodic part and a group language part, which is motived by the same result from unary languages, we can derive our decomposition result.

Theorem 11. *If L is a commutative regular language, then it is a finite union of commutative languages of the form $U \sqcup V$ where U is aperiodic and commutative and V is a commutative group-language.*

Using Theorem 6 and results from [13] and [2], we can refine our state complexity results for aperiodic languages. The method of proof is to look closely at automata for the operations on unary languages, which is done in the above mentioned articles. We apply those results to the unary projection languages and combine them appropriately to get an automaton for the result, a similar method is used in the proof of Theorem 6.

Theorem 12. *Let U and V be aperiodic commutative languages with index vectors $(i_1, \ldots, i_k)$ and $(j_1, \ldots, j_k)$. Then the state complexity of union and intersection is at most $\prod_{l=1}^{k}(\max\{i_l, j_l\} + 1)$. The state complexity of the shuffle $U \sqcup V$ is at most $\prod_{l=1}^{k}(i_l + j_l + 1)$.*

As for aperiodic languages by Theorem 10 the period is one, using Lemma 9 we get the next corollary to Theorem 12.

Corollary 10. *If U, V are aperiodic commutative languages, then the state complexity of union and intersection is at most $(\max\{\mathrm{sc}(U), \mathrm{sc}(V)\})^{|\Sigma|}$. The state complexity of the shuffle $U \sqcup V$ is at most $(\mathrm{sc}(U) + \mathrm{sc}(V) - 1)^{|\Sigma|}$.*

---

[4] Note that the notions of periodic and aperiodic languages appearing in this article are not meant to be related in dichotomous way.

5 Conclusion

We have examined the state complexity of shuffle, intersection and union for commutative regular languages and have given a tight bound for shuffle that is far better than the general bound given in [1]. We used the minimal commutative automaton from [7] introduced with the help of the Nerode right-congruence relation, investigated it further and introduced the index and period vector of some commutative regular language. We believe this to be useful notions also for other and future questions related to commutative regular languages.

In the course of our investigations we discovered two strict subclasses, the periodic languages from [4], and those of the form $L = \pi_1(L) \sqcup \ldots \sqcup \pi_k(L)$, which are properly contained in each other as shown by Example 4, and investigated their properties and the relation of the minimal commutative automaton to the general minimal automaton.

Acknowledgement. I thank my supervisor, Prof. Dr. Henning Fernau, for giving valuable feedback, discussions and research suggestions concerning the content of this article. I also thank the anonymous reviewers whose comments improved the presentation of this article. I also thank an anonymous reviewer of a previous version, whose feedback ultimately led to a new approach and stronger results.

References

1. Brzozowski, J., Jirásková, G., Liu, B., Rajasekaran, A., Szykuła, M.: On the state complexity of the shuffle of regular languages. In: Câmpeanu, C., Manea, F., Shallit, J. (eds.) DCFS 2016. LNCS, vol. 9777, pp. 73–86. Springer, Cham (2016). https://doi.org/10.1007/978-3-319-41114-9_6
2. Brzozowski, J.A., Liu, B.: Quotient complexity of star-free languages. Int. J. Found. Comput. Sci. **23**(6), 1261–1276 (2012)
3. Brzozowski, J.A., Simon, I.: Characterizations of locally testable events. Discrete Math. **4**(3), 243–271 (1973)
4. Ehrenfeucht, A., Haussler, D., Rozenberg, G.: On regularity of context-free languages. Theor. Comput. Sci. **27**, 311–332 (1983)
5. Gao, Y., Moreira, N., Reis, R., Yu, S.: A survey on operational state complexity. J. Autom. Lang. Comb. **21**(4), 251–310 (2017)
6. Ginsburg, S., Spanier, E.H.: Bounded regular sets. Proc. Am. Math. Soc. **17**, 1043–1049 (1966)
7. Cano Gómez, A., Álvarez, G.I.: Learning commutative regular languages. In: Clark, A., Coste, F., Miclet, L. (eds.) ICGI 2008. LNCS (LNAI), vol. 5278, pp. 71–83. Springer, Heidelberg (2008). https://doi.org/10.1007/978-3-540-88009-7_6
8. Jirásková, G., Masopust, T.: State complexity of projected languages. In: Holzer, M., Kutrib, M., Pighizzini, G. (eds.) DCFS 2011. LNCS, vol. 6808, pp. 198–211. Springer, Heidelberg (2011). https://doi.org/10.1007/978-3-642-22600-7_16
9. Maslov, A.N.: Estimates of the number of states of finite automata. Dokl. Akad. Nauk SSSR **194**(6), 1266–1268 (1970)
10. McNaughton, R.: The loop complexity of pure-group events. Inf. Control **11**(1/2), 167–176 (1967)

11. McNaughton, R., Papert, S.A.: Counter-Free Automata. MIT Research Monograph, vol. 65. The MIT Press, Cambridge (1971)
12. Pighizzini, G.: Unary language concatenation and its state complexity. In: Yu, S., Păun, A. (eds.) CIAA 2000. LNCS, vol. 2088, pp. 252–262. Springer, Heidelberg (2001). https://doi.org/10.1007/3-540-44674-5_21
13. Pighizzini, G., Shallit, J.: Unary language operations, state complexity and Jacobsthal's function. Int. J. Found. Comput. Sci. 13(1), 145–159 (2002)
14. Yu, S., Zhuang, Q., Salomaa, K.: The state complexities of some basic operations on regular languages. Theor. Comput. Sci. 125(2), 315–328 (1994)

Algebraic Systems Motivated by DNA Origami

James Garrett, Nataša Jonoska[⊠], Hwee Kim, and Masahico Saito

Department of Mathematics and Statistics, University of South Florida,
4202 E. Fowler Ave, Tampa, FL 33620, USA
jonoska@mail.usf.edu

Abstract. We initiate an algebraic approach to study DNA origami structures. We identify two types of basic building blocks and describe a DNA origami structure by their composition. These building blocks are taken as generators of a monoid, called the origami monoid, and motivated by the well studied Temperley-Lieb algebras, we identify a set of relations that characterize the origami monoid. We present several observations about Green's relations for the origami monoid and study the relations to a direct product of Jones monoids, which is a morphic image of an origami monoid.

Keywords: DNA origami · Temperley-Lieb algebra ·
Rewriting system

1 Introduction

In the past few decades, bottom-up assemblies at the nano scale have introduced new materials and molecular scaffoldings producing structures that have wide ranging applications (e.g. [5,8]), even materials that seem to violate standard chemistry behavior (e.g. [14]). "DNA origami", introduced by Rothemund [11] in 2006, significantly facilitated the construction of $\sim$100 $\times$ 100 nm 2D DNA nanostructures. The method typically involves combining an M13 single-stranded cyclic viral molecule called *scaffold* with 200–250 short *staple strands* to produce about 100 nm diameter 2D shapes [11], and more recently also to produce a variety of 3D constructs (e.g. [4]). Figure 1 (left) shows a schematic of an origami structure, where the thick black line represents a portion of the cyclic vector plasmid outlining the shape, and the colored lines are schematics of the short strands that keep the cyclic molecule folded in the shape. Because the chemical construction of DNA origami is much easier than previous methods, this form of DNA nanotechnology has become popular, with perhaps 300 laboratories in the world today focusing on it.

Although numerous laboratories around the world are successful in achieving various shapes with DNA origami, theoretical understanding and characterizations of these shapes is still lacking. With this paper we propose an algebraic

M. Ćirić et al. (Eds.): CAI 2019, LNCS 11545, pp. 164–176, 2019.
https://doi.org/10.1007/978-3-030-21363-3_14

system to describe and investigate DNA origami structures. The staple strands usually have 2–4 segments of about 8 bases joining 2–3 locations (folds) of the scaffold. All cross-overs between two staple strands and between two neighboring folds of the scaffold are antiparallel. We divide the DNA origami structure to local scaffold-staples interactions and to such local interactions we associate a generator of a monoid which we call an *origami monoid*. The origami monoid we present here is closely related to the Jones monoid [2,3] which is a monoid variant of the well studied Temperley-Lieb algebra [1]. We show that a DNA origami structure can be associated to an element of an origami monoid and propose a set of rewriting rules that are plausible for DNA segments to conform in DNA origami. The number of generators of an origami monoid depends on the number of parallel folds of the scaffold in the DNA origami. We observe that a direct product of two Jones monoids is a surjective image of an origami monoid, and we study the structure of the origami monoids through Green's relations. We characterize the origami monoids for small number of scaffold folds and propose several conjectures for general origami monoids.

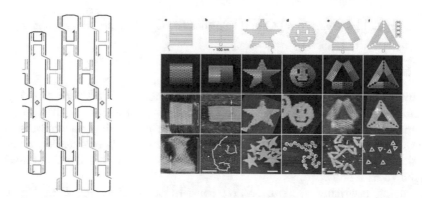

Fig. 1. (Left) A schematic figure of DNA origami structure with scaffold in black and staples in color (edited from [11]). (Right) Various shapes made by DNA origami, from [11]. Reprinted by permission from Springer Nature Customer Service Centre GmbH: Springer Nature (Folding DNA to create nanoscale shapes and patterns, Paul W. K. Rothemund), [4561420919172] 2006

2   Preliminaries

2.1   Jones Monoids

Temperley-Lieb algebras have been used in many fields, particularly in physics and knot theory (see, for example, [1,2,7,9]). The Jones monoid $\mathcal{J}_n$ is derived from the Temperley-Lieb algebras and is defined with generators and relations as follows [2,9]. The monoid $\mathcal{J}_n$ is generated by h_i, $i = 1, \ldots, n-1$, and has relations

(B) $h_i h_j h_i = h_i$ for $|i-j| = 1$, (C) $h_i h_i = h_i$ (D) $h_i h_j = h_j h_i$ for $|i-j| \geq 2$.

The elements of $\mathcal{J}_n$ may be represented as planar diagrams with non-crossing lines connecting n points at the top and n points at the bottom of the diagrams. The diagram for the generator h_i is depicted in Fig. 2(A) [9]. For each h_i, parallel vertical lines connect the top jth and bottom jth points ($j \neq i, i+1$) of the diagram for all but the ith and $(i+1)$st points, while the top ith and $(i+1)$st points are connected, and the bottom ith and $(i+1)$st points are connected. Multiplication of two elements is represented by concatenation of diagrams, placing the diagram of the first element on top of the second, and removing closed loops. The diagramatic representation of the monoid relations are depicted in Fig. 2(B), (C) and (D). More details can be found in [2,9].

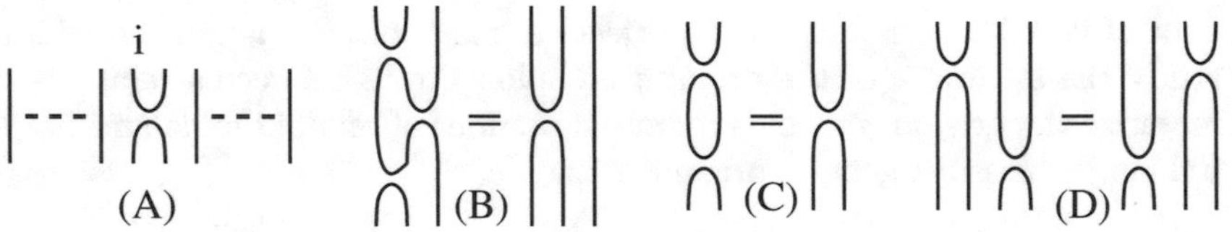

Fig. 2. The generators (A) and relations (B, C, D) of the Jones monoid

2.2 String Rewriting Systems

An alphabet Σ is a non-empty finite set of symbols. A word over Σ is a finite sequence of elements (symbols) from Σ, and Σ^* is the set of all words over Σ. This set includes the empty string, the word containing no symbols, often written as 1. A word u is called a *factor of a word* v if there exist words x and y, which may be empty, such that $v = xuy$. Note that this is also sometimes referred to as a subword.

A string rewriting system, (Σ, R) consists of an alphabet Σ and a set of rewriting rules, R, which is a binary relation on Σ^*. An element (x, y) of R is called a rewriting rule, and is written $x \to y$. We extend R to factors of words $\xrightarrow{}$, where for any $s, t \in \Sigma^*$, $s \xrightarrow{R} t$ if there exist $x, y, u, v \in \Sigma^*$ such that $s = uxv$, $t = uyv$, and $x \to y$. We also write $s \to t$ for simplicity if no confusion arises.

If there is a sequence of words $u = x_1 \to x_2 \to \cdots \to x_n = v$ in a rewriting system (Σ^*, R), we write $u \to_* v$. An element $x \in \Sigma^*$ is *confluent* if for all $y, z \in \Sigma^*$ such that $x \to_* y$ and $x \to_* z$, there exists $w \in \Sigma^*$ such that $y \to_* w$ and $z \to_* w$. If all words in Σ^* are confluent, then (Σ^*, R) is called *confluent*. In particular, if R is symmetric, then the system (Σ^*, R) is confluent.

2.3 Monoids and Green's Relations

A monoid is a pair $(M, \cdot)$ where M is a set and $\cdot$ is an associative binary operation on M that has an identity element 1. The set Σ^* is a (free) monoid generated by Σ with word concatenation as the binary operation, and the empty string as the identity element. Presentations of monoids are defined from the free monoid in a manner similar to presentations of groups. Rewriting systems define the monoid relations by taking the equivalence closure of the rewriting rules, which makes the rewriting system confluent.

For a monoid M, the *principal left (resp. right) ideal* generated by $a \in M$ is defined by $Ma = \{xa \mid x \in M\}$ (resp. aM), and the *principal two-sided ideal* s MaM. Green's relations $\mathscr{L}$, $\mathscr{R}$, and $\mathscr{J}$ are defined for $a, b \in M$ by $a\mathscr{L}b$ if $Ma = Mb$, $a\mathscr{R}b$ if $aM = bM$ and $a\mathscr{J}b$ if $MaM = MbM$. Green's $\mathscr{H}$ relation is defined by $a\mathscr{H}b$ if $a\mathscr{L}b$ and $a\mathscr{R}b$. Green's $\mathscr{D}$ relation is defined by $a\mathscr{D}b$ if there is c such that $a\mathscr{L}c$ and $c\mathscr{R}b$. The equivalence classes of $\mathscr{L}$ are called $\mathscr{L}$-classes, and similarly for the other relations. In a finite monoid, $\mathscr{D}$ and $\mathscr{J}$ coincide. The $\mathscr{D}$-classes can be represented in a matrix form called *egg boxes*, where the rows represent $\mathscr{R}$-classes, columns $\mathscr{L}$-classes, and each entry is an $\mathscr{H}$-class. See [10] for more details.

Example 1. In [3], $\mathscr{D}$-classes are obtained for Jones and related monoids. Here we include an example of a $\mathscr{D}$-class of $\mathfrak{J}_3$, which has a $\mathscr{D}$-class consisting of the identity element and another class of the (2×2)-matrix below, where each element is a an $\mathscr{H}$-class, which in this case are singletons:

$$\begin{bmatrix} h_1 & h_1 h_2 \\ h_2 h_1 & h_2 \end{bmatrix}$$

where rows $\{h_1, h_1 h_2\}$, $\{h_2 h_1, h_2\}$, are the $\mathscr{R}$-classes and columns $\{h_1, h_2 h_1\}$, $\{h_1 h_2, h_2\}$ are the $\mathscr{L}$-classes in this $\mathscr{D}$-class. For instance, we see that multiplying h_1 and $h_1 h_2$ by h_i to the right gives rise to the same right ideal.

3   Origami Monoid $\mathcal{O}_n$

3.1   Generators

Here we identify simple building blocks in DNA origami structures. With each block type we associate a generator of a monoid and derive string rewriting systems to describe DNA structures. We have two motivations for our choices. (1) In Fig. 1 (left), one notices repeated patterns of simple building blocks whose concatenation builds a larger structure. One type of these patterns is a cross-over by the staple strands, and the other is a cross-over of the scaffold strand. Thus, a natural approach to describe DNA origami structures symbolically is to associate generators of an algebraic system to simple building blocks, and to take multiplication in the system to be presented as concatenation of the blocks. (2) In knot theory, a knot diagram is decomposed into basic building blocks of crossings or tangles. For the Kauffman bracket version of the Jones polynomial [7], for example, whose generators resemble building blocks observed in Fig. 1 (left), are used.

For a positive integer n we define a monoid $\mathcal{O}_n$, where n represents the number of vertical double stranded DNA strands, that is, n is the number of parallel folds of the scaffold. For the structure in Fig. 1, $n = 6$. The generators of

O_n are denoted by α_i (corresponds to anti-parallel staple strands cross-over) and β_i (corresponds to antiparallel scaffold strand cross-over) for $i = 1, \ldots, n-1$, as depicted in Fig. 3. The subscript i represents the position of the left scaffold corresponding to α_i and β_i, respectively, by starting at 1 from the leftmost scaffold strand fold and counting right (Fig. 4).

Because DNA is chemically oriented, and the strands in the double stranded DNA are oppositely oriented, we define an orientation within the building blocks corresponding to generators. Because parallel scaffold strands are obtained by folding of the scaffold, consecutive scaffold strands run in alternating directions, while staple strands run in the opposite direction to the scaffold, and for convention we take that the first scaffold runs in an upwards direction. In this way, the direction of the scaffold/staple strands for any particular α_i or β_i depends entirely on the parity of i, as shown in Fig. 3.

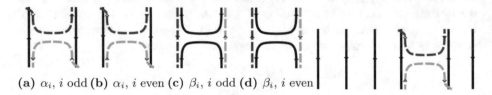

(a) α_i, i odd (b) α_i, i even (c) β_i, i odd (d) β_i, i even

Fig. 3. The generators identified

Fig. 4. α_4 in the context of a 6-fold stranded structure

Figure 4 shows a diagram corresponding to α_4 as an example of the "full picture" of one of these generators. For the sake of brevity, we neglect to draw the extra scaffold and staple strands in most diagrams, but it may be helpful to imagine them when we describe their concatenation. In addition, we often use α_i and β_i to refer to the corresponding diagrams. As in Fig. 4, parallel scaffolds in generator diagrams do not have counterpart parallel staples.

3.2   Concatenation as a Monoid Operation

To justify modeling DNA origami structures by words over the generators we make a correspondence between concatenations of generators α_i, β_i and concatenations of DNA segments. For a natural number $n \geq 2$, the set of generators of the monoid O_n is the set $\Sigma_n = \{\alpha_1, \alpha_2, \ldots, \alpha_{n-1}, \beta_1, \beta_2, \ldots, \beta_{n-1}\}$. For a product of two generators x_i and y_j in Σ_n, we place the diagram of the first generator above the second, lining up the scaffold strings of the two generators, and then we connect each respective scaffold string. If the two generators are adjacent, that is, if for indices i and j it holds $|i - j| \leq 1$, then we also connect their staples as described below. Otherwise, if $|i - j| \geq 2$, no staple connection is performed and the concatenation is finished.

We define a convention of connecting staples for adjacent generators, which is motivated by the manner in which staples connect in Fig. 1. Note how the staples of α-type protrude "outside" of the scaffold in Fig. 3. We refer to these ends of a staple as an "extending staple-ends", and all other staple ends as "non-extending staple-ends". We connect staples everywhere *except* when two non-extending staple-ends would have to cross a scaffold to connect (recall that the scaffold strands are connected first), as can be seen in Figs. 6 and 7.

Our choice of coloring staples in the figure is arbitrary, and we re-color staples in the same color if they get connected when we concatenate generators. By exhausting all possibilities, one can see that under our convention of connection, the staples remain short by concatenation without joining more than three scaffold folds. Note that concatenation of three or more generators is associative because generators can be connected in an associative manner following the rules described above.

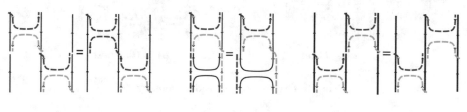

Fig. 5. $\alpha_i\alpha_{i+1}$, i odd   **Fig. 6.** $\alpha_i\beta_i$, i odd   **Fig. 7.** $\alpha_i\alpha_{i-1}$, i odd

3.3   Relations in $\mathcal{O}_n$

The rewriting rules (which generate the relations within the monoids) are motivated by similarity between the DNA origami structures as seen in Fig. 1 (left) and the diagrams of Jones monoids in Fig. 2. It is deemed that the relations of Jones monoids simplify the DNA origami structure, and may be useful for designing efficient and more solid structures by the rewriting rules proposed below. The figures in this section are for justifying feasibility of corresponding DNA structures, and to represent the rewriting system diagrammatically.

Rewriting Rules. For $\Sigma_n = \{\alpha_1, \alpha_2, \ldots, \alpha_{n-1}, \beta_1, \beta_2, \ldots, \beta_{n-1}\}$, we establish a set of rewriting rules that allows simplification of the DNA structure description. Define a string rewriting system (Σ_n, R) as follows.

To ease the notation, we define a bar on Σ_n by $\overline{\alpha_i} = \beta_i$ and $\overline{\beta_i} = \alpha_i$, and extend this operation to the free monoid by defining $\overline{w}$ for a word w by applying bar to each letter of w. Let $\gamma \in \{\alpha, \beta\}$ and $i \in \{1, \ldots, n-1\}$, then we have:

(1) (Idempotency) $\qquad\qquad \gamma_i\gamma_i \to \gamma_i$
(2) (Left Jones relation) $\quad\; \gamma_i\gamma_{i+1}\gamma_i \to \gamma_i$
(3) (Right Jones relation) $\; \gamma_i\gamma_{i-1}\gamma_i \to \gamma_i$
(4) (Inter $-$ commutation) $\quad \gamma_i\overline{\gamma_j} \to \overline{\gamma_j}\gamma_i$, for $|i-j| \geq 1$
(5) (Intra $-$ commutation) $\quad \gamma_i\gamma_j \to \gamma_j\gamma_i$, for $|i-j| \geq 2$

The rules are extended to Σ_n^* as described in Sect. 2.2.

The rewriting rules are inspired by Jones monoids, and they are also reflected in the reality of the diagrams of DNA origami, as shown in Figs. 8 and 9. Specifically, a pattern in the left of Fig. 8(a) has a small staple circle, which is deemed to be simplified by the right side. Staple strands are holding the scaffold in a certain position (obtained to the right of the arrow) and the cyclic staple only reinforces the structure. The small circle of a scaffold in Fig. 8(b) left cannot form in DNA origami, and therefore is simplified to the structure on the right of the arrow.

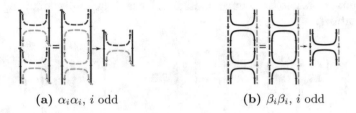

(a) $\alpha_i\alpha_i$, i odd   (b) $\beta_i\beta_i$, i odd

Fig. 8. Two examples of idempotency

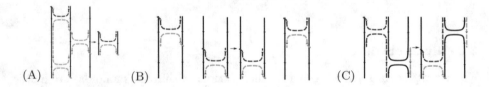

(A)   (B)   (C)

Fig. 9. Examples of (A) Jones relation, (B) Inter-commutation, and (C) Intra-commutation

Deriving Additional Rewriting Rules by Substitution. Since DNA origami structure has no internal scaffold loops, applying rewriting rules similar to (1)–(5) to concatenations of generators, that is, products of α's and β's, is plausible for DNA origami structures. We extend these rules to more general substitution rules for our specific case of generators α_i and β_i by considering other γ's, for instance $\gamma = \alpha\beta$. The composition diagrams show that such substitution rules describe the DNA origami staples/scaffold structure in the way we proposed above (see Fig. 10), while these new structures produce rules that cannot be derived from the listed ones in (1)–(5). Therefore we consider rewriting rules for concatenations of generators α's and β's. Furthermore, we focus on concatenations of generators with the same or 'neighboring' indexes, because only for these generators can the ends of the staples connect. However, α_i and β_j ($i \neq j$) can commute freely (by the inter-communication rule (4)), so we also do not need to consider substitutions such as $\gamma = \alpha_i\beta_{i+1}$. Further, we observe that by setting $\gamma \in \{\alpha_i\beta_i, \beta_i\alpha_i\}$, the idempotency rule (1) holds as seen in Fig. 10.

Therefore there are only four cases to consider: $\gamma \in \{\alpha_i\beta_i, \beta_i\alpha_i, \alpha_i\beta_i\alpha_i, \beta_i\alpha_i\beta_i\}$ and check the plausibility of corresponding DNA diagrams.

First, consider $\gamma \in \{\alpha\beta, \beta\alpha\}$, where γ_i indicates $\alpha_i\beta_i$. Then substituting γ into rewriting rules (1), (2), and (3) gives us new rewriting rules (1a), (2a), and (3a). For example, (1a) consists of $\alpha_i\beta_i\alpha_i\beta_i \to \alpha_i\beta_i$ and $\beta_i\alpha_i\beta_i\alpha_i \to \beta_i\alpha_i$. Note that a provisional rewriting rule (5a) could easily be obtained by the rewriting rule (5), so we do not consider it as a new rule. We also do not add rewriting rule (4a) since it conflicts with the structure of the scaffold, as shown in Fig. 11. Notice that the scaffold strand at the top left is connected to the second strand only on the left side of the figure, and on the right hand side of the figure it is connected to the third strand. Next we consider $\gamma \in \{\alpha\beta\alpha, \beta\alpha\beta\}$, which gives us rewriting rules (1b), (2b) and (3b). Similarly as before, rules (4b) and (5b) are not added, (4b) because of incompatible staple strands, and (5b) because it can be derived from (5). In addition, (1b) can also be derived from (1) and (1a), so it is not considered as a new rule. In the end, we are left with 10 rewriting rules which we use to define the general rewriting rules and the monoids.

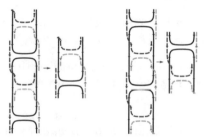

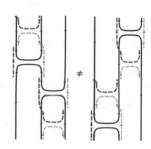

Fig. 10. Substitution of $\alpha\beta$ and $\beta\alpha$ (resp.) into the first rewriting rule (i odd)

Fig. 11. Substitution of $\gamma = \alpha\beta$ into rewriting rule (4) for i odd.

Definition 1. The *origami monoid* $\mathcal{O}_n$ is the monoid with a set of generators Σ_n and relations generated by the rewriting rules (1) through (5), (1a), (2a), (3a), (2b), (3b).

4   Monoid Structure of $\mathcal{O}_n$

In this section, we present computational results on Green's $\mathscr{D}$-classes and compare them to those for the Jones monoids obtained in [3]. For comparison, we use the monoid epimorphism from $\mathcal{O}_n$ to the product $\mathcal{J}_n \times \mathcal{J}_n$ defined below.

Let $\mathcal{J}_n$ be the Jones monoid of degree n with generators h_i, $i = 1, \ldots, n-1$. We denote the submonoid of $\mathcal{O}_n$ generated by αs (resp. βs), by $\mathcal{O}_n^\alpha$ (resp. $\mathcal{O}_n^\beta$). An equivalent description for $\mathcal{O}_n^\alpha$ is the set of all words consisting of only αs (plus the empty word), and similarly for $\mathcal{O}_n^\beta$. Let $\mathcal{O}_n^{\alpha\beta} = [\mathcal{O}_n \setminus (\mathcal{O}_n^\alpha \cup \mathcal{O}_n^\beta)] \cup \{1\}$.

Lemma 1. $\mathcal{O}_n^{\alpha\beta}$ is a submonoid of $\mathcal{O}_n$.

Proof. The left and right hand sides of each rewriting rule show that rewriting a word by these rules does not change the absence, or existence of at least one α in the word, and similarly for β. Thus multiplication of two words in $\mathcal{O}_n^{\alpha\beta}$ does not remove α's or β's from the product, hence the product remains in $\mathcal{O}_n^{\alpha\beta}$. □

Let $p_\alpha : \mathcal{O}_n \to \mathcal{J}_n$ be the epimorphism defined by 'projections' $p_\alpha(\alpha_i) = h_i$ and $p_\alpha(\beta_i) = 1$, for all $i = 1, \ldots, n-1$, and let p_β be defined similarly for βs. Define $p : \mathcal{O}_n \to \mathcal{J}_n \times \mathcal{J}_n$ by $p(x) = (p_\alpha(x), p_\beta(x))$ for $x \in \mathcal{O}_n$. Since the monoid relations of $\mathcal{O}_n$ hold under p, we have the following:

– $\mathcal{O}_n^\alpha \cong \mathcal{O}_n^\beta \cong \mathcal{J}_n$.
– The map $p : \mathcal{O}_n \to \mathcal{J}_n \times \mathcal{J}_n$ is a surjective monoid morphism.

In particular, it follows that the order of $\mathcal{O}_n$ is at least $|\mathcal{J}_n|^2$.

4.1 Orders of Origami Monoids

For $n = 2$ we can determine the order of $\mathcal{O}_2$ as follows.

Lemma 2. *Every non-empty element of $\mathcal{O}_2$ can be represented by the rewriting rules as one of the following words:* α_1, β_1, $\alpha_1\beta_1$, $\beta_1\alpha_1$, $\alpha_1\beta_1\alpha_1$, *or* $\beta_1\alpha_1\beta_1$.

Proof. Since $\Sigma_2 = \{\alpha_1, \beta_1\}$, we list the words of length 3 or less exhaustively. After applying rewriting rules to these words, they reduce to those words listed in the statement.

Now consider a word w with length greater than 3. We show that w can be reduced to a word with length 3 or less. If $\alpha_1\alpha_1$ or $\beta_1\beta_1$ are factors of w, we reduce them to α_1 or β_1, respectively. Repeating this process, we may assume that w is an alternating sequence of α_1 and β_1. Since $\alpha_1\beta_1$ and $\beta_1\alpha_1$ are idempotent, w reduces to a word of length less than 4. □

It is known that the elements of the Jones monoid $\mathcal{J}_n$ are in bijection with the linear chord diagrams obtained from the arcs of the diagrams representing them, and the total number of such chord diagrams is equal to the Catalan number $C_n = \dfrac{1}{n+1}\dbinom{2n}{n}$ [2]. Thus the numbers of elements of $\mathcal{J}_n$ for $n = 2, \ldots, 6$ are 2, 5, 14, 42, 429, respectively. GAP computations show that the number of non-identity elements in $\mathcal{O}_3$, $\mathcal{O}_4$, $\mathcal{O}_5$ and $\mathcal{O}_6$ are 44, 293, 2179, 19086 respectively [12]. This sequence of integers is not listed in the OEIS [13] list of sequences. We observe that the orders of origami monoids are much larger. In fact it is not apparent from the definition whether they are all finite. Thus we conjecture the following.

Conjecture 1. The order of $\mathcal{O}_n$ is finite for all n.

4.2   Green's Classes

We have the following observations for Green's classes of $\mathcal{O}_n$ for general n.

Lemma 3. *Let $x \in \mathcal{O}_n^\alpha$, $y \in \mathcal{O}_n^{\alpha\beta}$ be nonempty words and let D_x and D_y be the $\mathcal{D}$-classes containing x and y, respectively. Then $D_x \neq D_y$.*

Proof. By Lemma 1, if $y\mathcal{L}a$, then $a \in \mathcal{O}_n^{\alpha\beta}$, and if $a\mathcal{R}b$, then $b \in \mathcal{O}_n^{\alpha\beta}$. Thus we cannot have $y\mathcal{D}x$.

Corollary 1. *The conclusion of Lemma 3 holds for $x \in \mathcal{O}_n^\beta$, $y \in \mathcal{O}_n^{\alpha\beta}$ and $x \in \mathcal{O}_n^\alpha$, $y \in \mathcal{O}_n^\beta$.*

Remark 1. It follows from the definition of p that every $\mathcal{D}$-class of $\mathcal{O}_n$ maps into a $\mathcal{D}$-class of $\mathcal{J}_n \times \mathcal{J}_n$, and by Lemma 1.4 Ch. 5 in [6] the map is also onto. Also, if $\mathcal{O}_n$ is finite, then each $\mathcal{D}$-class of $\mathcal{J}_n \times \mathcal{J}_n$ is an image of a $\mathcal{D}$-class of $\mathcal{O}_n$ by p. We conjecture that there is in fact a one-to-one correspondence between the $\mathcal{D}$-classes of $\mathcal{O}_n$ and those of $\mathcal{J}_n \times \mathcal{J}_n$. We show that this observation is true for $n \leq 6$.

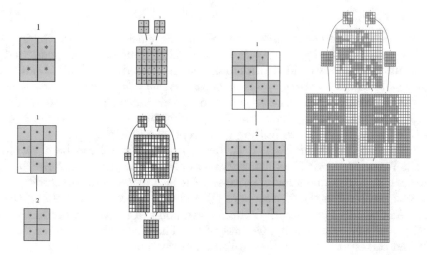

Fig. 12. $\mathcal{D}$-classes of $\mathcal{J}_n$ (left) and $\mathcal{O}_n$ (right) for $n = 3$ (top left), $n = 4$ (bottom left), and $n = 5$ (right)

4.3   Green's Classes for $n \leq 6$

Green's relations for $\mathcal{J}_n$ have been studied in [7]. We show results of GAP computations that determine $\mathcal{D}$-classes of origami monoids $\mathcal{O}_n$ for $n \leq 6$; the structures are presented in Figs. 12 and 13. Shaded squares represent $\mathcal{H}$-classes which contain an idempotent. We note that for $n \leq 6$, every $\mathcal{H}$-class of $\mathcal{O}_n$ is singleton, so each square in the figure represents precisely one element of $\mathcal{O}_n$.

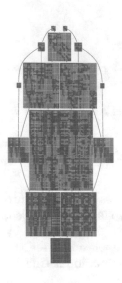
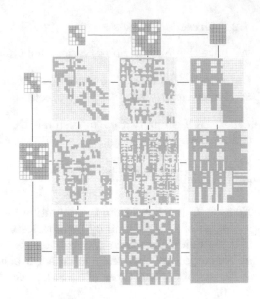

Fig. 13. $\mathscr{D}$-classes of $\mathcal{O}_6$ (left) and re-arranged and resized to fit the grid (right)

For $n \leq 6$, since $\mathcal{O}_n$ is finite, the $\mathscr{J}$ and $\mathscr{D}$ relations coincide. A preorder $\leq_{\mathscr{D}}$ is defined on $\mathcal{O}_n$ by $a \leq_{\mathscr{D}} b$ if the two-sided principal ideal generated by a is a subset of the two-sided principal ideal generated by b. This condition is equivalent to the existence of $x, y \in \mathcal{O}_n$ such that $xby = a$. Since any two elements of a $\mathscr{D}$-class generate the same principal ideal, this preorder may be extended to the set of $\mathscr{D}$-classes of $\mathcal{O}_n$ such that $D \leq_{\mathscr{D}} D'$ if for $a \in D$ and $b \in D'$, $a \leq_{\mathscr{D}} b$. The lines between $\mathscr{D}$-classes in the figures represent this preorder.

The relations between $\mathcal{O}_n$ and $\mathcal{J}_n$ described in Sect. 4.2 can be observed in Fig. 12. We omit the $\mathscr{D}$-class consisting of only the empty word from the diagrams, which is maximal among $\mathscr{D}$-classes. For each n, two copies of the $\mathscr{D}$-classes of $\mathcal{J}_n$ can be found as the $\mathscr{D}$-classes of $\mathcal{O}_n^\alpha$ and $\mathcal{O}_n^\beta$, respectively, in the $\mathscr{D}$-classes of $\mathcal{O}_n$. As described in Remark 1, these correspond to the direct product of one identity and one non-identity $\mathscr{D}$-class of $\mathcal{J}_n$. The other $\mathscr{D}$-classes are those of $\mathcal{O}_n^{\alpha\beta}$, and correspond to the direct product of two non-identity $\mathscr{D}$-classes of $\mathcal{J}_n$. Which pair of $\mathscr{D}$-classes of $\mathcal{J}_n$ correspond to which $\mathscr{D}$-class of $\mathcal{O}_n$ can be better seen in Fig. 13.

In Fig. 13, we arrange the $\mathscr{D}$-classes of $\mathcal{O}_6$ to better illustrate the relation between the $\mathscr{D}$-classes of $\mathcal{J}_n$, although the same process may be applied to other n. On the right, the preorder of the $\mathscr{D}$-classes remains, applying left-to-right as well as top-to-bottom. The $\mathscr{D}$-classes along the top row and left column are the $\mathscr{D}$-classes of $\mathcal{O}_6^\alpha$ and $\mathcal{O}_6^\beta$ respectively, which as previously described are isomorphic to $\mathcal{J}_n$. For any $\mathscr{D}$-class of $\mathcal{O}_n^{\alpha\beta}$, the $\mathscr{D}$-classes which it maps onto are greater in the preorder. Thus the grid of $\mathscr{D}$-classes may be thought of as a table, with the row and column of any entry determining the image of the $\mathscr{D}$-class by p_α and p_β, respectively. Since rewriting rules are equivalent for α and β, the $\mathscr{D}$-classes are symmetric up to switching rows and columns. This can be easily seen in the $\mathscr{D}$-classes in the upper right and lower left corners. However, the rows and columns of any $\mathscr{D}$-class may be ordered arbitrarily, and are automated by GAP, making the symmetry non-obvious for other $\mathscr{D}$-classes.

5   Concluding Remarks

In this paper, motivated by similarity to Temperley-Lieb algebras, we introduced an algebraic system that describes DNA origami structures. Generators in this system are defined such that they mimic basic building blocks of DNA origami. Following the structural properties of the DNA origami, we established rewriting rules, as well as monoids whose elements conform to the relations obtained from these rules. To each DNA origami structure we can associate an element from an appropriate monoid. For example, the structure in Fig. 1 corresponds to the element represented by the word $\alpha_1 \alpha_3 \alpha_5 \beta_2 \beta_4$. We hope that such representations of DNA origami may provide a tool for distinguishing constructs.

The monoids introduced here are connected to Jones monoids, and we provide several conjectures with the goal of relating them to known monoids. For example, from our findings for $n \leq 6$, we conjecture that $\mathcal{O}_n$ are finite for all n, and $\mathscr{H}$-classes are singletons. We also provide conjectures relating to the $\mathscr{D}$-classes of $\mathcal{O}_n$ and $\mathcal{J}_n$ under the morphism p. Specifically, we conjecture that the $\mathscr{D}$-classes of $\mathcal{O}_n$ are in one-to-one correspondence with the $\mathscr{D}$-classes of $\mathcal{J}_n \times \mathcal{J}_n$.

Acknowledgment. This work is partially supported by NIH R01GM109459, and by NSF's CCF-1526485 and DMS-1800443. This research was also partially supported by the Southeast Center for Mathematics and Biology, an NSF-Simons Research Center for Mathematics of Complex Biological Systems, under National Science Foundation Grant No. DMS-1764406 and Simons Foundation Grant No. 594594.

References

1. Abramsky, S.: Temperley-Lieb algebra: from knot theory to logic and computation via quantum mechanics. CoRR abs/0910.2737 (2009)
2. Borisavljević, M., Došen, K., Petrić, Z.: Kauffman monoids. J. Knot Theory Ramifications **11**(2), 127–143 (2002)
3. Dolinka, I., East, J.: The idempotent-generated subsemigroup of the Kauffman monoid. Glasgow Math. J. **59**(3), 673–683 (2017)
4. Douglas, S.M., Dietz, H., Liedl, T., Högberg, B., Graf, F., Shih, W.M.: Self-assembly of DNA into nanoscale three-dimensional shapes. Nature **459**, 414–418 (2009)
5. Geim, A.K., Novoselov, K.S.: The rise of graphene. Nature Mater. **6**, 183–191 (2007)
6. Grillet, P.A.: Semigroups: An Introduction to the Structure Theory. CRC Press, Boca Raton (1995)
7. Kauffman, L.H.: Knots and Physics. World Scientific, Singapore (2001)
8. Kim, H., et al.: Water harvesting from air with metal-organic frameworks powered by natural sunlight. Science **356**(6336), 430–434 (2017)

9. Lau, K.W., FitzGerald, D.G.: Ideal structure of the Kauffman and related monoids. Commun. Algebra **34**(7), 2617–2629 (2006)
10. Pin, J.E.: Varieties of Formal Languages. North Oxford Academic Publishers, Oxford (1986)
11. Rothemund, P.W.K.: Folding DNA to create nanoscale shapes and patterns. Nature **440**(7082), 297–302 (2006)
12. Gap - groups, algorithms, and programming, version 4.10.0. https://www.gap-system.org
13. The on-line encyclopedia of integer sequences, id:a047974. http://www.research.att.com/~njas/sequences/A047974
14. Zhang, W., et al.: Unexpected stable stoichiometries of sodium chlorides. Science **342**(6165), 1502–1505 (2013)

Algebraic Models for Arbitrary Strength Covering Arrays over v-ary Alphabets

Ludwig Kampel[1], Dimitris E. Simos[1(✉)], Bernhard Garn[1], Ilias S. Kotsireas[2], and Evgeny Zhereshchin[2]

[1] SBA Research, 1040 Vienna, Austria
{lkampel,dsimos,bgarn}@sba-research.org
[2] CARGO Lab, Wilfrid Laurier University, Waterloo, ON, Canada
ikotsire@wlu.ca, zher0340@mylaurier.ca

Abstract. Extending our previous work [7], we introduce a novel technique to model and compute arbitrary strength covering arrays over v-ary alphabets, using methods arising from linear algebra commutative algebra and symbolic computation. Concrete instances of covering arrays for given parameters then appear as points in varieties as they occur in solutions of multivariate polynomial equation systems. To solve these systems we apply polynomial solvers based on the theory of Gröbner bases and exhaustive search using serial and parallel programming techniques.

Keywords: Covering arrays · Algebraic models · Solvers

1 Introduction

In recent years, covering arrays (CAs) are applied in a branch of automated software testing called combinatorial testing [12]. Traditional applications of especially orthogonal and covering arrays lie in the field of Design of Experiments, however, the requirements of customizability of combinatorial testing led to the primary usage of covering arrays. Especially for applications in software testing the defining property of covering arrays, coverage of all t-tuples in subarrays, has been shown particularly beneficial; see [13]. Closely related NP-hard problems, as in [6,15,16], suggest that also the problem of finding *optimal covering arrays* is a hard combinatorial optimization problem, however, its actual complexity remains unknown [9].

Several approaches for the construction of CAs have been introduced in the literature so far. Amongst them are greedy heuristics, metaheuristics, combinatorial constructions and exact methods, see [17] and references therein. Exact methods can be based on backtracking, constraint programming [8] or on computational algebra formalisms as in [7]. For a survey of CA generation methods, the interested reader might have a look at [4] or [17].

© Springer Nature Switzerland AG 2019
M. Ćirić et al. (Eds.): CAI 2019, LNCS 11545, pp. 177–189, 2019.
https://doi.org/10.1007/978-3-030-21363-3_15

Structure. The work presented in this paper builds upon the work presented in [7]. In particular, in Sect. 3 the algebraic approach of modelling binary CAs of strength two in [7] is extended for CAs over arbitrary alphabets and for higher strengths. In Sect. 3.1, based on our algebraic model, we devise an algorithm for the construction of such CAs. Further, in Sect. 4 we discuss how the proposed algebraic modelling of CAs can be used when trying to extend a given CA with an additional column and show how the degrees of the appearing polynomials can be reduced. In Sect. 5 we list some experiments where we compute CAs arising as solutions of equation systems using different solving techniques. Last, Sect. 6 concludes the work and points out directions for future work.

2   Preliminaries

Notations. Throughout this paper we use the abbreviation $[v]$ for the integer interval $\{0, 1, \ldots, v - 1\}$, and by $\mathbb{Z}_v := \mathbb{Z}/(v\mathbb{Z})$ the ring of integers modulo v. In the scope of this paper we may identify the elements of $\mathbb{Z}_v$ with the elements of $[v]$. By $\binom{\{k\}}{t}$ we denote the set of all subsets of $\{1, \ldots, k\}$ of cardinality t. Further, with $\mathbf{1}$ we denote the vector having all entries equal to 1; the length of the vector is clear from the context. Additionally, for vectors $\mathbf{h} = (h_1, \ldots, h_N)^T$ we use the notation $\mathbf{prod}(\mathbf{h}) := \prod_{i=1}^{N} h_i$.

Although in general CAs can be over arbitrary alphabets, we restrict our attention to arrays over integer intervals $[v]$, as the properties we consider in this paper only depend on the size of the alphabets rather than their actual elements. The following definition follows the one given in [5], but is phrased slightly different.

Definition 1. A *covering array* $\mathsf{CA}(N; t, k, v)$ is an $N \times k$ array $(\mathbf{c}_1, \ldots, \mathbf{c}_k)$ with the following properties:

(i) For all $j \in [k]$ the values in the j-th column $\mathbf{c}_j$ belong to the set $\{0, \ldots, v-1\}$.
(ii) For each selection $\{\mathbf{c}_{j_1}, \ldots, \mathbf{c}_{j_t}\} \subseteq \{\mathbf{c}_1, \ldots, \mathbf{c}_k\}$ of t different columns, the subarray that is comprised by the columns $\mathbf{c}_{j_1}, \ldots, \mathbf{c}_{j_t}$ has the property that every t-tuple in $[v]^t$ appears at least once as a row.

The values t, k and v are also referred to as the CA parameters, and the parameter t is also called the *strength* of the CA.

The smallest integer N for which a $\mathsf{CA}(N; t, k, v)$ exists is called the *covering array number* (CAN) and is denoted as $\mathsf{CAN}(t, k, v)$. Arrays that meet this bound are called *optimal*. That this number is well defined and finite for all $1 \le t \le k$ and alphabet sizes v, follows from the fact that the full product $[v]^k$ always constitutes a CA for the respective parameters. The following notion formalizes the concept of t-tuples, and can be also found in [1].

Definition 2. For a given alphabet size v, we define a *v-ary t-way interaction* as a set of pairs $T = \{(p_1, u_1), \ldots, (p_t, u_t)\}$ with $1 \le p_1 < p_2 < \ldots p_t \le k$ and $u_i \in [v]\ \forall i = 1, \ldots, t$. Usually the underlying alphabet is clear from the context and we speak of *t-way interactions* for short.

We say an $N \times k$ array *covers* a t-way interaction $\{(p_1, u_1), \ldots, (p_t, u_t)\}$, if one of its rows has the entry u_i in position p_i for all $i \in \{1, \ldots, t\}$.

Aside from the problem of determining CANs, a variety of different problems arise in the realm of covering arrays. An extensive list can be found in [7]. Due to space limitations, in this paper we consider the following two problems, which correspond to Problems 2 and 9 of [7].

Problem 1 (Computational Existence). For given CA parameters and given $N \in \mathbb{N}$, construct one/all covering array(s) with exactly N rows, or terminate indicating there exists no such CA.

Problem 2 (Computational Factor Extension). Given a covering array and an alphabet size v, construct one/all new additional column(s) such that the extended matrix constitutes a covering array with the additional column (not adding any additional rows), or terminate indicating there exists no such a column.

3   Algebraic Models for CAs of Arbitrary Strength

Definition 3 (*v-ary t-way interaction distinguish property*). Let $(R, +, \cdot, 0, 1)$ be an integral domain with 1, and $a_1, \ldots, a_t \in R$. We say that $(R, a_1, \ldots, a_t)$ has the *v-ary t-way interaction distinguish property*, if and only if $\forall u_i \in [v]$, $\forall i = 1, \ldots, t$ the elements $u_1 a_1 + \ldots + u_t a_t \in R$, are pairwise different, where we interpret the natural numbers $u \in [v]$ embedded in R as $\underbrace{(1 + \ldots + 1)}_{u}$.

Remark. Notice that from the v-ary t-way interaction distinguish property, for $t \geq 1$, it immediately follows that $char(R) \geq v$ or $char(R) = 0$. Hence the set $[v] = \{0, 1, \ldots, v - 1\}$ of natural numbers is mapped injectively into R.

When the underlying alphabet $[v]$ is clear from the context, we also speak of the *t-way interaction distinguish property* for short. We can interpret the notion of the t-way interaction distinguish property as a special kind of linear independence, considering the appropriate algebraic structures. For that purpose, using the same notations as in Definition 3, we regard the integral domain R as a unitary $\mathbb{Z}_v$-module. Then $(R, a_1, \ldots, a_t)$ having the v-ary t-way interaction distinguish property is equivalent to the linear independence of $a_1, \ldots, a_t$. We give some examples of rings R and elements $a_1, \ldots, a_t$, such that $(R, a_1, \ldots, a_t)$ has the v-ary t-way interaction distinguish property:

1. For a ring S of characteristic $char(S) \geq v$ or $char(S) = 0$, let $R = S[x_1, \ldots, x_n]$ be the ring of all polynomials in the indeterminates x_i, and $a_i = x_i \; \forall i = 1, \ldots, n$.
2. Let $K \geq L$ be fields of characteristic $char(L) \geq v$ or $char(L) = 0$, $R = L$ and let $a_1, \ldots, a_t \in K$ be algebraically independent over L.

Lemma 1. *Let $(R, a_1, \ldots, a_t)$ have the v-ary t-way interaction distinguish property, then for any $(u_1, \ldots, u_t) \in [v]^t$ and $(x_1, \ldots, x_t) \in [v]^t$ we have:*

$$(x_1, \ldots, x_t) = (u_1, \ldots, u_t) \Leftrightarrow (x_1, \ldots, x_t) \cdot (a_1, \ldots, a_t)^T - \sum_{i=1}^{t} u_i a_i = 0.$$

Proof. The claim follows directly from the definition of the v-ary t-way interaction distinguish property.

In the following we will use elements $a_1, \ldots, a_t$ to select t different columns of a given matrix. For that purpose we need the following additional definition.

Definition 4. *Let $t \leq k \in \mathbb{N}$ and $C = \{c_1, \ldots, c_t\} \subseteq [k]$ with $|C| = t$. Further let $(R, +, \cdot, 0)$ be a ring and $a_1, \ldots, a_t \in R \setminus \{0\}$, then we define $\iota^C_{t,k}(a_1, \ldots, a_t)$ as the (column) vector of length k having entry a_i in position c_i, $\forall i = 1, \ldots, t$ and the entry 0 in all other positions.*

For example, for $C = \{2, 3, 6\}$ then $\iota^C_{3,6}(a_1, a_2, a_3) = (0, a_1, a_2, 0, 0, a_3)^T$.

Lemma 2. *Let R be an integral domain (with 1) and $(R, a_1, \ldots, a_t)$ have the v-ary t-way interaction distinguish property, M be a given $N \times k$ matrix $M = (\mathbf{m}_1, \ldots, \mathbf{m}_k)$ defined over R, $C = \{c_1, \ldots, c_t\}$ with $1 \leq c_1 < \ldots < c_t \leq k$ and $M_C = (\mathbf{m}_{c_1}, \ldots, \mathbf{m}_{c_t})$ be the matrix comprised of the t columns defined by C. Further consider a t-tuple $(u_1, \ldots, u_t) \in [v]^t$, then the following statements are equivalent:*

1. *The tuple $(u_1, \ldots, u_t)$ appears at least once as a row in the matrix M_C.*
2. *The vector $\mathbf{h} := (h_1, \ldots, h_N)^T := M \cdot \iota^C_{t,k}(a_1, \ldots, a_t) - \mathbf{1}(\sum_{i=1}^{t} u_i a_i)$ contains at least one component equal to zero.*
3. *$prod(\mathbf{h}) = \prod_{i=1}^{N} h_i = 0$.*

Proof. The equivalence of 1 and 2 follows from Lemma 1. The equivalence of 2 and 3 holds since R is an integral domain.

Equations as in item 2 of Lemma 2 are formulated in such a way that they are semantically equivalent to the appearance of a t-way interaction in an array. Considering these equations for all t-way interactions and all selections of t columns, Lemma 2 leads to the main result of this section.

Theorem 1. *Let R be a ring and $(R, a_1, \ldots, a_t)$ have the v-ary t-way interaction distinguish property. Then for a matrix $M \in \mathbb{Z}_v^{N \times k}$ the following statements are equivalent :*

1. *M is a $\mathsf{CA}(N; t, k, v)$*
2. *$\forall C \in \binom{\{k\}}{t}$, $\forall (u_1, \ldots, u_t) \in [v]^t$:*

$$prod(M \cdot \iota^C_{t,k}(a_1, \ldots, a_t) - \mathbf{1} \cdot (u_1, \ldots, u_t) \cdot (a_1, \ldots, a_t)^T) = 0. \tag{1}$$

Proof. The assertion of the theorem follows immediately considering that the equivalence of 2 and 3 of Lemma 2 holds for all $C \subseteq \{1, \ldots, k\}$ with $|C| = t$ and all $(u_1, \ldots, u_t) \in [v]^t$.

Based on this algebraic characterization of CAs, we can now describe the previously mentioned computational or decisional problems for CAs as related problems found in multivariate polynomial algebra. In particular, through our algebraic modelling the problem(s) of constructing and computing covering arrays can be formulated as instances of algebraic equation systems, where each solution of the system, provided existence, corresponds to a covering array.

Corollary 1. *Let R be a ring and $(R, a_1, \ldots, a_t)$ have the v-ary t-way interaction distinguish property, and $X := (x_{i,j})$ be an $N \times k$ array of variables. Then any solution to the following system of equations in the unknowns $x_{i,j}$ yields a* CA$(N; t, k, v)$:

1. $\forall i \in \{1, \ldots, N\}, \forall j \in \{1, \ldots, k\}$

$$\prod_{r=0}^{v-1} (x_{i,j} - r) = 0. \tag{2}$$

2. $\forall C \in \binom{\{k\}}{t}, \ \forall(u_1, \ldots, u_t) \in [v]^t:$

$$\mathbf{prod}(X \cdot \iota_{t,k}^C(a_1, \ldots, a_t) - \mathbf{1} \cdot (u_1, \ldots, u_t) \cdot (a_1, \ldots, a_t)^T) = 0. \tag{3}$$

Following the terminology of [7], we call the Eq. (2) of Corollary 1 the *domain equations* and the Eq. (3) the *coverage equations*.

3.1 An Algebraic Algorithm for Searching CAs

Provided the derived algebraic characterization (Theorem 1, Corollary 1) of CAs, it is possible to interpret these combinatorial structures as elements in varieties corresponding to ideals in polynomial rings over fields. In [7] an algorithm was presented that addresses Problem 1 in the binary case for strength $t = 2$, interpreting the appearing polynomials as elements of $R = \mathbb{Q}[x_1, \ldots, x_\gamma, a_1, a_2]$, depending on the binary 2-way interaction distinguish property of (R, a_1, a_2).

With the results presented in this paper, a natural way to generalize this algorithm to the case of CA$(N; t, k, v)$ for arbitrary t and v is possible by interpreting appearing polynomials as elements of $R = \mathbb{Q}[x_1, \ldots, x_\gamma, a_1, \ldots, a_t]$ and relying on the t-way interaction distinguish property of $(R, a_1, \ldots, a_t)$. Before we describe such an algorithm, we address how the replacement of the indeterminates a_i can reduce the number of symbolic variables. For example, in Algorithm 3 of [7] the indeterminates a_i were replaced by random elements of $\mathbb{Q}$. In this work, we show how we can choose values for the a_i while still ensuring the t-way interaction distinguish property.

Lemma 3. *Let $R = \mathbb{Q}[x_1, \ldots, x_\gamma]$ and let $v \in \mathbb{N}$ with $v \geq 2$, then $(R, 1, v^1, \ldots, v^{t-1})$ has the v-ary t-way interaction distinguish property.*

Proof. We have to show that all elements $u_t v^{t-1} + \ldots u_2 v^1 + u_1 v^0$, for $u_i \in [v]$ $\forall i = 1, \ldots, t$ are pairwise different. Certainly this holds, as the elements of this set are exactly the natural numbers in $\{0, \ldots, v^t - 1\}$ and $(u_t, \ldots, u_1)$ corresponds to their base v representation.

Summarizing briefly, Algorithm 1 initializes an $N \times k$ array of symbolic variables, and generates all *coverage equations* (lines 4–9) and *domain equations* (lines 10–13) for this matrix according to Eqs. (2) and (3). Provided the previous Lemma 3 we can interpret the appearing polynomials as elements of $\mathbb{Q}[x_1, \ldots, x_{Nk}]$ using the elements $a_i = v^{i-1}$ for $i = 1, \ldots, t$ providing the t-way interaction distinguish property. This system of multivariate polynomial equations then is fed to an external SOLVE procedure. In Sect. 5 we will describe how such a procedure can be instantiated. Depending on the instantiation of this SOLVE procedure, the respective version of Problem 1 for searching *all* or *one* CA(s) is targeted.

Algorithm 1. ALGEBRAICSEARCHCAS

1: INPUT: N, t, k, v
Require: $t \le k$
2: Create a symbolic $N \times k$ array X containing variables $x_1, \ldots, x_{Nk}$
3: $EQall := \emptyset$
4: **for** $C \in \binom{\{k\}}{t}$ **do**   ▷ Add coverage equations
5:     **for** $\mathbf{u} \in [v]^t$ **do**
6:         $EQ := \boldsymbol{prod}(X \cdot \iota_{t,k}^C(a_1, \ldots, a_t) - \mathbf{1} \cdot (u_1, \ldots, u_t) \cdot (v^0, \ldots, v^{t-1})^T) = 0$
7:         add EQ to $EQall$
8:     **end for**
9: **end for**
10: **for** $i = 1, \ldots, Nk$ **do**   ▷ Add domain equations
11:     $EQ := \prod_{j=0}^{v-1}(x_i - j) = 0$
12:     add EQ to $EQall$
13: **end for**
14: Interpret $EQall$ as subset of $\mathbb{Q}[x_1, ..., x_{Nk}]$
15: $V = $ SOLVE$(EQall)$ ▷ Call external solver
16: **if** $V \neq \emptyset$ **then**
17:     return V;
18: **else** print "No CA exists";
19: **end if**

---

4   An Algebraic Model for Column Extensions of CAs

Similar to Sect. 3, in this section we devise a model such that Problem 2 can be treated as a problem of computational algebra. When extending a CA, with one column it is sufficient to ensure that in all subarrays comprised by t columns, involving the newly added column, all t-way interactions are covered, to guarantee that the defining properties of Definition 1 hold. Note that this technique of iteratively extending an existing CA with a column, followed by possible row extensions, is applied in the widely used IPO strategy for CA construction (see [14]). We illustrate this by the following example.

Example 1. Consider the following $\mathsf{CA}(9; 2, 3, 3)$:

$$M = \begin{pmatrix} 0\,0\,0\,1\,1\,1\,2\,2\,2 \\ 0\,1\,2\,0\,1\,2\,0\,1\,2 \\ 0\,1\,2\,1\,2\,0\,2\,0\,1 \end{pmatrix}^T .$$

Next, we will interpret the problem of extending M by one column as a problem of finding solutions to the unknowns $x_1, \ldots, x_9$, such that

$$\overline{M} = \begin{pmatrix} 0 & 0 & 0 & 1 & 1 & 1 & 2 & 2 & 2 \\ 0 & 1 & 2 & 0 & 1 & 2 & 0 & 1 & 2 \\ 0 & 1 & 2 & 1 & 2 & 0 & 2 & 0 & 1 \\ x_1 & x_2 & x_3 & x_4 & x_5 & x_6 & x_7 & x_8 & x_9 \end{pmatrix}^T \tag{4}$$

is again a CA. One such solution is e.g. $(x_1, \ldots, x_9) = (0, 1, 2, 2, 0, 1, 1, 2, 0)$.

Theorem 2. *Consider the ring $R = \mathbb{Q}[x_1, \ldots, x_N]$ and elements $a_1, \ldots, a_t$ such that $(R, a_1, \ldots, a_t)$ has the v-ary t-way interaction distinguish property. Given an $N \times k$ matrix M that is a $\mathsf{CA}(N; t, k, v)$, the following assertions are equivalent:*

1. *There exists a vector $\mathbf{c} \in [v]^{N \times 1}$ such that the horizontal extension $(M|\mathbf{c})$ of M by the column $\mathbf{c}$ is a $\mathsf{CA}(N; t, k+1, v)$.*
2. *The system of equations in the unknowns $x_1, \ldots, x_N$ consisting of the equations in (5) and (6) has a non trivial solution.*

$$\forall i \in \{1, \ldots, N\} : \prod_{j=0}^{v-1} (x_i - j) = 0. \tag{5}$$

$$\forall C \in \binom{\{k\}}{t-1} \forall (u_1, \ldots, u_t) \in \prod_{i \in C} [v_i] \times [v_{k+1}] :$$

$$\boldsymbol{prod}((M|\mathbf{x}) \cdot \iota_{t,k+1}^{C \cup \{k+1\}}(a_1, \ldots, a_t) - \mathbf{1} \sum_{i=1}^{t} u_i a_i) = 0. \tag{6}$$

Note that in (6) we only consider those subsets of $\{1, \ldots, k, k+1\}$ having cardinality t that contain the element $k + 1$.

Proof. If there exists a vector $\mathbf{c} = (c_1, \ldots, c_N) \in [v]^{N \times 1}$ such that $(M|\mathbf{c})$ is a $\mathsf{CA}(N; t, k+1, v)$, then $\mathbf{x} := \mathbf{c}$ obviously satisfies the equations in (5). From Theorem 1 we also get that x satisfies all equations in (6) when substituting k with $k + 1$ in Theorem 1. Conversely assume $\mathbf{x} = (x_1, \ldots, x_N) \in [v]^{N \times 1}$ is a solution to the system of equations given by (5) and (6). Since M is an $\mathsf{CA}(N; t, k, v)$, from Theorem 1 we get that $\boldsymbol{prod}(M \cdot \iota_{t,k}^C(a_1, \ldots, a_t) - \mathbf{1} \sum_{i=1}^{t} u_i a_i) = 0$ holds for all $C \in \binom{\{k\}}{t}$ and $(u_1, \ldots, u_t) \in [v]^t$. Together with (6) we have that

$\mathbf{prod}((M|\mathbf{x}) \cdot \iota^C_{t,k+1}(a_1,\ldots,a_t) - \mathbf{1}\sum_{i=1}^t u_i a_i) = 0$ holds for the remaining $C \in \binom{\{k+1\}}{t}$ and $(u_1,\ldots,u_t) \in [v]^t$. Corollary 1 then ensures that $(M|\mathbf{x})$ is a CA.

Remark 1 (Reduction of appearing Degree). Taking a closer look at the linear factors, e.g. the r-th factor $(m_{r,1},\ldots,m_{r,k},x_r) \cdot \iota^{C \cup \{k+1\}}_{t,k+1}(a_1,\ldots,a_t) - \sum_{i=1}^t u_i a_i = \sum_{i=1}^{t-1}(m_{r,c_i} - u_i)a_i + (x_r - u_t)a_t$, of the polynomials appearing in (6), we can see that some of them can never evaluate to zero, independent of the choice of x_r. This is due to the t-way interaction distinguish property, which ensures that $\sum_{i=1}^{t-1}(m_{r,c_i} - u_i)a_i + (x_r - u_t)a_t = 0$ if and only if $(m_{r,c_1},\ldots,m_{r,c_{t-1}},x_r) = (u_1,\ldots,u_t)$. Thus, $(m_{r,c_1},\ldots,m_{r,c_{t-1}}) = (u_1,\ldots,u_{t-1})$ is a necessary condition so that there exists a value for x_r, such that the r-th factor evaluates to zero. Therefore we can significantly reduce the degrees of the polynomials appearing in the coverage equations, as in (6), when using the result of Theorem 2 for the computation of CAs. We make this explicit by providing a small example.

Example 2. Continuing Example 1, we consider the matrix $\overline{M}$, as given in (4), as a matrix over $GF(3)[a_1,a_2]$, the ring of polynomials in the indeterminates a_1, a_2 over the finite field with three elements, and consider the coverage equation for $C = \{1,4\}$ and $(u_1,u_2) = (1,0)$.

$$\mathbf{prod}((M|\mathbf{x}) \cdot (a_1,0,0,a_2)^T - (a_1,a_1)) =$$
$$(0 \cdot a_1 + x_1 \cdot a_2 - a_1)(0 \cdot a_1 + x_2 \cdot a_2 - a_1)(0 \cdot a_1 + x_3 \cdot a_2 - a_1) \cdot$$
$$(1 \cdot a_1 + x_4 \cdot a_2 - a_1)(1 \cdot a_1 + x_5 \cdot a_2 - a_1)(1 \cdot a_1 + x_6 \cdot a_2 - a_1) \cdot$$
$$(2 \cdot a_1 + x_7 \cdot a_2 - a_1)(2 \cdot a_1 + x_8 \cdot a_2 - a_1)(2 \cdot a_1 + x_9 \cdot a_2 - a_1) = 0. \quad (7)$$

Due to the 3-ary 2-way interaction distinguish property of $GF(3)[a_1,a_2]$ we have e.g. $(0a_1 + x_1 a_2 - a_1) \neq 0$ for any value of $x_1 \in GF(3)$ or $(2a_1 + x_9 a_2 - a_1) \neq 0$ for any value of $x_9 \in GF(3)$. Hence Eq. (7) is equivalent to

$$(a_1 + x_4 a_2 - a_1)(a_1 + x_5 a_2 - a_1)(a_1 + x_6 a_2 - a_1) = 0, \quad (8)$$

reducing the degree of the polynomial in this coverage equation from 9 to 3. A similar reduction of the degrees can be done for the other coverage equations. The combinatorial interpretation, or reason, for this reduction of the degrees, is that, again considering the above example, the 2-way interaction $\{(1,1),(4,0)\}$ can only be covered by the 4-th, 5-th or 6-th row when extending M with one column.

Similar to Algorithm 1, based on Theorem 2 one can formulate an algorithm that treats Problem 2 i.e. an algorithm that finds all possible column extensions to a given CA when they exist. Due to the simplicity of this algorithm and space limitations we do not include it in this paper.

5   Experiments Using Gröbner Bases and Supercomputing

We employed two methodologies to solve the systems of polynomial equations arising from our algebraic modelling for CAs. Firstly, Gröbner bases (GB) computations in Maple and Magma, and secondly, exhaustive search using C and parallel programming using C/MPI (Message Passing Interface). Each of these different solving implementations have been used as a means to instantiate the SOLVE procedure in Algorithm 1.

5.1 Solving Using Gröbner Bases

For the Gröbner bases computations, we simply encode the system in Maple and Magma format and computed lexicographical and total degree Gröbner bases. If the result of the (reduced) Gröbner bases computation is equal to $\{1\}$, then we know that the system does not have any solutions [2]. If the result of the (reduced) Gröbner bases computation is not equal to $\{1\}$, then we use the actual basis to recover some solution of the system. We observed that in general, Maple and Magma are able to successfully compute Gröbner bases for systems of polynomial equations arising from CA constructions, for up to 20 binary and 10 ternary variables. We give a related example below:

Example 3 (Column extension of a covering array). Continuing Example 1, recall that

$$M = \begin{pmatrix} 0\,0\,0\,1\,1\,1\,2\,2\,2 \\ 0\,1\,2\,0\,1\,2\,0\,1\,2 \\ 0\,1\,2\,1\,2\,0\,2\,0\,1 \end{pmatrix}^T \quad \text{and} \quad \overline{M} = \begin{pmatrix} 0 & 0 & 0 & 1 & 1 & 1 & 2 & 2 & 2 \\ 0 & 1 & 2 & 0 & 1 & 2 & 0 & 1 & 2 \\ 0 & 1 & 2 & 1 & 2 & 0 & 2 & 0 & 1 \\ x_1 & x_2 & x_3 & x_4 & x_5 & x_6 & x_7 & x_8 & x_9 \end{pmatrix}^T .$$

We generate the system of domain and coverage equations, where the 9 polynomials in the domain equations have degree 3, and the degree of the polynomials in the 27 coverage equations can be reduced from 9 to 3, when applying the reduction of degrees as described in Remark 1. This system has 6 solutions, namely $\{(0, 1, 2, 2, 0, 1, 1, 2, 0), (0, 2, 1, 1, 0, 2, 2, 1, 0), (1, 0, 2, 2, 1, 0, 0, 2, 1),$ $(1, 2, 0, 0, 1, 2, 2, 0, 1), (2, 0, 1, 1, 2, 0, 0, 1, 2), (2, 1, 0, 0, 2, 1, 1, 0, 2)\}$. Note the nice linear equalities: $x_1 = x_5 = x_9$ and $x_3 = x_4 = x_8$ revealed by the GB and reflected in the corresponding positions above, for the 6 solutions.

5.2 Exhaustive Search Using Supercomputing

For systems with more than 20 binary or 10 ternary variables we designed a serial C program to perform exhaustive search. The program uses the ranking and unranking functions described in [11] to efficiently enumerate all combinations of values for the binary or the ternary variables. Obviously this approach readily generalizes to quaternary variables and beyond. For each combination of the variables generated, we solve the equations incrementally, i.e. we look at the equations as constraints that must be satisfied simultaneously and proceed by first checking whether the first equation is satisfied then secondly, by checking whether the second equation is satisfied and so forth. If a particular generated combination of values satisfies all equations then it is a solution and we collect all solutions found in a result file for post-processing. We found it beneficial to use a meta-programming approach, i.e. a bash script that parses the systems of polynomial equations arising from CA constructions automatically and generates the corresponding serial C program, without any intervention by hand. Using meta-programming allows us to produce massive amounts of bug-free and reliable C code with minimal effort.

For systems where our serial C program approach is insufficient, to either produce solutions or verify that no solutions exist, we parallelize our automatically generated C program using MPI. The parallelization is achieved with meta-programming again, in order to make efficient use of the ranking and unranking functions to distribute the workload among the parallel processors. We run our generated C/MPI code on the heterogeneous cluster known as "graham"[1], operated by Compute Canada at the University of Waterloo. Given that the system of polynomial equations arising from CA constructions exhibit a very precise structure and symmetries, we use this structure to distribute the computation not only at the level of variables, but at the level of equations as well. More precisely, by construction, the equations are divided in groups of r equations, where the number of variables featured in each group is a function of r and is significantly smaller than the total number of variables γ. This clearly suggests a two-phase approach to solve the original system:

1. Solve each group of equations independently and in parallel.
2. Look for common solutions among the solutions of all the groups.

The first phase of the above two-phase approach is reminiscent of the selection of subsets of clauses when applying resolution to large CNFs. Subsequently, we revised the first phase, by amalgamating one or more groups together, which has the advantage that fewer solutions are generated, and at the same time may prove insolvability of the system, if one or more groups of equations do not possess any solutions. Using our meta-programming bash script, we are able to run multiple experiments, to determine optimal cut-off points, as far as the number of groups of equations that can be solved independently, with the aim to keep the sizes of the generated solutions files small enough for the second-phase processing. We give below a related example using this approach:

Example 4 (Computation of optimal CAs). In this example we want to show how an optimal $\mathsf{CA}(9; 2, 3, 3)$ can be computed based on Corollary 1 using exhaustive search techniques. Therefore, we initialize a 9×3 array X of symbolic variables

$$X = \begin{pmatrix} x_{1,1} \ x_{1,2} \ x_{1,3} \ x_{1,4} \ x_{1,5} \ x_{1,6} \ x_{1,7} \ x_{1,8} \ x_{1,9} \\ x_{2,1} \ x_{2,2} \ x_{2,3} \ x_{2,4} \ x_{2,5} \ x_{2,6} \ x_{2,7} \ x_{2,8} \ x_{2,9} \\ x_{3,1} \ x_{3,2} \ x_{3,3} \ x_{3,4} \ x_{3,5} \ x_{3,6} \ x_{3,7} \ x_{3,8} \ x_{3,9} \end{pmatrix}^T ,$$

generate all 27 domain equations of degree 3 and the 27 coverage equations of degree 9, according to Eqs. (2) and (3), where we use $(a_1, a_2) = (3, 1)$ based on Lemma 3:

---

[1] https://docs.computecanada.ca/wiki/Graham.

1. $\forall i \in \{1, \ldots, 9\}, \forall j \in \{1, 2, 3\}$

$$x_{i,j}(x_{i,j} - 1)(x_{i,j} - 2) = 0.$$

2. $\forall C \in \binom{\{3\}}{2}, \ \forall (u_1, u_2) \in \{0, 1, 2\}^2$:

$$\mathbf{prod}(X \cdot \iota_{2,3}^{C}(3, 1) - \mathbf{1} \cdot (u_1, u_2) \cdot (3, 1)^T) = 0.$$

Any solution of this system yields an optimal $\mathsf{CA}(9; 2, 3, 3)$, one of which is e.g.

$$\begin{pmatrix} x_{1,1} \ x_{1,2} \ x_{1,3} \ x_{1,4} \ x_{1,5} \ x_{1,6} \ x_{1,7} \ x_{1,8} \ x_{1,9} \\ x_{2,1} \ x_{2,2} \ x_{2,3} \ x_{2,4} \ x_{2,5} \ x_{2,6} \ x_{2,7} \ x_{2,8} \ x_{2,9} \\ x_{3,1} \ x_{3,2} \ x_{3,3} \ x_{3,4} \ x_{3,5} \ x_{3,6} \ x_{3,7} \ x_{3,8} \ x_{3,9} \end{pmatrix}^T = \begin{pmatrix} 0 \ 0 \ 0 \ 1 \ 1 \ 1 \ 2 \ 2 \ 2 \\ 0 \ 1 \ 2 \ 0 \ 1 \ 2 \ 0 \ 1 \ 2 \\ 0 \ 1 \ 2 \ 1 \ 2 \ 0 \ 2 \ 0 \ 1 \end{pmatrix}^T.$$

5.3   Initial Experiments for Computation of Optimal CAs

Last, in Table 1 we list some initial experiments of ours, aiming for reconstruction and, if possible, updating the values (upper bounds) of $\mathsf{CAN}(3, k, 2)$ and $\mathsf{CAN}(2, k, 3)$. Based on the algebraic model for CAs presented in Sect. 3, we used an implementation of Algorithm 1. For the computation of the precise value of $\mathsf{CAN}(t, k, v)$, the input values t, k, v to Algorithm 1 are specified according to the respective CA instance (given in the first column). Whereas the input value N, determining the number of rows of the target CA to be constructed, is set to either the exact value of $\mathsf{CAN}(t, k, v)$, or $(\mathsf{CAN}(t, k, v) - 1)$. In case of $N < \mathsf{CAN}(t, k, v)$ our implementation returned "No CA exists", as expected, and for $N = \mathsf{CAN}(t, k, v)$ we found solutions for all cases documented above.

For example, we computed all optimal $\mathsf{CA}(8; 3, 4, 2)$ and 3022997 optimal $\mathsf{CA}(9; 2, 3, 3)$. Note that we report the number of all solutions, i.e. we do distinguish between equivalent CAs. We also proved that there do not exist $\mathsf{CA}(7; 3, 4, 2)$ and $\mathsf{CA}(8; 2, 3, 3)$ arising from our algebraic models. Note that the GB computations have been carried out in both, Maple and Magma.

Table 1. In column **# Vars** we list the number of unknowns in the respective equation systems; in column **# Sols** we list the exact number of solutions, i.e. CAs for the respective instance, except for entries $\geq x$, which indicate that at least x solutions were found. Column **CAN** lists the exact values of CAN and also the size of the retrieved CA solutions.

CA instance	Solver	# Vars	# Sols	CAN	Reference
$\mathsf{CA}(4; 2, 3, 2)$	GB	12	48	4	[10]
$\mathsf{CA}(5; 2, 3, 2)$	GB	15	1440	4	[10]
$\mathsf{CA}(5; 2, 4, 2)$	GB	20	1920	5	[10]
$\mathsf{CA}(8; 3, 4, 2)$	GB	32	80640	8	[3]
$\mathsf{CA}(9; 2, 3, 3)$	C/MPI	27	$\geq 3 \cdot 10^6$	5	[3]

6 Conclusion and Future Work

We presented a way to model CAs of arbitrary strengths as solutions of multivariate polynomial equation systems, and solved these systems with algebraic solvers based on Gröbner bases or supercomputing. Even though the presented modelling is only considered for CAs over v-ary alphabets, a generalization for other classes of covering arrays (e.g. mixed level or variable strength) might be possible. We plan to investigate such extensions of our modelling as part of future work. Moreover, we believe that our approach is capable of computing the CAN for larger CA instances, especially when used in conjunction with symmetry breaking in the equation systems and solving these systems with the aid of supercomputing.

Acknowledgements. This research was carried out partly in the context of the Austrian COMET K1 program and publicly funded by the Austrian Research Promotion Agency (FFG) and the Vienna Business Agency (WAW). Kotsireas and Zhereshchin are supported by an NSERC grant.

References

1. Bryce, R.C., Colbourn, C.J.: A density-based greedy algorithm for higher strength covering arrays. Softw. Test. Verif. Reliab. **19**(1), 37–53 (2009)
2. Buchberger, B.: Bruno Buchberger's PhD thesis 1965: an algorithm for finding the basis elements of the residue class ring of a zero dimensional polynomial ideal. J. Symb. Comput. **41**, 475–511 (2006)
3. Bush, K.A., et al.: Orthogonal arrays of index unity. Ann. Math. Stat. **23**(3), 426–434 (1952)
4. Colbourn, C.J.: Combinatorial aspects of covering arrays. Le Mathematiche **59**(1, 2), 125–172 (2004)
5. Colbourn, C.J., Dinitz, J.H.: Handbook of Combinatorial Designs. CRC Press, Boca Raton (2006)
6. Danziger, P., Mendelsohn, E., Moura, L., Stevens, B.: Covering arrays avoiding forbidden edges. In: Yang, B., Du, D.-Z., Wang, C.A. (eds.) COCOA 2008. LNCS, vol. 5165, pp. 296–308. Springer, Heidelberg (2008). https://doi.org/10.1007/978-3-540-85097-7_28
7. Garn, B., Simos, D.E.: Algebraic modelling of covering arrays. In: Kotsireas, I.S., Martínez-Moro, E. (eds.) ACA 2015. PROMS, vol. 198, pp. 149–170. Springer, Cham (2017). https://doi.org/10.1007/978-3-319-56932-1_10
8. Hnich, B., Prestwich, S.D., Selensky, E., Smith, B.M.: Constraint models for the covering test problem. Constraints **11**(2), 199–219 (2006)
9. Kampel, L., Simos, D.E.: A survey on the state of the art of complexity problems for covering arrays. Theoret. Comput. Sci. (2018, to appear)
10. Kleitman, D.J., Spencer, J.: Families of k-independent sets. Discrete Math. **6**(3), 255–262 (1973)
11. Kreher, D.L., Stinson, D.R.: Combinatorial Algorithms: Generation, Enumeration, and Search, 1st edn. CRC Press, Boca Raton (1998)
12. Kuhn, D., Kacker, R., Lei, Y.: Introduction to Combinatorial Testing. Chapman & Hall/CRC Innovations in Software Engineering and Software Development Series. Taylor & Francis, London (2013)

13. Kuhn, R., Kacker, R., Lei, Y., Hunter, J.: Combinatorial software testing. Computer **42**(8), 94–96 (2009)
14. Lei, Y., Tai, K.C.: In-parameter-order: a test generation strategy for pairwise testing. In: Proceedings of the Third IEEE International High-Assurance Systems Engineering Symposium (Cat. No. 98EX231), pp. 254–261 (1998)
15. Nayeri, P., Colbourn, C.J., Konjevod, G.: Randomized post-optimization of covering arrays. Eur. J. Comb. **34**(1), 91–103 (2013)
16. Seroussi, G., Bshouty, N.H.: Vector sets for exhaustive testing of logic circuits. IEEE Trans. Inf. Theory **34**(3), 513–522 (1988)
17. Torres-Jimenez, J., Izquierdo-Marquez, I.: Survey of covering arrays. In: 2013 15th International Symposium on Symbolic and Numeric Algorithms for Scientific Computing, pp. 20–27 (2013)

The Precise Complexity of Finding Rainbow Even Matchings

Martin Loebl$^{(\boxtimes)}$

Department of Applied Mathematics, Charles University, Prague, Czech Republic
loebl@kam.mff.cuni.cz

Abstract. A progress in complexity lower bounds might be achieved by studying problems where a very precise complexity is conjectured. In this note we propose one such problem: Given a planar graph on n vertices and disjoint pairs of its edges $p_1, \ldots, p_g$, perfect matching M is RAINBOW EVEN MATCHING (REM) if $|M \cap p_i|$ is even for each $i = 1, \ldots, g$. A straightforward algorithm finds a REM or asserts that no REM exists in $2^g \times \mathrm{poly}(n)$ steps and we conjecture that no deterministic or randomised algorithm has complexity asymptotically smaller than 2^g. Our motivation is also to pinpoint the curse of dimensionality of the MAX-CUT problem for graphs embedded into orientable surfaces: a basic problem of statistical physics.

Keywords: Matching · Max cut · Exponential time hypothesis · Ising partition function

1 Introduction

Given a graph $G = (V, E)$, a set of edges $M \subseteq E$ is called *perfect matching* if the graph (V, M) has degree one at each vertex. In this paper we introduce and study the following matching problems which, as far as we know, were not studied before.

Given a graph $G = (V, E)$ and disjoint pairs of its edges $p_1, \ldots, p_g$, we say that a perfect matching M is a RAINBOW EVEN MATCHING (REM) if $|M \cap p_i|$ is even for each $i = 1, \ldots, g$. For example, let C be a cycle of length 8 consisting of consecutive edges $e_1, e_2, \ldots, e_8$. If $g \geq 1$ and $p_1 = \{e_1, e_2\}$ then there is no REM and if $g = 3$ and $p_1 = \{e_1, e_3\}, p_2 = \{e_2, e_4\}, p_3 = \{e_5, e_6\}$ then both perfect matchings of C are REM. We consider the following problems:

1. **Decision Rainbow Even Matching problem (DREM):** Given a planar graph G on n vertices and disjoint pairs of edges $p_1, \ldots, p_g$, decide if there is a REM.

The author was partially supported by the H2020-MSCA-RISE project CoSP- GA No. 823748.

M. Ćirić et al. (Eds.): CAI 2019, LNCS 11545, pp. 190–201, 2019.
https://doi.org/10.1007/978-3-030-21363-3_16

2. **Enumeration Rainbow Even Matching problem (EREM):** Given a planar graph G on n vertices and disjoint pairs of edges $p_1, \ldots, p_g$, calculate the number of REMs.
3. If an integer weight function is given on the edge-set of the graph G then DREM has a natural weighted version, denoted by **OptDREM**, to find the maximum total weight of a REM, and EREM is turned into the problem denoted by **GenREM** to find the generating function of weighted REMs.

There is a straightforward algorithm of complexity $2^g\text{poly}(n)$ to solve Opt-DREM: For each $S \subset \{1, \ldots, g\}$ we find a maximum weight extension of the set $\cup_{i \in S} p_i$ into a perfect matching by edges of $E \setminus \cup_{i \leq g} p_i$. The weighted perfect matching algorithm does it.

There is also a straightforward algorithm of complexity $2^g\text{poly}(n)$ to solve GenREM: For each $S \subset \{1, \ldots, g\}$ we calculate the generating function of the REMs which contain all edges of $\cup_{i \in S} p_i$ and no edge of $\cup_{i \notin S} p_i$. This can be done by the *method of Kasteleyn orientations* briefly introduced in Subsect. 1.3.

Main Contribution

- We propose that the above standard algorithms are in fact optimal. Our Frustration Conjecture 1 below states that up to a polynomial factor the *precise complexity* of OptDREM with edge-weight in $\{-1, 0, 1\}$ is 2^g. This is more tight complexity specification than the *Strong Exponential Time Hypothesis*.
- We show that refuting the Frustration Conjecture 1 implies that in the class of graphs where the crossing number is equal to the genus, the complexity of the MAX-CUT problem is smaller than the *additive determinantal complexity* of cuts enumeration. At present, no natural class of embedded graphs with this property is known.

1.1 The Exponential Time Hypothesis

The *Exponential Time Hypothesis (ETH)* is an unproven computational hardness assumption that was formulated by Impagliazzo and Paturi [7]. For each k let s_k be the infimum of reals s for which there exists an algorithm solving k–SAT in time $O(2^{sn})$, where n is the number of variables. ETH states that for each $k > 2$, $s_k > 0$. We note that 2–SAT can be solved in polynomial time. In the same paper [7], the authors prove the *Sparsification Lemma* which implies that ETH is equivalent to a potential strengthening of ETH where the k–SAT instances have the number of clauses bounded from above by $c_k n$ for some constant c_k; n denotes the number of the variables.

The ETH was strengthened by Impagliazzo, Paturi and Zane [8] to the *Strong Exponential Time Hypothesis (SETH)*: For all $d < 1$ there is a k such that k-SAT cannot be solved in $O(2^{dn})$ time. No Sparsification Lemma is known for SETH.

Both ETH and SETH have a very natural role: they are used to argue that known algorithms are probably optimal.

I believe that the method of Kasteleyn orientations provides optimal algorithms for the MAX-CUT problem in the classes of embedded graphs.

My motivation for introducing REM has been to pinpoint this 'curse of dimensionality' by a problem formulated with no reference to the geometry.

Conjecture 1 (Frustration Conjecture). No deterministic or randomised algorithm can solve OptDREM with the edge-weights from $\{-1, 0, 1\}$ in asymptotically less than 2^g steps.

1.2   Justification for the Frustration Conjecture

An exponential lower bound for DREM is simply implied by ETH, see Corollary 1. Next, Theorem 2 connects the Frustration Conjecture 1 to the additive determinantal complexity of cuts enumeration.

A well-established way to approach matching problems is to determine whether some specific coefficient of the generating function of the perfect matchings (with suitable substitutions) is non-zero. This can be achieved because of the *Isolation Lemma*, see [13], by calculating a single Pfaffian of a matrix where the entries are monomials in possibly more than one variable. The Pfaffian is a determinant type expression which can be computed with essentially the same complexity as that of the determinant (of the same matrix). The complexity of calculating the determinant of matrices with polynomial entries essentially depends on the number of the variables.

After many failed attempts to use this machinery to disprove the Frustration Conjecture I am convinced that this approach will not beat the 2^g lower bound. However, I do not have at present a general theorem of this nature, only some partial results.

We can reduce, in a simple way suggested by Bruno Loff, (1 in 3)-SAT to DREM showing DREM is NP-complete.

Theorem 1. *DREM is an NP-complete problem.*

Proof. The reduction of (1 in 3)-SAT to DREM is best explained by an example. If the input of (1 in 3)-SAT is $(x_1 \lor \neg x_2 \lor x_3) \land (x_2 \lor x_1 \lor x_4)$ where the first clause is denoted by C_1 and the second clause by C_2 then the input graph for the corresponding DREM is depicted in Fig. 1, with $g = 2$ and $p_1^1 = \{e_1^1, e_2^1\}, p_1^2 = \{e_1^2, e_2^2\}$. This simply generalises.

Let $x_1, \ldots, x_n$ be the variables and let $C_1, \ldots, C_m$ be the clauses of a (1 in 3)-SAT input. (1) With each clause C_j we associate a copy $S(j)$ of the star with three leaves. (2) Let $x(i, j)$ denote the appearance of variable x_i in clause C_j. (3) If $x(i, j)$ is equal to x_i then let $P(i, j)$ be a copy of the path of three edges. (4) If $x(i, j)$ is equal to $\neg x_i$ then let $P(i, j)$ be a copy of the path of five edges. (5) Let variables $x_{i_1}, x_{i_2}, x_{i_3}$ appear in clause C_j. Then we identify the two leaves of $P(i_1, j)$ ($P(i_2, j), P(i_3, j)$ respectively) with two leaves of $S(j)$ as indicated in Fig. 1. (6) Finally we specify $g = 3m - n$ disjoint pairs of edges: Let $i \leq n$ and let $x(i, j_1), \ldots, x(i, j_k)$ be all the appearances of variable x_i. For $l = 1, \ldots, k$ let

$e(i, j_l)$ be an edge adjacent to the middle edge of $P(i, j_l)$. For each $i \leq n$ and $l \in \{1, \ldots, k-1\}$ we will have edge-pair $p_l^i = \{e(i, j_l), e(i, j_{l+1})\}$.

(7) These pairs assure the following: Let M be a REM and $1 \leq i \leq n$. Then the middle edge of each $P(i, j)$ belongs to M or the middle edge of no $P(i, j)$ belongs to M. This simply implies that there is a REM iff there is a (1 in 3)-satisfying assignment.

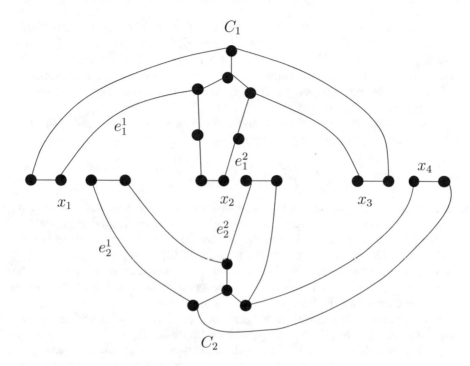

Fig. 1. An example of the DREM input graph for an instance of the (1 in 3)-SAT.

Corollary 1. *Let D be the infimum of reals d for which there exists an algorithm solving DREM in time $O(2^{dg})$, where g is the number of the input pairs of edges. Let us assume that the Exponential time hypothesis holds. Then $D > 0$.*

Proof. We first note that 3-SAT with n variables and m clauses can be reduced to (1 in 3)-SAT with $n + 6m$ variables and $5m$ clauses by a construction of Schaefer [14]. By the discussion in Sect. 1.1 we can assume that $m \leq c_3 n$. After this reduction we use the construction of the proof of Theorem 1.

1.3   Kasteleyn Orientations and Optimisation by Enumeration

Let me state a curious phenomenon: There is a strongly polynomial algorithm to solve the MAX-CUT problem in the planar graphs based on a reduction to the weighted perfect matching problem, see e.g. [10].

For the graphs of fixed genus $g \geq 1$ the situation is different: There is a weakly polynomial algorithm by Galluccio and Loebl ([4]; see also [5,6]); it was implemented several times and applied in extensive statistical physics calculations (see [12]). Recently other related algorithms based on the Valiant's theory [16] of holographic algorithms appeared (see [1,3]). All presently known approaches are of enumeration nature even for the class of the toroidal square grids. The weakly polynomial **optimisation by enumeration method** of [4] is as follows:

1. Let $G = (V, E)$ be a graph. A set of edges $E' \subseteq E$ is called *even* if each degree of the graph (V, E') is even. A set of edges $C \subseteq E$ is called an *edge-cut* of G, if there is a $V' \subseteq V$ so that $C = \{e \in E : |e \cap V'| = 1\}$. The Max-Cut problem, one of the basic optimisation problems, asks for the maximum size of an edge-cut in the input graph G, or, if weights on the edges are given, for the maximum total weight of an edge-cut.

2. If a weight-function $w : E \to \mathbf{R}$ and a set S of subsets of E are given then the *generating function of S* is defined as

$$\mathcal{F}(G, w, x) = \sum_{A \in S} \prod_{e \in A} x^{w(e)}.$$

3. The generating function of the edge-cuts is simply equivalent to the Ising partition function of the same graph, and it can be computed from the generating function of the even sets by a theorem of Van der Waender (for the definitions, theorems and their proofs see e.g. [10]).

4. The generating function of the even sets can be computed by the Fisher construction described in Sect. 2.3 as the generating function of the perfect matchings of a modified graph.

5. The seminal technical proposition was formulated by Kasteleyn [9] and proved by Galluccio, Loebl [4] and independently by Tesler [15]:

    The generating function of perfect matchings of a graph of genus g can be efficiently written as a linear combination of 2^{2g} Pfaffians. Pfaffians are determinant type expressions that can be computed efficiently by a variant of the Gaussian elimination. Cimasoni and Reshetikhin [2] provided a beautiful interpretation of the formula which then became known as the Arf invariant formula.

6. Summarising, the weakly polynomial algorithm solving the Max-Cut problem for the graphs of genus g by Galluccio and Loebl consists in calculating 2^{2g} Pfaffians and produces the complete generating function of the edge-cuts of the embedded graph.

1.4   Additive Determinantal Complexity

A recent result of Loebl and Masbaum [11] indicates that this might be optimum for the cuts enumeration. It is shown by Loebl and Masbaum in [11] that, if we want to enumerate the edge-cuts of each possible size of an input graph G of genus g, then in a strongly restricted setting called *additive determinantal complexity* the number of the Pfaffian calculations cannot be smaller than 2^{2g}.

This leads to a question: Is there an algorithm for solving the MAX-CUT problem in (a natural subclass of) the embedded graphs, whose complexity *beats* the additive determinantal complexity of the cuts enumeration? At present no such algorithm for a natural subclass of embedded graphs is known.

I believe that the answer to this question is NO and in Theorem 2 below we present a partial result. We show that the Frustration Conjecture implies that for the class of embedded graphs where the crossing number is equal to the genus, there is no algorithm to solve the MAX-CUT problem whose complexity beats the additive determinantal complexity bound. The proof of Theorem 2 is included in Sect. 2.

Theorem 2. *Let G be a graph with n vertices and embedded to the plane with g crossings. One can efficiently construct planar graph G' with edge-weights in $\{-1, 0, 1\}$ and a set of $2g$ disjoint pairs of edges of G' so that finding the maximum size of an edge-cut in G is polynomial time reducible to determining the maximum weight of a REM in G'.*

2   Edge-Cuts in Embedded Graphs

Let $G = (V, E)$ be a graph. A set of edges $E' \subseteq E$ is called *even* if each degree of the graph (V, E') is even. A set of edges $C \subseteq E$ is called an *edge-cut* of G, if there is a $V' \subseteq V$ so that $C = \{e \in E : |e \cap V'| = 1\}$. The MAX-CUT problem, one of the basic optimisation problems, asks for the maximum size of an edge-cut in the input graph G, or, if weights on the edges are given, for the maximum total weight of an edge-cut.

2.1   Surfaces

We recall the following standard description of a genus g surface S_g with one boundary component (we follow [10,11]). (We reserve the notation Σ_g for a closed surface of genus g.)

Definition 1. A 1-*highway* (see Fig. 2) is a surface $\bar{S}_g$ which consists of a base polygon R_0 and bridges $R_1, \ldots, R_{2g}$, where

- R_0 is a convex $4g$-gon with vertices $a_1, \ldots, a_{4g}$ numbered clockwise.
- Each R_{2i-1} is a rectangle with vertices $x(i, 1), \ldots, x(i, 4)$ numbered clockwise and glued to R_0. Edges $[x(i, 1), x(i, 2)]$ and $[x(i, 3), x(i, 4)]$ of R_{2i-1} are identified with edges $[a_{4(i-1)+1}, a_{4(i-1)+2}]$ and $[a_{4(i-1)+3}, a_{4(i-1)+4}]$ of R_0, respectively.
- Each R_{2i} is a rectangle with vertices $y(i, 1), \ldots, y(i, 4)$ numbered clockwise and glued to R_0. Edges $[y(i, 1), y(i, 2)]$ and $[y(i, 3), y(i, 4)]$ of R_{2i-1} are identified with edges $[a_{4(i-1)+2}, a_{4(i-1)+3}]$ and $[a_{4(i-1)+4}, a_{4(i-1)+5}]$ of R_0, respectively. (Here, indices are considered modulo $4g$.)

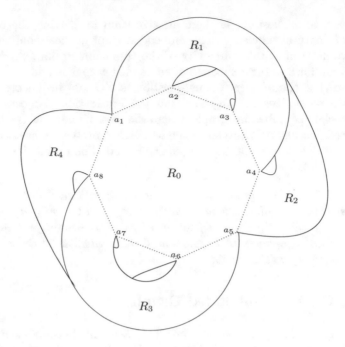

Fig. 2. A 1-highway.

Before proceeding, we point out a simple fact that we will soon exploit: the boundary of a 1-highway is isotopic to the boundary of a disk.

Now assume the graph G is embedded into a closed orientable surface Σ_g of genus g. We think of Σ_g as S_g union an additional disk δ glued to the boundary of S_g. By an isotopy of the embedding, we may assume that G does not meet the disk δ and that, moreover, all vertices of G lie in the interior of R_0.

We may also assume that the intersection of G with any of the rectangular bridges R_i consists of disjoint straight lines connecting the two sides of R_i which are glued to the base polygon R_0.

Next, follows the standard analogous description of a genus g surface S_g with more than one boundary component.

Definition 2. A *highway surface* S_g is obtained from a 2-sphere Z with h disjoint polygons $R_0^1, \ldots, R_0^h$ specified, and h disjoint 1-highway surfaces $\bar{S}_{g_1}^1, \ldots, \bar{S}_{g_h}^h$, where $g = g_1 + \ldots + g_h$, by first identifying the base polygon of each $\bar{S}_{g_i}^i$ with the polygon R_0^i, and then by deletion of the interiors of these polygons R_0^i ($i = 1, \ldots, h$).

Now assume the graph G is embedded into a closed orientable surface Σ_g of genus g. We again think of Σ_g as S_g union h additional disks δ_i ($i = 1, \ldots, h$), glued to the h boundaries of S_g. By an isotopy of the embedding, we may assume that G does not meet the disks δ_i's and that, moreover, no vertex of G lies in a bridge. We may also assume that the intersection of G with any of the rectangular bridges R_i^j consists of disjoint straight lines connecting the two sides of R_i^j which are glued to the base sphere Z.

2.2 Local Non-planarity

We note that each embedding of a graph G into Σ_g defines its *geometric dual*, usually denoted by G^*, as follows: the vertices of G^* are the faces of the embedding of G and for each edge e of G there is an edge e^* of G^* connecting the faces which have e on their boundary. For example, each toroidal square grid is self-dual. We note that a dual can have loops and multiple edges.

We consider simultaneous embeddings of the graph and its geometric dual into Σ_g.

Definition 3. Let $G = (V, E)$ be a graph. A *simultaneous embedding* of G into Σ_g consists of (1) an embedding N of graph G, and (2) an embedding N^* of the geometric dual $G^* = (V^*, E^*)$ of N. In addition, we require that (a) G is the geometric dual of N^*, (b) each vertex of G^* (of G respectively) is embedded in the face of N (N^* respectively) it represents, (c) each pair of dual edges e, e^* intersects exactly once, and N, N^* have no other intersections, and (d) both N, N^* are embeddings into $S_g \subseteq \Sigma_g$.

For a collection of edges $S \subseteq E$ we denote by $S^* \subseteq E^*$ the collection of dual edges e^* such that $e \in S$.

Since a simultaneous embedding of G into Σ_g is by definition a subset of $S_g \subseteq \Sigma_g$, we will also call it *simultaneous embedding into S_g*.

We may also assume that the intersection of G with any of the rectangular bridges R_i^j consists of disjoint straight lines connecting the two sides of R_i^j which are glued to the base sphere Z.

Definition 4. We recall that the intersection of an embedding of G in S_g with any of the rectangular bridges R_i^j of S_g consists of disjoint straight lines connecting the two sides of R_i^j which are glued to the base sphere.

Let G be embedded in S_g. An even set $E' \subset E$ of the edges of G of which crosses each bridge of S_g by an even number of disjoint straight lines will be called *admissible*.

A simultaneous embedding of G into S_g is called *even* if it holds that $C \subseteq E$ is an edge-cut of G if and only if $C^* \subseteq E^*$ is an admissible even set of the embedding of G^*.

A basic example of an even simultaneous embedding is a toroidal square grid and its geometric dual.

Definition 5. We say that a simultaneous even embedding of a graph G into some S_g is *restricted* if E^* intersects each bridge by at most 2 disjoint straight lines.

Definition 6. We say that graph G belongs to class $\mathcal{C}_g$ if G is drawn to the plane with exactly g edge-crossings and for each crossing there is a planar disc where the drawing looks as depicted in Fig. 3.

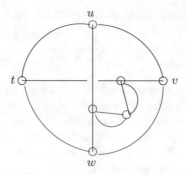

Fig. 3. Illustration for Definition 6

Theorem 3. *If $G \in \mathcal{C}_g$, then G admits a restricted even simultaneous embedding into S_g.*

Proof. We consider the simultaneous local embedding of the graph G as described in Fig. 4. The embedding is clearly restricted. We need to show that the embedding is even.

We first observe that the set $\delta(v)$ of the edges of G incident with any vertex v of G satisfies that $\delta^*(v)$ intersects each bridge in an even number of segments. Since each edge-cut of G is the symmetric difference of some sets $\delta(v), v \in V$, we get: If C is an edge-cut of G, then C^* is admissible.

In order to prove that the embedding is even we need to show that each admissible set C^* of dual edges is a symmetric difference of faces of G^*; this implies that C is an edge-cut of G. We can assume that C^* has empty intersection with the bridges (depicted in Fig. 4).

Consider the pair of bridges in Fig. 4. There is a face F_1 of G^* with exactly 2 edges on the vertical bridge and no edge on the horizontal bridge, and also a face F_2 of G^* where the role of the two bridges is exchanged. We can use the symmetric difference of C^* with F_1 or F_2 to produce a new even set C_0^* which has empty intersection with each bridge. Moreover, if C_0^* is a symmetric difference of faces of G^* then so is C^*.

It follows that C^* is an even subset of an embedded planar subgraph of G^*. For the planar graphs, the boundaries of faces generate all even sets of edges by the symmetric difference operation. Hence the proof is finished if we show that each face F of this planar subgraph is a symmetric difference of the faces of G^*.

Indeed, such F is either a face of G^* itself, or it looks like the square of Fig. 4 comprised of edges depicted as thick lines, which is the symmetric difference of the dual faces encircling the three unlabelled vertices of G of Fig. 4.

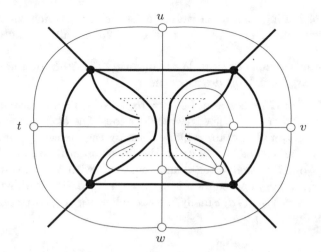

Fig. 4. Simultaneous embedding of graph $G \in \mathcal{C}_g$ near a crossing. There is one pair of bridges; the boundaries of the vertical bridge are depicted by dotted lines and the boundaries of the horizontal bridge are not depicted to simplify the presentation. The edges of G are depicted by normal lines and the dual edges are depicted by thick lines.

2.3   Proof of Theorem 2

We show that for a graph $G = (V, E)$ with n vertices and embedded to the plane with g edge crossings one can efficiently construct a planar graph $H = (W, E')$ with edge-weights in $\{-1, 0, 1\}$ and with $2g$ specified disjoint pairs of its edges so that the maximum size of an edge-cut in G is equal to the maximum weight of a REM in H. The construction goes as follows:

Step 1. We subdivide each edge of G near to each crossing; if $e \in E$ got subdivided into edges $e_1, \ldots, e_k$ which form the path $(e_1, \ldots, e_k)$ then we let the weight of e_1 equal to 1 and the weight of $e_2, \ldots, e_k$ equal to -1. The resulting weighted graph will be denoted by G_1. We note that the MAX-CUT problem in G is reduced to the weighted MAX-CUT problem in G_1.

Step 2. We add, for each edge crossing of G_1, the four edges of weight zero forming a 4–cycle (denoted by $uvwt$ in Fig. 4) and further one new vertex which we connect by four edges of weight zero to the two vertices near to this crossing added in Step 1 so that the resulting graph, which we denote by G_2, is in $\mathcal{C}_g$. We note that G_2 is uniquely determined and the weighted MAX-CUT problem in G_1 is reduced to the weighted MAX-CUT problem in G_2.

Step 3. We use Theorem 3. Let G_2^* be the dual from the restricted simultaneous even embedding of G_2 into S_g. The weight of each edge e^* of G_2^* is defined to be equal to the weight of the corresponding edge e of G_2. We specify $2g$ pairs $p_1, \ldots, p_{2g}$ of edges of G_2^*:

   Each pair consists of the two edges embedded on one of the $2g$ bridges of S_g (see Fig. 4). We note that the weighted MAX-CUT problem for G_2 is reduced to the problem of finding maximum weight even set of G_2^* which contains an even number of elements of each pair $p_i, i = 1, \ldots, 2g$. Finally we note that G_2^* is planar.

Step 4: Fisher's construction. We transform G_2^* into H by the Fisher's construction (see e.g. book [10]) described next.

Definition 7. Let G be a graph. Let $\sigma = (\sigma_v)_{v \in V(G)}$ be a choice, for every vertex v, of a linear ordering of the edges incident to v. The *blow-up*, or Δ-*extension*, of (G, σ) is the graph G^σ obtained by performing the following operation one by one for each vertex v. Let $e_1, \ldots, e_d$ be the linear ordering σ_v and let $e_i = vu_i$, $i = 1, \ldots, d$. We delete the vertex v and replace it with a path consisting of $6d$ new vertices $v_1, \ldots, v_{6d}$ and edges $v_i v_{i+1}, i = 1, \ldots, 6d - 1$. To this path, we add edges $v_{3j-2} v_{3j}, j = 1, \ldots, 2d$. Finally, we add edges $v_{6i-4} u_i$ corresponding to the original edges $e_1, \ldots, e_d$.

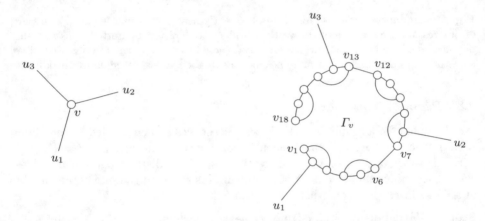

Fig. 5. For a node v with the neighborhood illustrated in (a) the associated gadget Γ_v is depicted in (b).

The subgraph of G^σ spanned by the $6d$ vertices $v_1, \ldots, v_d$ that replaced a vertex v of the original graph will be called a *gadget* and denoted by Γ_v. The edges of G^σ which do not belong to a gadget are in natural bijection with the edges of G. By abuse of notation, we will identify an edge of G with the corresponding edge of G^σ. Thus $E(G^\sigma)$ is the disjoint union of $E(G)$ and the various $E(\Gamma_v)$ $(v \in V(G))$ (Fig. 5).

It is important to note that different choices of linear orderings σ_v at the vertices of G may lead to non-isomorphic graphs G^σ. Nevertheless, one always has the following:

Lemma 1. *There is a natural bijection between the set of even subsets of G and the set of perfect matchings of G^σ. More precisely, every even set $E' \subseteq E(G)$ uniquely extends to a perfect matching $M \subset E(G^\sigma)$, and every perfect matching of G^σ arises (exactly once) in this way.*

It follows that if we set the weights of the edges of the gadgets of $(G_2^*)^\sigma$ equal to zero, we get that the value of the MAX-CUT problem for G is equal to the max REM of $H = (G_2^*)^\sigma$. This finishes the proof of Theorem 2.

Acknowledgement. This project initially started as a joint work with Marcos Kiwi. I would like to thank Marcos for many helpful discussions.

References

1. Bravyi, S.: Contraction of matchgate tensor networks on non-planar graphs. Contemp. Math. **482**, 179–211 (2009). Advances in quantum computation
2. Cimasoni, D., Reshetikhin, N.: Dimers on surface graphs and spin structures I. Commun. Math. Phys. **275**(1), 187–208 (2007)
3. Curticapean, R., Xia, M.: Parametrizing the permanent: genus, apices, evaluation mod 2^k. In: 2015 IEEE 56th Annual Symposium on Foundations of Computer Science, pp. 994–1009. IEEE Computer Society (2015)
4. Galluccio, A., Loebl, M.: On the theory of Pfaffian orientations II. T-joins, k-cuts, and duality of enumeration. Electron. J. Comb. **6**, Article Number R7 (1999)
5. Galluccio, A., Loebl, M., Vondrak, J.: A new algorithm for the Ising problem: partition function for finite lattice graphs. Phys. Rev. Lett. **84**, 5924 (2000)
6. Galluccio, A., Loebl, M., Vondrak, J.: Optimization via enumeration: a new algorithm for the max cut problem. Math. Program. **90**, 273–290 (2001)
7. Impagliazzo, R., Paturi, R.: On the complexity of k-SAT. J. Comput. Syst. Sci. **62**(2), 367–375 (2001)
8. Impagliazzo, R., Paturi, R., Zane, F.: Which problems have strongly exponential complexity? J. Comput. Syst. Sci. **63**(4), 512–530 (2001)
9. Kasteleyn, P.: Graph theory and crystal physics. In: Harary, F. (ed.) Graph Theory and Theoretical Physics. Academic Press, London (1967)
10. Loebl, M.: Discrete Mathematics in Statistical Physics. Vieweg and Teubner, Verlag (2010)
11. Loebl, M., Masbaum, G.: On the optimality of the Arf invariant formula for graph polynomials. Adv. Math. **226**, 332–349 (2011)
12. Lukic, J., Galluccio, A., Marinari, E., Martin, O.C., Rinaldi, G.: Critical thermodynamics of the two-dimensional $+/-$J Ising spin glass. Phys. Rev. Lett. **92**(11), 117202 (2004)
13. Mulmuley, K., Vazirani, U.V., Vazirani, V.V.: Matching is as easy as matrix inversion. Combinatorica **7**(1), 105–113 (1987)
14. Schaefer, J.: The complexity of satisfiability problems. In: 10th Annual ACM Symposium on Theory of Computing, STOC 1978, Proceedings, pp. 216–226 (1978)
15. Tesler, G.: Matchings in graphs on non-orientable surfaces. J. Comb. Theory Ser. B **78**, 198–231 (2000)
16. Valiant, L.: Holographic algorithms (extended abstract). In: 45th Annual IEEE Symposium on Foundations of Computer Science, FOCS 2004. IEEE Computer Society (2004)

New Cryptcodes for Burst Channels

Daniela Mechkaroska, Aleksandra Popovska-Mitrovikj$^{(\boxtimes)}$, and Verica Bakeva

Faculty of Computer Science and Engineering, Ss. Cyril and Methodius University,
Skopje, Macedonia
{daniela.mechkaroska,aleksandra.popovska.mitrovikj,
verica.bakeva}@finki.ukim.mk

Abstract. RCBQ are cryptcodes proposed in 2007. After that, several papers for performances of these codes for transmission through a binary-symmetric and Gaussian channels, have been published. Also, for improving the performances of these codes several new algorithms have been defined. In this paper we proposed a new modification of existing cryptcodes obtaining new cryptcodes suitable for transmission in burst channels. For generating burst errors we use the model of Gilbert-Elliott burst channel. Experimental results for bit-error and packet-error probabilities obtained for different channel and code parameters are presented. Also, we made comparison of the results obtained with the old and the new algorithms for RCBQ.

Keywords: Cryptcoding · Gilbert-Elliott channel · SNR ·
Bit-error probability · Burst errors · Quasigroup

1 Introduction

Cryptcodes based on quasigroups, called Random Codes Based on Quasigroups (RCBQ) are defined in [2]. In this paper we consider performances of RCBQ for decoding data transmitted through a Gilbert-Elliott channel. In order to improve performances of these codes for correction of burst errors we propose a new modification of coding/decoding algorithms. There are several coding/decoding algorithms for RCBQ, but we consider Cut-Decoding and 4-Sets-Cut-Decoding (#3) algorithms proposed in [9,10]. In all algorithms for these codes in the process of coding/decoding an encryption/decryption algorithm is used and therefore these codes can correct some of the transmission errors and at the same time they encrypt the messages. A few similar combinations of error-correcting codes and cryptographic algorithms are proposed for cryptographic purposes [8,11,12]. Cryptographic properties of RCBQ are already investigated in several papers, for example [1,6,7]. Here, we consider only error-correction capabilities of RCBQ.

The rest of the paper is organized on the following way. The model of Gilbert-Elliott burst channel is described in Sect. 2. In Sect. 3 we explain Cut-Decoding and 4-Sets-Cut-Decoding algorithms for RCBQ and we define new algorithms (called Burst-Cut-Decoding and Burst-4-Sets-Cut-Decoding algorithms) for improving the performances of RCBQ for correction of burst errors.

© Springer Nature Switzerland AG 2019
M. Ćirić et al. (Eds.): CAI 2019, LNCS 11545, pp. 202–212, 2019.
https://doi.org/10.1007/978-3-030-21363-3_17

Experimental results obtained with Cut-Decoding, 4-Sets-Cut-Decoding, Burst-Cut-Decoding and Burst-4-Sets-Cut-Decoding algorithms are given in Sect. 4. At the end, we give some conclusions for the presented results.

2   Gilbert-Elliott Burst Model

Gilbert-Elliott Burst Model is a channel model introduced by E. Gilbert and E. O. Elliott. This model is based on a Markov chain with two states G (good or gap) and B (bad or burst). In good state the probability for incorrect transmission of a bit is small, and in bad state this probability is large. This model is widely used for describing burst error patterns in transmission channels. The model is shown in Fig. 1, where G represents the good state and B represents the bad state. The transmission probability from bad to good state is P_{BG} and this probability from good to bad state is P_{GB} [4,5].

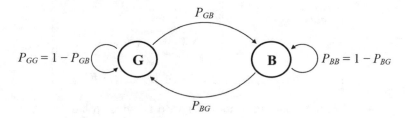

Fig. 1. Gilbert-Elliott burst model

We made experiments with two kinds of Gilbert-Eliott channel. In the first one, in each state the channel is binary symmetric with bit error probabilities $P_e(G)$ in a good state and $P_e(B)$ in a bad state. In the second one, the channels are Gaussian where SNR_G in a good state is high and SNR_B in a bad state is low.

3   Description of Coding/Decoding Algorithms

RCBQs are designed using algorithms for encryption and decryption from the implementation of TASC (Totally Asynchronous Stream Ciphers) by quasigroup string transformation [3]. These cryptographic algorithms use the alphabet Q and a quasigroup operation $*$ on Q together with its parastrophe $\backslash$.

3.1   Description of Coding

At first, let describe Standard coding algorithm for RCBQs proposed in [2]. The message $M = m_1 m_2 \ldots m_l$ (of $N_{block} = 4l$ bits where $m_i \in Q$ and Q is an alphabet of 4-bit symbols (nibbles)) is extended to message $L = L^{(1)} L^{(2)} \ldots L^{(s)} =$

$L_1L_2...L_m$ by adding redundant zero symbols. The produced message L has $N = 4m$ bits ($m = rs$), where $L_i \in Q$ and $L^{(i)}$ are sub-blocks of r symbols from Q. In this way we obtain (N_{block}, N) code with rate $R = N_{block}/N$. The codeword is produced after applying the encryption algorithm of TASC (given in Fig. 2) on the message L. For this purpose, a key $k = k_1k_2...k_n \in Q^n$ should be chosen. The obtained codeword of M is $C = C_1C_2...C_m$, where $C_i \in Q$.

Encryption	Decryption
Input: Key $k = k_1k_2 \ldots k_n$ and	**Input**: The pair
$L = L_1L_2 \ldots L_m$	$(a_1a_2 \ldots a_r, k_1k_2 \ldots k_n)$
Output: codeword	**Output**: The pair
$C = C_1C_2...C_m$	$(c_1c_2 \ldots c_r, K_1K_2 \ldots K_n)$
For $j = 1$ to m	For $i = 1$ to n
$\quad X \leftarrow L_j;$	$\quad K_i \leftarrow k_i;$
$\quad T \leftarrow 0;$	For $j = 0$ to $r - 1$
$\quad$ For $i = 1$ to n	$\quad X, T \leftarrow a_{j+1};$
$\quad\quad X \leftarrow k_i * X;$	$\quad temp \leftarrow K_n;$
$\quad\quad T \leftarrow T \oplus X;$	$\quad$ For $i = n$ to 2
$\quad\quad k_i \leftarrow X;$	$\quad\quad X \leftarrow temp \setminus X;$
$\quad k_n \leftarrow T$	$\quad\quad T \leftarrow T \oplus X;$
Output: $C_j \leftarrow X$	$\quad\quad temp \leftarrow K_{i-1};$
	$\quad\quad K_{i-1} \leftarrow X;$
	$\quad X \leftarrow temp \setminus X;$
	$\quad K_n \leftarrow T;$
	$\quad c_{j+1} \leftarrow X;$
	Output: $(c_1c_2 \ldots c_r, K_1K_2 \ldots K_n)$

Fig. 2. Algorithms for encryption and decryption

In Cut-Decoding algorithm, instead of using (N_{block}, N) code with rate R, we use together two $(N_{block}, N/2)$ codes with rate $2R$ and for coding we apply the encryption algorithm (given in Fig. 2) two times, on the same redundant message L using different parameters (different keys or quasigroups). We obtain the codeword of the message as concatenation of two codewords of $N/2$ bits. In 4-Sets-Cut-Decoding algorithm we use four $(N_{block}, N/4)$ codes with rate $4R$ and the codeword of the message is a concatenation of four codewords of $N/4$ bits.

3.2   Description of Decoding

The decoding in all algorithms for RCBQ is actually a list decoding and the speed of the decoding process depends on the list size (a shorter list gives faster decoding).

In Standard decoding algorithm for RCBQs, after transmission through a noisy channel, the codeword C will be received as a message $D = D^{(1)}$ $D^{(2)} \ldots D^{(s)} = D_1D_2 \ldots D_m$ where $D^{(i)}$ are blocks of r symbols from Q and $D_i \in Q$. The decoding process consists of four steps: (i) procedure for generating the sets with predefined Hamming distance, (ii) inverse coding algorithm, (iii) procedure for generating decoding candidate sets and (iv) decoding rule.

Let B_{max} be a given integer which denotes the assumed maximum number of bit errors that occur in a block during transmission. We generate the sets

$H_i = \{\alpha | \alpha \in Q^r, \quad H(D^{(i)}, \alpha) \leq B_{max}\}$, for $i = 1, 2, \ldots, s$, where $H(D^{(i)}, \alpha)$ is Hamming distance between $D^{(i)}$ and α. The decoding candidate sets S_0, S_1, $S_2, \ldots, S_s$ are defined iteratively. Let $S_0 = (k_1 \ldots k_n; \lambda)$, where λ is the empty sequence. Let S_{i-1} be defined for $i \geq 1$. Then S_i is the set of all pairs $(\delta, w_1 w_2 \ldots w_{4ri})$ obtained by using the sets S_{i-1} and H_i as follows (w_j are bits). For each element $\alpha \in H_i$ and each $(\beta, w_1 w_2 \ldots w_{4r(i-1)}) \in S_{i-1}$, we apply the inverse coding algorithm (i.e., algorithm for decryption given in Fig. 2) with input (α, β). If the output is the pair (γ, δ) and if both sequences γ and $L^{(i)}$ have the redundant zeros in the same positions, then the pair $(\delta, w_1 w_2 \ldots w_{4r(i-1)} c_1 c_2 \ldots c_r) \equiv (\delta, w_1 w_2 \ldots w_{4ri})$ ($c_i \in Q$) is an element of S_i.

In Cut-Decoding algorithm, after transmission through a noisy channel, we divide the outgoing message $D = D^{(1)} D^{(2)} \ldots D^{(s)}$ in two messages $D_1 = D^{(1)} D^{(2)} \ldots D^{(s/2)}$ and $D_2 = D^{(s/2+1)} D^{(s/2+2)} \ldots D^{(s)}$ with equal lengths and we decode them parallel with the corresponding parameters. In this decoding algorithm we make modification in the procedure for generating decoding candidate sets. Let $S_i^{(1)}$ and $S_i^{(2)}$ be the decoding candidate sets obtained in the ith iteration of both parallel decoding processes, $i = 1, \ldots, s/2$. Then, before the next iteration we eliminate from $S_i^{(1)}$ all elements whose second part does not match with the second part of an element in $S_i^{(2)}$, and vice versa. In the $(i+1)^{th}$ iteration the both processes use the corresponding reduced sets $S_i^{(1)}$ and $S_i^{(2)}$.

In [10], authors proposed 4 different versions of decoding with 4-Sets-Cut-Decoding algorithm. The best results are obtained using 4-Sets-Cut-Decoding algorithm#3 and here we use only this version. In this algorithm after transmitting through a noisy channel, we divide the outgoing message $D = D^{(1)} D^{(2)} \ldots D^{(s)}$ in four messages D^1, D^2, D^3 and D^4 with equal lengths and we decode them parallel with the corresponding parameters. Similarly, as in Cut-Decoding algorithm, in each iteration of the decoding process we reduce the decoding candidate sets obtained in the four decoding processes, as follows. Let $S_i^{(1)}$, $S_i^{(2)}$, $S_i^{(3)}$ and $S_i^{(4)}$ be the decoding candidate sets obtained in the ith iteration of four parallel decoding processes, $i = 1, \ldots, s/4$. Let $V_j = \{w_1 w_2 \ldots w_{r \cdot a \cdot i} | (\delta, w_1 w_2 \ldots w_{r \cdot a \cdot i}) \in S_i^{(j)}\}$, $j = 1, 2, 3, 4$ and $V = V_1 \cap V_2 \cap V_3 \cap V_4$. If $V = \emptyset$ then $V = (V_1 \cap V_2 \cap V_3) \cup (V_1 \cap V_2 \cap V_4) \cup (V_1 \cap V_3 \cap V_4) \cup (V_2 \cap V_3 \cap V_4)$.

Before the next iteration we eliminate from $S_i^{(j)}$ all elements whose second part is not in V, $j = 1, 2, 3, 4$.

The decoding rule is following. After the last iteration, if all reduced sets $S_{s/2}^{(1)}$, $S_{s/2}^{(2)}$ in Cut-Decoding (or $S_{s/4}^{(1)}$, $S_{s/4}^{(2)}$, $S_{s/4}^{(3)}$, $S_{s/4}^{(4)}$ in 4-Sets-Cut-Decoding) have only one element with a same second component then this component is the decoded message L. In this case, we say that we have a *successful decoding*. If the decoded message is not the correct one then we have an *undetected-error*. If the reduced sets obtained in the last iteration have more than one element then we have a *more-candidate-error*. If we obtain $S_i^{(1)} = S_i^{(2)} = \emptyset$ in some iteration of Cut-Decoding or $S_i^{(1)} = S_i^{(2)} = S_i^{(3)} = S_i^{(4)} = \emptyset$ in some iteration of 4-Sets-Cut-Decoding algorithm, then the process will finish (a *null-error* appears). But, if we obtain at least one nonempty decoding candidate set in an iteration then the decoding continues with the nonempty sets (the reduced sets are obtained by intersection of the non-empty sets only).

4   New Cryptcodes for Burst Channels

In experiments with burst channels explained in Sect. 2, we do not obtain good results using Cut-Decoding and 4-Sets-Cut-Decoding algorithms and therefore here we propose new algorithms for coding/decoding called Burst-Cut-Decoding and Burst-4-Sets-Cut-Decoding algorithms. It is known that interleaving and deinterleaving are useful for handling burst errors in a communication system. So, in the new algorithms, we include an interleaver in coding algorithm and the corresponding deinterleaver in the decoding algorithm. Namely, in the process of coding before the concatenation of two (or four) codewords we apply the interleaving on each codeword, separately. The interleaver rearranges (by rows) m nibbles of a codeword in a matrix of order $(m/k) \times k$. The output of the interleaver is a mixed message obtained reading the matrix by columns. Then, after transmission of a concatenated message through a burst channel we divide the outgoing message D in two (or four) messages with equal length and before the parallel decoding we apply deinterleaving on each messages, separately. The coding/decoding process in the new algorithms is schematically presented on Fig. 3.

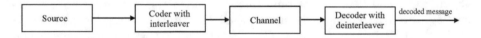

Fig. 3. Coding/decoding process in the new algorithms

5   Experimental Results

In this section we present experimental results obtained with RCBQ for transmission through a burst channel. For simulation of the channel we use the Gilbert-Elliott model explained in Sect. 2. We compare the values of packet-error probability (PER) and bit-error probability (BER) obtained with Cut-Decoding, 4-Sets-Cut-Decoding, Burst-Cut-Decoding and Burst-4-Sets-Cut-Decoding algorithms. We consider codes $(72, 288)$ with rate 1/4 using Cut-Decoding algorithm and Burst-Cut-Decoding algorithm, and also $(72, 576)$ with rate 1/8 using all algorithms (Cut-Decoding, 4-Sets-Cut-Decoding, Burst-Cut-Decoding and Burst-4-Sets-Cut-Decoding). In the experiments we use the following code parameters.

- For code $(72, 288)$ in Cut-Decoding and Burst-Cut-Decoding algorithm, the parameters are:
    - redundancy pattern: 1100 1110 1100 1100 1110 1100 1100 1100 0000 for rate 1/2 and two different keys of 10 nibbles.

– For code $(72, 576)$, the code parameters are:
  - in Cut-Decoding/Burst-Cut-Decoding - redundancy pattern: 1100 1100 1000 00001100 1000 1000 0000 1100 1100 1000 0000 1100 1000 1000 0000 0000 0000, for rate 1/4 and two different keys of 10 nibbles,
  - in 4-Sets-Cut-Decoding/Burst-4-Sets-Cut-Decoding - redundancy pattern: 1100 1110 1100 1100 1110 1100 1100 1100 0000 for rate 1/2 and four different keys of 10 nibbles.

For all experiments we use $B_{max} = 4$ and the same quasigroup on Q given in Table 1.

Table 1. Quasigroup of order 16 used in the experiments

*	0	1	2	3	4	5	6	7	8	9	a	b	c	d	e	f
0	3	c	2	5	f	7	6	1	0	b	d	e	8	4	9	a
1	0	3	9	d	8	1	7	b	6	5	2	a	c	f	e	4
2	1	0	e	c	4	5	f	9	d	3	6	7	a	8	b	2
3	6	b	f	1	9	4	e	a	3	7	8	0	2	c	d	5
4	4	5	0	7	6	b	9	3	f	2	a	8	d	e	c	1
5	f	a	1	0	e	2	4	c	7	d	3	b	5	9	8	6
6	2	f	a	3	c	8	d	0	b	e	9	4	6	1	5	7
7	e	9	c	a	1	d	8	6	5	f	b	2	4	0	7	3
8	c	7	6	2	a	f	b	5	1	0	4	9	e	d	3	8
9	b	e	4	9	d	3	1	f	8	c	5	6	7	a	2	0
a	9	4	d	8	0	6	5	7	e	1	f	3	b	2	a	c
b	7	8	5	e	2	a	3	4	c	6	0	d	f	b	1	9
c	5	2	b	6	7	9	0	e	a	8	c	f	1	3	4	d
d	a	6	8	4	3	e	c	d	2	9	1	5	0	7	f	b
e	d	1	3	f	b	0	2	8	4	a	7	c	9	5	6	e
f	8	d	7	b	5	c	a	2	9	4	e	1	3	6	0	f

For new burst algorithms, we made experiments for different values of k (number of columns in the interleaver matrix), i.e., for all divisors of 36. Namely, the number of nibbles of two (or four) concatenated codewords in Cut-Decoding/Burst-Cut-Decoding (or 4-Sets-Cut-Decoding/Burst-4-Sets-Cut-Decoding) algorithms is 36. Best results are obtained for $k = 9$. Further on, we will present only results for this value of k.

In Subsect. 5.1. we present experimental results for Gilbert-Elliott model with Binary Symmetric Channels for different values of probability of bit error and transition probabilities. The experimental results for different values of SNR and transition probabilities in Gilbert-Elliott model with Gaussian channels, are given in Subsect. 5.2.

5.1   Experiments for Gilbert-Elliott with BSC Channels

In all experiments for Gilbert-Elliott model with Binary Symmetric Channels we use the value of bit-error probabilities in the good state $P_e(G) = 0.01$ and some different values of bit error probabilities in the bad state $P_e(B) \in \{0.2, 0.16, 0.13, 0.1\}$.

In Table 2, we give experimental results for bit-error probabilities BER_{cut} and packet-error probabilities PER_{cut} (obtained with Cut-Decoding algorithm) and the corresponding probabilities BER_{b-cut} and PER_{b-cut} (obtained with Burst-Cut-Decoding algorithm) for code with rate 1/4, and following combinations of transition probabilities from good to good state P_{GG} and from bad to bad state P_{BB}:

- $P_{GG} = 0.8$ and $P_{BB} = 0.8$
- $P_{GG} = 0.5$ and $P_{BB} = 0.5$
- $P_{GG} = 0.2$ and $P_{BB} = 0.8$
- $P_{GG} = 0.8$ and $P_{BB} = 0.2$

Table 2. Experimental results with $R = 1/4$

$P_e(B)$	PER_{cut}	PER_{b-cut}	BER_{cut}	BER_{b-cut}
$P_{GG} = 0.8$		$P_{BB} = 0.8$		
0.1	0.14069	0.06818	0.10453	0.05030
0.13	0.34014	0.18173	0.26055	0.13535
0.16	0.58078	0.32466	0.45424	0.24698
0.2	0.81271	0.51087	0.67081	0.40820
$P_{GG} = 0.5$		$P_{BB} = 0.5$		
0.1	0.13464	0.04665	0.09799	0.03376
0.13	0.32711	0.12283	0.24417	0.08970
0.16	0.56365	0.23394	0.43088	0.17073
0.2	0.82358	0.43581	0.65848	0.33149
$P_{GG} = 0.2$		$P_{BB} = 0.8$		
0.1	0.20470	0.14177	0.14853	0.10266
0.13	0.46781	0.34619	0.35206	0.25621
0.16	0.73048	0.59497	0.57155	0.45788
0.2	0.942468	0.84951	0.79485	0.68579
$P_{GG} = 0.8$		$P_{BB} = 0.2$		
0.1	0.05709	0.00921	0.04201	0.00560
0.13	0.14292	0.01555	0.10386	0.01066
0.16	0.27728	0.03333	0.20678	0.02316
0.2	0.48898	0.07250	0.37152	0.05260

Analyzing the results in Table 2, we can conclude that for all values of $P_e(B)$ and for all values of P_{GG} and P_{BB} the results for BER_{b-cut} are better than the corresponding results of BER_{cut}. Namely,

- for $P_{GG} = 0.8$ and $P_{BB} = 0.8$, BER_{b-cut} is from 1.6 to 2.5 times better than BER_{cut};
- for $P_{GG} = 0.5$ and $P_{BB} = 0.5$, BER_{b-cut} is from 1.9 to 2.8 times better than BER_{cut};
- for $P_{GG} = 0.2$ and $P_{BB} = 0.8$, BER_{b-cut} is about 1.2 times better than BER_{cut};
- for $P_{GG} = 0.8$ and $P_{BB} = 0.2$, BER_{b-cut} is from 7.5 to 10 times better than BER_{cut}.

The same conclusions can be derived for comparison of PER_{b-cut} and PER_{cut}.

In Table 3, we give experimental results for code with rate 1/8, with same combinations of transition probabilities P_{GG} and P_{BB} as for the code with rate 1/4. There, BER_{cut}, BER_{b-cut}, BER_{4sets} and $BER_{b-4sets}$ are bit-error probabilities obtained with Cut-Decoding, Burst-Cut-Decoding, 4-Sets-Cut-Decoding and Burst-4-Sets-Cut-Decoding algorithm, correspondingly. Also, PER_{cut}, PER_{b-cut}, PER_{4sets} and $PER_{b-4sets}$ are corresponding packet-error probabilities.

Table 3. Experimental results with $R = 1/8$

$P_e(B)$	BER_{cut}	BER_{b-cut}	BER_{4sets}	$BER_{b-4sets}$	PER_{cut}	PER_{b-cut}	PER_{4sets}	$PER_{b-4sets}$
	$P_{GG} = 0.8$	$P_{BB} = 0.8$						
0.1	0.09183	0.04224	0.01796	0.00601	0.15790	0.07445	0.03578	0.01252
0.13	0.23069	0.11561	0.08060	0.02616	0.37334	0.19549	0.16957	0.05429
0.16	0.41820	0.22276	0.20798	0.07405	0.62600	0.35476	0.41101	0.15207
0.2	0.64522	0.39519	0.45947	0.20217	0.85671	0.56624	0.75302	0.34288
	$P_{GG} = 0.5$	$P_{BB} = 0.5$						
0.1	0.08336	0.02759	0.01619	0.00412	0.14804	0.04989	0.03333	0.00784
0.13	0.21774	0.07662	0.07504	0.01353	0.36182	0.13364	0.16330	0.02887
0.16	0.20663	0.15975	0.20663	0.04268	0.41748	0.27016	0.41733	0.09245
0.2	0.63793	0.31083	0.47000	0.13001	0.87975	0.49294	0.79255	0.27095
	$P_{GG} = 0.2$	$P_{BB} = 0.8$						
0.1	0.13132	0.09052	0.03269	0.01683	0.22753	0.16028	0.07099	0.03729
0.13	0.32319	0.23406	0.13834	0.08086	0.52584	0.39026	0.29586	0.17713
0.16	0.54950	0.43033	0.36372	0.22914	0.80105	0.66229	0.66993	0.45636
0.2	0.77847	0.67042	0.66579	0.50899	0.97227	0.89840	0.95542	0.82632
	$P_{GG} = 0.8$	$P_{BB} = 0.2$						
0.1	0.03391	0.00320	0.00576	0.00044	0.05889	0.00583	0.01094	0.00079
0.13	0.0926	0.00866	0.02041	0.00124	0.16503	0.01684	0.04133	0.00230
0.16	0.18494	0.02014	0.06019	0.00256	0.31617	0.03578	0.12802	0.00518
0.2	0.34766	0.04408	0.16918	0.00803	0.55465	0.07913	0.34569	0.01569

From the results in Table 3, we can conclude that for all values of $P_e(B)$ and for all values of P_{GG} and P_{BB}, the results for BER_{b-cut} are better than the corresponding results of BER_{cut} and the results for $BER_{b-4sets}$ are better than the corresponding results of BER_{4sets}. Also, if we compare the results for burst algorithms, we can conclude that Burst-4-Sets-Cut-Decoding algorithm gives from 2 to 7 times better results than Burst-Cut-Decoding algorithm depending of

the channel parameters. The same conclusions can be derived for corresponding
packet-error probabilities.

5.2   Experiments for Gilbert-Elliott with Gaussian Channels

In this subsection, we presented the experimental results for Gilbert-Elliott
model with Gaussian channels with $SNR_G = 4$ and for different values of
$SNR_B \in \{-3, -2, -1\}$ and the same transition probabilities as in the experiments with binary symmetric channels. First, in Table 4 we give experimental
results for code with rate $1/4$, where we use the same notations as previously.

Table 4. Experimental results with $R = 1/4$

SNR_B	PER_{cut}	PER_{b-cut}	BER_{cut}	BER_{b-cut}
	$P_{GG} = 0.8$	$P_{BB} = 0.8$		
-3	0.56322	0.32366	0.44058	0.24605
-2	0.34886	0.17727	0.26411	0.13127
-1	0.16467	0.08179	0.12212	0.05911
	$P_{GG} = 0.5$	$P_{BB} = 0.5$		
-3	0.55235	0.23135	0.41796	0.16945
-2	0.32754	0.12348	0.24303	0.08867
-1	0.15243	0.05609	0.10993	0.03993
	$P_{GG} = 0.2$	$P_{BB} = 0.8$		
-3	0.73127	0.57783	0.57146	0.44170
-2	0.46867	0.35419	0.35223	0.26129
-1	0.22775	0.16561	0.16732	0.12115
	$P_{GG} = 0.8$	$P_{BB} = 0.2$		
-3	0.27282	0.03513	0.20121	0.02449
-2	0.15056	0.01980	0.10891	0.01393
-1	0.06732	0.00986	0.04852	0.00589

From Table 4, we can see that the results obtained with the new Burst-Cut-
Decoding algorithm are from 2 to 8 times better than the corresponding results
obtained with the old Cut-Decoding algorithm.

In Table 5, we give experimental results for bit-error probabilities and packet-
error probabilities for codes with rate $1/8$. The notations are previously given.
From this table, we can make similar conclusions for rate $1/8$ as for rate $1/4$.
Namely, we can conclude that for all values of SNR, results for BER and
PER obtained with the new algorithms are better than the corresponding
results obtained with the old versions of the algorithms. Also, Burst-4-Sets-
Cut-Decoding algorithm gives from 2 to 8 times better results than Burst-Cut-
Decoding algorithm.

Table 5. Experimental results with $R = 1/8$

SNR_B	BER_{cut}	BER_{b-cut}	BER_{4sets}	$BER_{b-4sets}$	PER_{cut}	PER_{b-cut}	PER_{4sets}	$PER_{b-4sets}$
$P_{GG} = 0.8,$	$P_{BB} = 0.8$							
-3	0.40468	0.21957	0.202399	0.07334	0.61031	0.34835	0.40048	0.15092
-2	0.23806	0.11587	0.08246	0.02539	0.38256	0.19693	0.17461	0.05457
-1	0.10796	0.04945	0.02266	0.00871	0.17821	0.08640	0.04860	0.01771
$P_{GG} = 0.5$	$P_{BB} = 0.5$							
-3	0.38576	0.15412	0.20372	0.04329	0.60865	0.26116	0.40710	0.09454
-2	0.21698	0.07825	0.07806	0.01490	0.37028	0.13637	0.16748	0.03204
-1	0.09497	0.03024	0.02002	0.00385	0.16402	0.05501	0.04169	0.00835
$P_{GG} = 0.2$	$P_{BB} = 0.8$							
-3	0.53986	0.42327	0.35535	0.22462	0.78650	0.65300	0.65797	0.44758
-2	0.33082	0.24201	0.15007	0.08345	0.53269	0.40408	0.31820	0.18130
-1	0.15119	0.10676	0.03686	0.02265	0.26101	0.18613	0.08208	0.04910
$P_{GG} = 0.8$	$P_{BB} = 0.2$							
-3	0.17904	0.02085	0.05540	0.00328	0.30645	0.03809	0.11880	0.00604
-2	0.09442	0.01037	0.02074	0.00134	0.16345	0.01886	0.04428	0.00252
-1	0.04009	0.00503	0.00703	0.00060	0.07063	0.00907	0.01404	0.00115

6 Conclusion

In this paper we define two new algorithms called Burst Cut-Decoding and Burst-4-Sets-Cut-Decoding algorithm for improving the performances for transmission through a burst channel. In the new algorithms, we include an interleaver in coding algorithm and the corresponding deinterleaver in the decoding algorithm. In this way, we obtain better results for packet-errors and bit-error probabilities than with the old Cut-Decoding and 4-Sets-Cut-Decoding algorithm. As further work, we will investigate performances of proposed algorithms for other fading channels.

Acknowledgment. This research was partially supported by Faculty of Computer Science and Engineering at "Ss Cyril and Methodius" University in Skopje.

References

1. Dimitrova, V., Markovski, J.: On quasigroup pseudo random sequence generators. In: Proceedings of 1st Balkan Conference in Informatics, Thessaloniki, Greece, pp. 393–401 (2003)
2. Gligoroski, D., Markovski, S., Kocarev, L.: Error-correcting codes based on quasigroups. In: Proceedings of 16th International Conference on Computer Communications and Networks, pp. 165–172 (2007)
3. Gligoroski, D., Markovski, S., Kocarev, L.: Totally asynchronous stream ciphers + redundancy = cryptocoding. In: Aissi, S., Arabnia, H.R. (eds.) Proceedings of the International Conference on Security and Management, SAM 2007, pp. 446–451. CSREA Press, Las Vegas (2007)

4. Knag, J., Stark, W., Hero, A.: Turbo codes for fading and burst channels. In: IEEE Theory Mini Conference, pp. 40–45 (1998)

5. Labiod, H.: Performance of Reed Solomon error-correcting codes on fading channels. In: IEEE International Conference on Personal Wireless Communications (Cat. No. 99TH8366), Jaipur, India, pp. 259–263 (1999)

6. Markovski, S., Gligoroski, D., Bakeva, V.: Quasigroup string processing: part 1. Contrib. Sect. Nat. Math. Biotech. Sci. **20**(1–2), 13–28 (1999)

7. Markovski, S., Gligoroski, D., Kocarev, L.: Unbiased random sequences from quasigroup string transformations. In: Gilbert, H., Handschuh, H. (eds.) FSE 2005. LNCS, vol. 3557, pp. 163–180. Springer, Heidelberg (2005). https://doi.org/10.1007/11502760_11

8. Mathur, C.N., Narayan, K., Subbalakshmi, K.P.: High diffusion cipher: encryption and error correction in a single cryptographic primitive. In: Zhou, J., Yung, M., Bao, F. (eds.) ACNS 2006. LNCS, vol. 3989, pp. 309–324. Springer, Heidelberg (2006). https://doi.org/10.1007/11767480_21

9. Popovska-Mitrovikj, A., Markovski, S., Bakeva, V.: Increasing the decoding speed of random codes based on quasigroups. In: Markovski, S., Gusev, M. (eds.) ICT Innovations 2012, Web Proceedings, ISSN 1857–7288, pp. 93–102 (2012)

10. Popovska-Mitrovikj, A., Markovski, S., Bakeva, V.: 4-Sets-cut-decoding algorithms for random codes based on quasigroups. Int. J. Electron. Commun. (AEU) **69**(10), 1417–1428 (2015)

11. Hwang, T., Rao, T.R.N.: Secret error-correcting codes (SECC). In: Goldwasser, S. (ed.) CRYPTO 1988. LNCS, vol. 403, pp. 540–563. Springer, New York (1990). https://doi.org/10.1007/0-387-34799-2_39

12. Zivic, N., Ruland, C.: Parallel joint channel coding and cryptography. Int. J. Electr. Electron. Eng. **4**(2), 140–144 (2010)

Zeroing Neural Network Based on the Equation $AXA = A$

Marko D. Petković$^{(\boxtimes)}$ and Predrag S. Stanimirović

Faculty of Sciences and Mathematics, University of Niš,
Višegradska 33, 18000 Niš, Serbia
dexterofnis@gmail.com, pecko@pmf.ni.ac.rs

Abstract. Zeroing Neural Network (ZNN) design arising from different error monitoring functions (or Zeroing functions) defined on the basis of Penrose matrix equations are considered. New Zeroing function based on the Penrose equation $AXA = A$ and initiated ZNN design for computing the time-varying pseudoinverse are defined and investigated. Also, an explicit form of defined model is proposed. Illustrative simulation results are given to verify theoretical results.

Keywords: Zeroing neural network · Moore-Penrose inverse · Dynamic equation · Convergence

AMS subject classifications: 68T05 · 15A09

1 Introduction

For any matrix $A \in \mathbb{C}^{n \times n}$, its range and the null space are denoted by $\mathcal{R}(A)$ and $\mathcal{N}(A)$, respectively.

The initial point in the investigation and computation of generalized inverses of $A \in \mathbb{C}_r^{m \times n}$ are Penrose equations with respect to unknown matrix X:

$$AXA = A \tag{1}$$

$$XAX = X \tag{2}$$

$$(AX)^* = AX \tag{3}$$

$$(XA)^* = XA. \tag{4}$$

The left inverse A_L^{-1} of $A \in \mathbb{C}_m^{m \times n}$ satisfies $A_L^{-1} A = I$, while the right inverse A_R^{-1} of $A \in \mathbb{C}_n^{m \times n}$ satisfies $AA_R^{-1} = I$.

Recently, a number of nonlinear and linear recurrent neural network (RNN) models have been developed for the purpose of numerical evaluation of the matrix inverse and generalized inverses. RNN models dedicated to find zeros of equations

The authors gratefully acknowledge support from the Research Project 174013 of the Ministry of Education, Science and Technological Development, Republic of Serbia.

© Springer Nature Switzerland AG 2019
M. Ćirić et al. (Eds.): CAI 2019, LNCS 11545, pp. 213–224, 2019.
https://doi.org/10.1007/978-3-030-21363-3_18

or to minimize nonlinear functions are frequently used in computing generalized inverses [1,3,5,6]. These models represent optimization networks. Optimization RNN models can be divided in two classes: *Gradient Neural Networks (GNN)* and *Zeriong (or Zhang) Neural Networks* (ZNN).

The GNN models are aimed to solving time-invariant problems, while the ZNN models are able to solve time-varying problems [14]. The first step in defining both the GNN and ZNN design is to define an appropriate error function $E(t)$ on the basis of the matrix equation which is currently being solved. The error matrix $E(t)$ in a GNN design is defined by replacing the unknown matrix from the considered problem by the time-varying matrix $V(t)$ which will be approximated during the time $t \geq 0$. The error function $E(t)$ in ZNN models for solving matrix algebra problems represents a complex matrix-valued error-monitoring function, called the Zeroing (or Zhang) function (ZF).

The goal function of a GNN dynamics is the scalar function which is defined by the Frobenius norm of $E(t)$:

$$\varepsilon(t) = \frac{\|E(t)\|_F^2}{2}, \quad \|E\|_F = \sqrt{\mathrm{Tr}(E^{\mathrm{T}} E)}.$$

The GNN dynamic evolution is based on direct proportionality between the time derivative $\dot{V}(t)$ and the negative gradient of the goal function $\varepsilon(t)$:

$$\dot{V}(t) = \frac{\mathrm{d}V(t)}{\mathrm{d}t} = -\gamma \mathcal{F}\left(\frac{\partial \varepsilon(t)}{\partial V}\right), \quad V(0) = V_0. \tag{1.1}$$

Here, $V(t)$ is the matrix of activation state variables, $t \in [0, +\infty)$ is the time and $\mathcal{F}(\cdot) : \mathbb{R}^{m \times n} \to \mathbb{R}^{m \times n}$ denotes an odd and monotonically increasing activation function. Larger values of the scaling parameter γ enable faster convergence.

The linear ZNN design assumes application of defined ZF by the dynamical implicit evolution of the form

$$\dot{E}(t) = \frac{\mathrm{d}E(t)}{\mathrm{d}t} = -\gamma \mathcal{F}(E(t)), \tag{1.2}$$

where $\dot{E}(t)$ is the time derivative of $E(t)$ and $\gamma \in \mathbb{R}$ is a positive scalar used to scale the convergence rate. The design parameter γ should be as large as the hardware permits [1].

A comparison of zeroing neural network and gradient neural network evolution was considered in [11].

GNN model for finding the constant matrix inversion was investigated in [4,5,9]. Various GNN dynamical systems for computing generalized inverses of rank-deficient matrices were designed in [6].

ZNN models for solving online solution to complex-valued time-varying matrix inversion problem were considered in [7,10,12]. Different ZNN models for computing the Moore-Penrose inverse of online time-varying full-rank matrix were generalized, investigated and analyzed in [8]. Liao and Yhang in [1] proposed five different complex ZFs and, accordingly developed and investigated five complex ZNN models for computing the time-varying complex matrix pseudoinverse.

Our goal in the present paper is to investigate GNN and ZNN models arising from different error monitoring functions $E(t)$ defined on the basis of the matrix Eqs. (1)–(4). New ZF based on the Penrose equation (1) and initiated ZNN design for computing the time-varying pseudoinverse are defined. Convergence of defined dynamical system is considered and illustrative simulation results are presented.

The remainder of the manuscript is organized as follows. Preliminaries, subject, underlying motivation and related results are presented in Sect. 2. A new explicit ZNN model for the pseudoinverse computation is defined in Sect. 3. Convergence analysis of defined ZNN design in a constant matrix case is considered in Sect. 4. Section 5 exposes simulation numerical examples.

2 Preliminaries, Subject, Motivation and Related Results

Zhang and Guo in [13] proposed five complex fundamental error-monitoring functions for computing the time-varying Moore-Penrose inverse:

$$
\begin{aligned}
E_1(t) &= A(t)V(t) - I, \\
E_2(t) &= V(t)A(t) - I, \\
E_3(t) &= V(t)A(t)A(t)^* - A^*(t), \\
E_4(t) &= A(t)^*A(t)V(t) - A^*(t), \\
E_5(t) &= A(t) - V(t)^\dagger.
\end{aligned}
\tag{2.1}
$$

It is important to mention that the error functions $E_1(t)$ and $E_2(t)$ are usable for computing the inverse of a nonsingular matrix. A ZNN design arising from $E_1(t)$ and $E_2(t)$ and appropriate for computing the Moore-Penrose inverse was presented in [13]. This extension is based on the Tikhonov regularization principle. Also, the error functions $E_3(t)$ and $E_4(t)$ are usable for computing the left and right inverse. These models can be used in computing the Moore-Penrose inverse using the Tikhonov regularization, which leads to the following error functions:

$$
\begin{aligned}
E_3(t) &= V(t)(A(t)A(t)^* + \lambda I) - A(t)^*, \\
E_4(t) &= (A(t)^*A(t) + \lambda I)V(t) - A(t)^*,
\end{aligned}
\tag{2.2}
$$

where $\lambda > 0$ is a small real parameter.

Our intention is to consider ZF and initiated ZNN design based on the Penrose equation (1):

$$
E(t) = A(t) - A(t)V(t)A(t).
\tag{2.3}
$$

Let us mention that the GNN design aimed to the computation of the Moore-Penrose inverse and arising on the error matrix (2.3) was investigated in [3]. In the present article we investigate the ZNN model based on the matrix equation $AXA = A$, i.e., on the basis of the ZF (2.3).

The main contributions of the paper are emphasized as follows.

(i) New ZF of the form (2.3) is proposed staring from the Penrose equation (1). Accordingly, new ZD for computing the time-varying complex matrix Moore-Penrose inverse is proposed.
(ii) Defined ZNN model is explicit.
(iii) Illustrative simulation results are presented during the computation of the Moore-Penrose inverse.

3   ZNN Model for the Pseudoinverse Computation

Assume that $A(t) \in \mathbb{R}^{m \times n}$ is the time-varying matrix where $t \in [0, +\infty)$ and the entries of $A(t)$ are differentiable functions. The time derivative of $A(t)$ is denoted by $\dot{A}(t)$. The goal is to construct the zeroing neural network for computing the Moore-Penrose inverse $A(t)^{\dagger}$.

In order to do that, we need to assume that the range $\mathcal{R}(A(t))$ and the null space $\mathcal{N}(A(t))$ do not depend on time t, i.e., that there exists subspaces T and S such that $\mathcal{R}(A(t)) = T$ and $N(A(t)) = S$ for all $t \geq 0$. Denote by $T^{\perp}$ and $S^{\perp}$ the orthogonal complements of T and S respectively. Then $\mathcal{R}(A(t)^{\dagger}) = S^{\perp}$ and $\mathcal{N}(A(t)^{\dagger}) = T^{\perp}$. The following proposition gives the property of the range and null space of the time derivative matrix $\dot{A}(t)$.

Proposition 1 [2, Proposition 5.1]. *Let $A(t) \in \mathbb{R}^{m \times n}$ has differentiable entries, its range and null space $\mathcal{R}(A(t)) = T$ and $\mathcal{N}(A(t)) = S$ are constant for every $t > 0$. The range and the null space of the time derivative $\dot{A}(t)$ satisfy $\mathcal{R}(\dot{A}(t)) \subset \mathcal{R}(A(t)) = T^{\perp}$ and $\mathcal{N}(\dot{A}(t)) \supset \mathcal{N}(A) = S^{\perp}$.*

Here the idea is to define the ZNN evolution design based on the underlying error function (2.3) arising from the Penrose equation (1). It is known that, if $\mathcal{R}(V(t)) \subseteq S^{\perp}$ and $\mathcal{N}(V(t)) \supseteq T^{\perp}$ then $V(t) = A(t)^{\dagger}$ is the unique solution of the first Penrose equation (1), i.e. of the Eq. (2.3). As a consequence, in the rest of this section, it will be assumed that the conditions $\mathcal{R}(V(t)) \subseteq S^{\perp}$ and $\mathcal{N}(V(t)) \supseteq T^{\perp}$ are valid. Note that the matrix $E(t)$ also satisfies similar conditions, i.e., $\mathcal{R}(E(t)) \subseteq T$ and $\mathcal{N}(E(t)) \supseteq S$.

In order to preserve range and the null space conditions for the matrix $E(t)$, we will further investigate the linear ZNN design

$$\dot{E}(t) = -\gamma E(t). \tag{3.1}$$

The last step of defining the ZNN evolution requires the expansion of the design formula (3.1). The time derivative $\dot{E}(t)$ is obtained directly from the defining Eq. (2.3):

$$\dot{E}(t) = \dot{A}(t) - \dot{A}(t)V(t)A(t) - A(t)\dot{V}(t)A(t) - A(t)V(t)\dot{A}(t). \tag{3.2}$$

Now, combining (2.3), (3.1) and (3.2), the following implicit dynamic evolution equation of the ZNN model can be defined:

$$\begin{aligned}\dot{A}(t) - \dot{A}(t)V(t)A(t) &- A(t)\dot{V}(t)A(t) - A(t)V(t)\dot{A}(t) \\ &= -\gamma\left(A(t) - A(t)V(t)A(t)\right).\end{aligned} \tag{3.3}$$

In order to obtain the explicit dynamical system, one needs to isolate the term with $\dot{V}(t)$ on the left side of the equation, i.e.:

$$A(t)\dot{V}(t)A(t) = \\ \dot{A}(t) - \dot{A}(t)V(t)A(t) - A(t)V(t)\dot{A}(t) + \gamma\left(A(t) - A(t)V(t)A(t)\right). \tag{3.4}$$

Since the range and null space assumptions on $A(t)$ and $V(t)$ imply $V(t)A(t)A(t)^{\dagger} = V(t) = A(t)^{\dagger}A(t)V(t)$, multiplying the previous equation by $A(t)^{\dagger}$ both from left and right yields to the following explicit dynamic evolution equation:

$$\dot{V}(t) = A(t)^{\dagger}\dot{A}(t)A(t)^{\dagger} - A(t)^{\dagger}\dot{A}(t)V(t) - V(t)\dot{A}(t)A(t)^{\dagger} \\ + \gamma A(t)^{\dagger}\left(A(t) - A(t)V(t)A(t)\right)A(t)^{\dagger}. \tag{3.5}$$

The problem with (3.5) is the presence of the Moore-Penrose inverse $A(t)^{\dagger}$ on the right-hand side. To solve that problem, one needs to replace all occurrences of $A(t)^{\dagger}$ by $V(t)$ on the right-hand side of (3.5). In such a way, we finally obtain the following explicit dynamical system for computing the Moore-Penrose inverse $A(t)^{\dagger}$ of the time-varying matrix $A(t)$:

$$\dot{V}(t) = -V(t)\dot{A}(t)V(t) + \gamma V(t)\left(A(t) - A(t)V(t)A(t)\right)V(t). \tag{3.6}$$

The first term in (3.6) vanishes if the matrix $A(t)$ is constant. The model (3.6) will be further denoted by **ZNN-EQ1**.

Due to the replacements performed in the last step of the construction process, the obtained model Eq. (3.6) and the initial (3.1) are no longer equivalent. Hence, one needs to prove its convergence. It is done in the following section.

4   Convergence Analysis in the Constant Matrix Case

Assume that $A(t)$ is constant. Therefore, in the rest of this section it will be simply denoted by A. The model **ZNN-EQ1** now reduces to

$$\dot{V}(t) = \gamma V(t)\left(A - AV(t)A\right)V(t). \tag{4.1}$$

Denote by $\rho(M)$ the spectral radius of the matrix M, by $\lambda_i(M)$ the i-th eigenvalue of the matrix M, and by O the zero matrix of the appropriate size. Also denote by $\mathcal{R}(A) = T$ and $\mathcal{N}(A) = S$ the range and the null space of A respectively.

The following is the main theorem which proves the convergence of the model **ZNN-EQ1** for the constant matrix A, given by (4.1).

Theorem 1. *Assume that the initial matrix $V(0)$ satisfies $\mathcal{R}(V(0)) = S^{\perp}$, $\mathcal{N}(V(0)) = T^{\perp}$, $\rho(AA^{\dagger} - AV(0)) < 1$ and $(AV(0))^{\mathrm{T}} = AV(0)$. Then $V(t)$ defined by (4.1) converges to the Moore-Penrose inverse $A^{\dagger}$ when $t \to +\infty$.*

Before we proceed to the proof of Theorem 1, we show the following two auxiliary lemmas.

Lemma 1. *Under the assumptions of Theorem 1, the activation state variables matrix $V(t)$ satisfies $V(t) = V(t)AA^\dagger$ and $A^\dagger AV(t) = V(t)$ for every $t \geq 0$.*

Proof. Denote by $\tilde{V}(t) = V(t)AA^\dagger$. Using the equality (4.1) and $AA^\dagger A = A$ we get

$$\dot{\tilde{V}}(t) = \dot{V}(t)AA^\dagger = \gamma V(t)(A - AV(t)AA^\dagger A)V(t)AA^\dagger$$
$$= \gamma V(t)(A - A\tilde{V}(t)A)\tilde{V}(t)$$
$$= \gamma V(t)(AA^\dagger A - AA^\dagger A\tilde{V}(t)A)\tilde{V}(t)$$
$$= \gamma V(t)AA^\dagger(A - A\tilde{V}(t)A)\tilde{V}(t)$$
$$= \gamma \tilde{V}(t)(A - A\tilde{V}(t)A)\tilde{V}(t).$$

In other words, matrix $\tilde{V}(t) = V(t)AA^\dagger$ also satisfies (4.1). The assumptions on $V(0)$ imply $\tilde{V}(0) = V(0)AA^\dagger = V(0)$. Now the uniqueness of the solution of (4.1) under prescribed initial value, we can conclude that $\tilde{V}(t) = V(t)$. The second equation $AA^\dagger V(t) = V(t)$ can be proved analogously.   □

Lemma 2. *Under the assumptions of Theorem 1, the matrix $AV(t)$ is symmetric, i.e. $(AV(t))^\mathrm{T} = AV(t)$ for every $t \geq 0$.*

Proof. Denote $W(t) = AV(t)$. Equation (4.1) directly implies:

$$\dot{W}(t) = A\dot{V}(t) = \gamma(AV(t))^2(I - AV(t)) = \gamma\left(W(t)^2 - W(t)^3\right)$$

It is evident that $W(t)^\mathrm{T}$ also satisfies the previous equation. Now using $W(0) = W(0)^\mathrm{T}$ (by assumption) and the same argument as in Lemma 1, one can conclude that $W(t) = W(t)^\mathrm{T}$ for all $t \geq 0$.   □

Now we are ready to prove Theorem 1.

Proof (Theorem 1). Consider the residual matrix $E_p(t) = AA^\dagger - AV(t)$. Lemma 1 implies $E_p(t)AA^\dagger = E_p(t)$, while $AA^\dagger E_p(t) = E_p(t)$ follows immediately. Now

$$E_p(t) = -A\dot{V}(t) = -\gamma AV(t)(A - AV(t)A)V(t) = -\gamma(AV(t))^2(I - AV(t))$$
$$= -\gamma(AA^\dagger - E_p(t))^2(I - AA^\dagger + E_p(t))$$

Furthermore, since

$$(AA^\dagger - E_p(t))(I - AA^\dagger + E_p(t)) = AA^\dagger E_p(t) - E_p(t)^2 = (AA^\dagger - E_p(t))E_p(t)$$

we obtain

$$\dot{E}_p(t) = -\gamma(AA^\dagger - E_p(t))^2 E_p(t). \tag{4.2}$$

Let $\epsilon_1(t) = \|E_p(t)\|_F^2/2 = \mathrm{Tr}(E_p(t)^\mathrm{T}E_p(t))/2$ where $\|\cdot\|_F$ is the Frobenius norm and $\mathrm{Tr}(\cdot)$ is the matrix trace. Its time derivative is given by

$$\dot{\epsilon}_1(t) = \mathrm{Tr}(\dot{E}_p(t)E_p(t)^\mathrm{T}) = -\gamma\mathrm{Tr}\left((AA^\dagger - E_p(t))^2 E_p(t)E_p(t)^\mathrm{T}\right)$$
$$= -\gamma\mathrm{Tr}\left((AA^\dagger - E_p(t))^\mathrm{T}(AA^\dagger - E_p(t))E_p(t)E_p(t)^\mathrm{T}\right)$$
$$= -\gamma\mathrm{Tr}\left(((AA^\dagger - E_p(t))E_p(t))^\mathrm{T}(AA^\dagger - E_p(t))E_p(t)\right).$$

We used the fact that the matrix $AA^\dagger - E_p(t) = AV(t)$ is symmetric (Lemma 2). Since $\epsilon_1(t) \leq 0$ for every $t \geq 0$, the Lyapunov stability theory implies that either $E_p(t) \to O$ or $E_p(t) \to AA^\dagger$ when $t \to +\infty$.

Suppose that $E_p(t) \to AA^\dagger$ when $t \to +\infty$. Then $\epsilon_1(t) < 0$ implies further $\|E_p(t)\|_F \leq \|E_p(0)\|_F$ and therefore $\sqrt{r} = \|A^\dagger A\|_F \leq \|E_p(0)\|_F$, wherein $r = \mathrm{rank}(A) = \mathrm{rank}(E_p(0))$. But, on the other hand, $\rho(E_p(0)) < 1$ implies

$$\|E_p(0)\|_F = \sqrt{\sum_{i=1}^{r} \lambda_i(E_p(0))^2} < \sqrt{r}$$

which is the contradiction.

Therefore, $E_p(t) \to O$ and also (using once again Lemma 1):

$$\|A^\dagger - V(t)\|_F = \|A^\dagger AA^\dagger - A^\dagger AV(t)\|_F \leq \|A^\dagger\|_F \|E_p(t)\|_F \to 0$$

when $t \to +\infty$. This completes the proof of Theorem 1.   □

5   Numerical Examples

We used Matlab Simulink implementation of the model **ZNN-EQ1** for testing purposes. The simulink is presented in Fig. 1. The solver `ode23s` was used in all simulations. One suitable choice for the initial matrix $V(0)$ satisfying the conditions of Theorem 1 is

$$V(0) = \alpha A^{\mathrm{T}}, \qquad \alpha = 2/\|A\|_F^2.$$

That same choice will be used in all subsequent numerical examples including once with the time-varying matrix $A(t)$.

Example 1. Consider the constant matrix

$$A = \begin{bmatrix} -200 & -100 & -200 \\ 110 & 10 & -10 \\ -204 & -84 & -156 \\ -234 & -90 & -162 \end{bmatrix}$$

having $\mathrm{rank}(A) = 2$, while its Moore-Penrose inverse is given by

$$A^\dagger = \begin{bmatrix} \frac{23275}{10011978} & \frac{275965}{40047912} & -\frac{17953}{20023956} & -\frac{29125}{13349304} \\ -\frac{13225}{10011978} & -\frac{38075}{20023956} & \frac{2965}{10011978} & \frac{209}{6674652} \\ -\frac{43025}{10011978} & -\frac{295055}{40047912} & \frac{9829}{20023956} & \frac{10823}{13349304} \end{bmatrix}.$$

We apply the **ZNN-EQ1** model to compute the Moore-Penrose inverse of the matrix A. Figure 2 shows the trajectories of each element of the state matrix $V(t)$ for $\gamma = 10^4$ inside the total simulation time $3 \cdot 10^{-3}$. Initial value is given by $V(0) = 2/\|A\|_F^2 \cdot A^{\mathrm{T}}$.

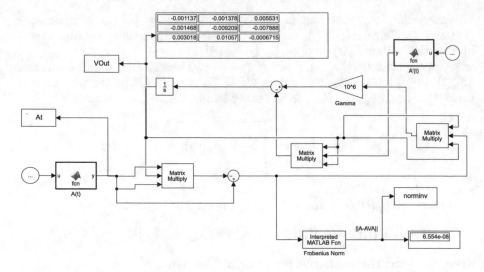

Fig. 1. Simulink implementation of the **ZNN-EQ1** model (3.6).

The evolution of the residual norm $\|AV(t)A - A\|_F$ for different values of the parameter γ is shown in Fig. 3. It can be seen that larger values of γ significantly decrease the required evolution time to obtain the Moore-Penrose inverse with desired precision.

Example 2. Consider the time-varying matrix $A(t)$ given by

$$A(t) = M_0 + M_1 \cos(t) + M_2 \sin(t)$$

where

$$M_0 = \begin{bmatrix} -15 & 0 & 21 \\ -30 & -18 & 60 \\ 60 & -45 & -39 \end{bmatrix}, \quad M_1 = \begin{bmatrix} 5 & 0 & -7 \\ 20 & -6 & -22 \\ 5 & -15 & 8 \end{bmatrix}, \quad M_2 = \begin{bmatrix} -5 & 0 & 7 \\ -10 & -6 & 20 \\ 20 & -15 & -13 \end{bmatrix}.$$

and $\omega > 0$ is the real parameter. It satisfies

$$\mathcal{R}(A(t)) = T = \text{span}\{(45, 164, 5), (-353, 0, 3177)\},$$
$$\mathcal{N}(A(t)) = S = \text{span}\{(7, 5, 5)\}.$$

We apply the model **ZNN-EQ1** on the matrix $A(t)$ taking $V(0) = 2/\|A(0)\|_F^2 \cdot A(0)^{\mathrm{T}}$ as the initial value and $\gamma = 10$. Figure 4 shows the trajectories of each element of the state matrix $V(t)$, as well as the exact value of the Moore-Penrose inverse $A(t)^\dagger$. It can be seen that the convergence is achieved in the steady-state regime.

The residual norm $\|A(t)V(t)A(t) - A(t)\|_F$ for different values of γ is shown in Fig. 5. It can be seen that the steady state residual norm decreases when γ is increasing. However, we see that for a large value of γ, the residual norm fluctuates heavily, which means that no significant benefit is obtained by increasing γ above the certain threshold.

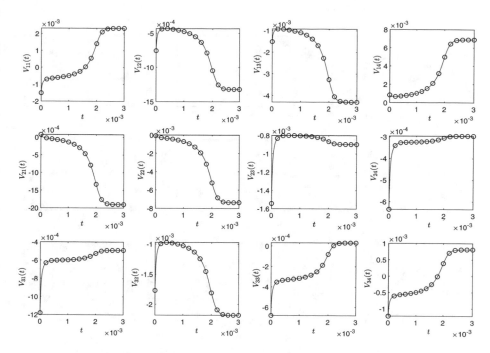

Fig. 2. Elementwise trajectories of the $V(t)$ of **ZNN-EQ1**, for the matrix A and $\gamma - 10^4$ in Example 1.

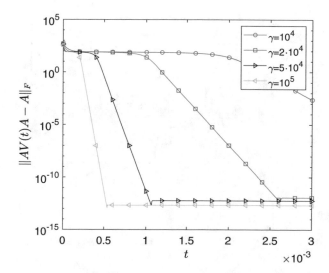

Fig. 3. Residual norm $\|AV(t)A - A\|_F$ as the function of time, for different values of the parameter γ. Model **ZNN-EQ1** is used for the matrix A in Example 1.

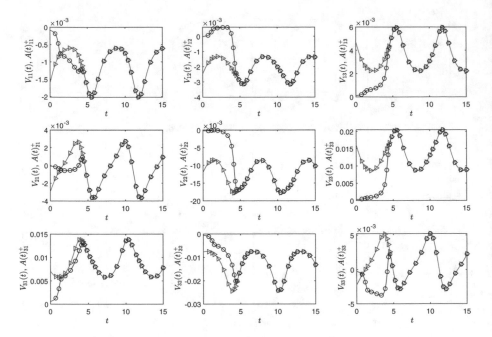

Fig. 4. Elementwise trajectories of the $V(t)$ of **ZNN-EQ1**, for the matrix $A(t)$ and $\gamma = 10$ in Example 2. Black and circled lines correspond to the elements of $V(t)$ while blue and triangled lines correspond to the elements of the exact value $A(t)^{\dagger}$.

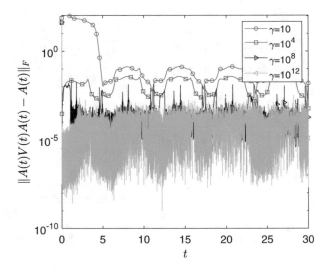

Fig. 5. Residual norm $\|A(t)V(t)A(t) - A(t)\|_F$ as the function of time, for different values of γ. Model **ZNN-EQ1** is used for the matrix $A(t)$ in Example 2.

6 Conclusion

New Zeroing Neural Network (ZNN) dynamical evolution design for approximating the Moore-Penrose inverse is defined and investigated. The proposed dynamical system is unique and original because of the fact that it is defined on the basis of the original error monitoring function arising from the Penrose equation (1). Moreover, after some approximations, an equivalent explicit for of the introduced model is derived. Convergence of defined ZNN design in the constant matrix case is considered. Simulation numerical examples are presented.

As the testing results suggest (on several matrices including one from Example 2), the model **ZNN-EQ1** is also convergent for non-rapidly changing time-varying matrices $A(t)$, under the assumption that $\mathcal{R}(A(t))$ and $\mathcal{N}(A(t))$ are constant, introduced in the Sect. 3. The formulation of the exact convergence conditions for the time-varying case is left for the further research.

References

1. Liao, B., Zhang, Y.: Different complex ZFs leading to different complex ZNN models for time-varying complex generalized inverse matrices. IEEE Trans. Neural Netw. Learn. Syst. **25**, 1621–1631 (2014)
2. Petković, M.D., Stanimirović, P.S., Katsikis, V.N.: Modified discrete iterations for computing the inverse and pseudoinverse of the time-varying matrix. Neurocomputing **289**, 155–165 (2018)
3. Stanimirović, P.S., Petković, M.D., Gerontitis, D.: Gradient neural network with nonlinear activation for computing inner inverses and the Drazin inverse. Neural Process. Lett. **48**, 109–133 (2018)
4. Wang, J.: A recurrent neural network for real-time matrix inversion. Appl. Math. Comput. **55**, 89–100 (1993)
5. Wang, J.: Recurrent neural networks for computing pseudoinverses of rank-deficient matrices. SIAM J. Sci. Comput. **18**, 1479–1493 (1997)
6. Wei, Y.: Recurrent neural networks for computing weighted Moore-Penrose inverse. Appl. Math. Comput. **116**, 279–287 (2000)
7. Zhang, Y., Ge, S.S.: A general recurrent neural network model for time-varying matrix inversion. In: 2003 Proceedings of 42nd IEEE Conference on Decision and Control, pp. 6169–6174, San Diego (2003)
8. Zhang, Y., Yang, Y., Tan, N., Cai, B.: Zhang neural network solving for time-varying full-rank matrix Moore-Penrose inverse. Computing **92**, 97–121 (2011)
9. Zhang, Y., Shi, Y., Chen, K., Wang, C.: Global exponential convergence and stability of gradient-based neural network for online matrix inversion. Appl. Math. Comput. **215**, 1301–1306 (2009)
10. Zhang, Y.: Design and analysis of a general recurrent neural network model for time-varying matrix inversion. IEEE Trans. Neural Netw. **16**(6), 1477–1490 (2005)
11. Zhang, Y., Chen, K.: Comparison on Zhang neural network and gradient neural network for time-varying linear matrix equation $AXB = C$ solving. IEEE International Conference on Industrial Technology (2008). https://doi.org/10.1109/ICIT. 2008.4608579
12. Zhang, Y., Yi, C., Ma, W.: Simulation and verification of Zhang neural network for online time-varying matrix inversion. Simul. Model. Pract. Theory **17**, 1603–1617 (2009)

13. Zhang, Y., Guo, D.: Zhang Functions and Various Models. Springer, Heidelberg (2015)
14. Zhang, Y., Ge, S.S.: Design and analysis of a general recurrent neural network model for time-varying matrix inversion. IEEE Trans. Neural Netw. **16**(6), 1477–1490 (2005)

An Application of Computer Algebra and Dynamical Systems

Predrag S. Stanimirović[1](✉), Yimin Wei[2], Dejan Kolundžija[1],
Juan Rafael Sendra[3], and Juana Sendra[4]

[1] Faculty of Sciences and Mathematics, University of Niš, Višegradska 33,
18000 Niš, Serbia
pecko@pmf.ni.ac.rs, sidejan@gmail.com
[2] School of Mathematical Sciences and Key Laboratory of Mathematics for
Nonlinear Sciences, Fudan University, Shanghai 200433, People's Republic of China
ymwei@fudan.edu.cn
[3] Dpto. de Física y Matemáticas, Universidad de Alcalá, Ap. Correos 20,
28871 Alcalá de Henares, Madrid, Spain
rafael.sendra@uah.es
[4] Dpto. Matemática Aplicada a las TIC, ETSIST, Universidad Politécnica de Madrid,
Madrid, Spain
juana.sendra@upm.es

Abstract. An algorithm for the symbolic computation of outer inverses of matrices is presented. The algorithm is based on the exact solution of the first order system of differential equations which appears in corresponding dynamical system. The domain of the algorithm are matrices whose elements are integers, rational numbers as well as one-variable or multiple-variable rational or polynomial expressions.

Keywords: Outer generalized inverse · Computer algebra ·
Mathematica

1 Introduction

According to the traditional notation, $\mathbb{C}_r^{m \times n}$ (resp. $\mathbb{R}_r^{m \times n}$) denotes the set of all complex (resp. real) $m \times n$ matrices of rank r. The identity matrix of an appropriate order is denoted by I, while O denotes an appropriate zero matrix.

P. S. Stanimirović Gratefully acknowledge support from the Research Project 174013 of the Serbian Ministry of Education, Science and Technological Development.
P. S. Stanimirović and Y. Wei are supported by the bilateral project between China and Serbia, "The theory of tensors, operator matrices and applications" (no. 4–5).
Y. Wei is supported by the National Natural Science Foundation of China under grant 11771099.
J. R. Sendra and J. Sendra are supported by the Spanish Ministerio de Economía y Competitividad, by the European Regional Development Fund (ERDF), under the MTM2017-88796-P.

© Springer Nature Switzerland AG 2019
M. Ćirić et al. (Eds.): CAI 2019, LNCS 11545, pp. 225–236, 2019.
https://doi.org/10.1007/978-3-030-21363-3_19

Furthermore, the notations A^*, $\mathcal{R}(A)$, rank(A), $\mathcal{N}(A)$ and $\sigma(A)$ stand for the conjugate transpose, the range, the rank, the null space and the spectrum of the matrix A, respectively. In addition, $\mathbb{R}[\mathcal{S}]$ (resp. $\mathbb{R}(\mathcal{S})$) denotes the polynomials (resp. rational functions) with real coefficients with respect to the unknown variables $\mathcal{S} = s_1, \ldots, s_k$. The set of $m \times n$ matrices with elements in $\mathbb{R}[\mathcal{S}]$ (resp. $\mathbb{R}(\mathcal{S})$) is denoted by $\mathbb{R}[\mathcal{S}]^{m \times n}$ (resp. $\mathbb{R}(\mathcal{S})^{m \times n}$).

The problem of pseudoinverses computation leads to the, so called, Penrose equations

(1) $AXA = A$ (2) $XAX = X$ (3) $(AX)^* = AX$ (4) $(XA)^* = XA.$

For any matrix A there exists a unique element in the set $A\{1, 2, 3, 4\}$, called the Moore-Penrose inverse of A, which is denoted by $A^\dagger$. The Drazin inverse of a square matrix $A \in \mathbb{C}^{n \times n}$, denoted by A^D, is the unique matrix $X \in \mathbb{C}^{n \times n}$ which fulfills the matrix equation (2) and two additional matrix equations:

$$(1^k) \quad A^{l+1}X = A^l, \ l \geq \text{ind}(A), \quad (5) \ AX = XA.$$

Here, the notation ind(A) denotes the index of a square matrix A and it is defined as ind$(A) = \min\{j|\ \text{rank}(A^j) = \text{rank}(A^{j+1})\}$. In the case ind$(A) = 1$, the Drazin inverse becomes the group inverse $X = A^\#$. Consider a subset $\mathfrak{E} \subset \{1, 2, 3, 41^k, 5\}$. We will say that the equation (i) is defined by $\mathfrak{E}$ in the case $i \in \mathfrak{E}$. The set of all matrices satisfying the equations defined by $\mathfrak{E}$ is denoted by $A\{\mathfrak{E}\}$. Any matrix in $A\{\mathfrak{E}\}$ is called the $\mathfrak{E}$-inverse of A and is denoted by $A^{\mathfrak{E}}$.

The outer generalized inverse $A^{(2)}_{T,S}$ of $A \in \mathbb{C}^{m \times n}$ is the matrix $X \in \mathbb{C}^{n \times m}$ which satisfies the Penrose equation (2) and has predefined range and null space:

$$XAX = X, \quad \mathcal{R}(X) = T, \quad N(X) = S. \tag{1.1}$$

If $A \in \mathbb{C}^{m \times n}_r$, T is a subspace of $\mathbb{C}^n$ of dimension $t \leq r$ and S is a subspace of $\mathbb{C}^m$ of dimension $m - t$, then A has a $\{2\}$-inverse X such that $\mathcal{R}(X) = T$ and $\mathcal{N}(X) = S$ if and only if $AT \oplus S = \mathbb{C}^m$, in which case X is unique and it is denoted by $A^{(2)}_{T,S}$.

The Moore-Penrose inverse $A^\dagger$ and the weighted Moore-Penrose inverse $A^\dagger_{M,N}$, the Drazin inverse A^D as well as the group inverse $A^\#$ can be derived by means of appropriate choices of T and S as follows (see, for example, [20, 26]):

$$
\begin{aligned}
A^\dagger &= A^{(2)}_{\mathcal{R}(A^*),\mathcal{N}(A^*)}, \quad A^\dagger_{M,N} = A^{(2)}_{\mathcal{R}(A^\sharp),\mathcal{N}(A^\sharp)}, \quad A^\sharp = N^{-1}A^*M \\
A^D &= A^{(2)}_{\mathcal{R}(A^k),\mathcal{N}(A^k)}, \quad k \geq \text{ind}(A), \quad A^\# = A^{(2)}_{\mathcal{R}(A),\mathcal{N}(A)}, \quad \text{ind}(A) = 1.
\end{aligned}
\tag{1.2}
$$

For other important properties of generalized inverses see [1, 20, 26].

2 Short Overview on Symbolic Computation Algorithms

Many numerical algorithms for computing generalized inverses lack numerical stability. In addition when rounding errors ate inherent, one has to identify some small quantities as being zero. It is therefore clear that cumulative round off errors should be totally eliminated. During the symbolic implementation, variables are stored in "exact" form or can be left "unassigned" (without numerical values) resulting in no loss of accuracy during the calculation [5]. Symbolic computation of generalized inverses is one of interesting applications of computer algebra. Moreover, algorithms presented for matrices in symbolic form are applicable to significantly wider class of matrices and to a wider set of problems, with respect to algorithms intended for constant matrices. Also, algorithms applicable to matrices of unassigned symbols can be used in the construction of test matrices and in the verification of some hypotheses.

Symbolic algorithms for various types of generalized inverses could be classified into several different categories.

- Algorithms based on the multiple-modulus residue arithmetic, aimed to error-free computation of reflexive generalized inverses and the Moore-Penrose inverse of a matrix having rational entries, were developed in [8] and [12], respectively.
- Various extensions of the Leverrier-Faddeev algorithm, applicable in computing generalized inverses of polynomial matrices, were investigated in [2,4–6].
- Several extensions of the Greville's partitioning method from [3], which are applicable to rational and polynomial matrices, were established in [10,11, 17,19].
- The algorithm based on the LDL^* factorization and aimed for computing $\{1,2,3\},\{1,2,4\}$ inverses and the Moore–Penrose inverse of a given rational matrix was developed in [15].
- An algorithm for the evaluation of the full-rank QDR decomposition and its application in developing a new method and algorithm for the symbolic computation of $A_{T,S}^{(2)}$ inverses of one-variable polynomial or rational matrices was proposed in [18].
- Yu and Wang in [28] introduced an algorithm for calculating $\{2\}$-inverses of a polynomial matrix with prescribed image and kernel. It is based on the finite algorithm for generalized inverse $A_{T,S}^{(2)}$ of a matrix A over an integral domain and the discrete Fourier transform. The algorithm proposed in [28] extends the algorithms from [7].
- Sendra et al. in [13] showed how to extend the computation of Drazin inverses over certain computable fields to the computation of Drazin inverses of matrices with rational functions as entries. In [14], the authors considered the computation of the Moore-Penrose inverse in a field with an involutory automorphism, having the property that all matrices over them have Moore-Penrose inverse.

Recently, a number of nonlinear and linear dynamical systems and initiated recurrent neural network models have been developed in order to numerically evaluate the inverse matrix and the pseudoinverse of full-row or full-column rank rectangular matrices (for more details, see [9,21,22]). Also, various recurrent neural networks for computing generalized inverses of rank-deficient matrices were designed in [23, 24]. The most general design evolution for solving the matrix equation $AXB = D$ was investigated in [16].

Although numerical computation clearly plays a more dominant role within dynamical systems than symbolic computation, the application of computer algebra techniques in this field is becoming more and more popular. In this paper, we outline an algorithm for the symbolic computation of outer inverses of integral and rational matrices using the approach based on the exact solution of dynamic state equations defined by the Gradient Neural Network (GNN) dynamical systems for computing generalized inverses. More precisely, the algorithm arises from the exact solution of systems of differential equations which appear in dynamic state equations included in the GNN evolution of generalized inverses.

There are very sophisticated numerical methods for ordinary differential equations and systems. Do we need analytic methods? There are several reasons, for example

- When available, a formula covers all cases and is accurate, saving the effort of multiple numerical integrations;
- one can vary parameters;
- one can use the result in all subsequent stages of calculation.

Our intention is to unify both the Recurrent Neural Network (RNN) and symbolic approaches in a single computational method. What can be expected? Usually, traditional numerical algorithms are of serial-processing nature and may not be efficient enough for online or real-time simulations, which must guarantee a response within specified time constraints. This deficiency can be overcome using the dynamical system approach and RNN models.

On the other hand, numerical algorithms for computing generalized inverses lack numerical stability and accumulate rounding errors. This deficiency can be overcome using the symbolic computation techniques.

The rest of the paper is organized as follows. Dynamic state equations and initiated dynamical models are defined in Sect. 3. Section 4 investigates symbolic computation of outer inverses based on finding exact solutions of underlying dynamic state equations. Section 5 presents a number of illustrative examples.

3   GNN Models for Computing Outer Inverses

The dynamics of the GNN models for solving a matrix equation $\mathcal{M}$ is defined on the usage of the error matrix $E(t)$. The error matrix is defined by replacing the unknown matrix in $\mathcal{M}$ by the time-varying matrix $V(t)$. The next step is the minimization of the function which is defined as the scalar-valued norm-based error function

$$\varepsilon(t) = \varepsilon(V(t)) = \frac{1}{2}\|E(t)\|_F^2, \tag{3.1}$$

where and $\|A\|_F := \sqrt{\mathrm{Tr}(A^\mathrm{T} A)}$ denotes the Frobenius norm of the matrix A and $\mathrm{Tr}(\cdot)$ denotes the trace of a matrix. The linear GNN design model is defined as the search towards the descent direction $-\frac{\partial \varepsilon(V(t))}{\partial V}$ of the goal function, as follows:

$$\dot{V}(t) = \frac{dV(t)}{dt} = -\gamma \frac{\partial \varepsilon(V(t))}{\partial V}. \tag{3.2}$$

The scaling real parameter γ in (3.2) represents an inductance parameter or the reciprocal of a capacitance parameter. Greater values of γ initiate faster convergence.

The GNN design corresponding to the matrix equation

$$\mathcal{M} := AXB = D$$

was investigated and applied in [16]. The error matrix for solving $\mathcal{M}$ is defined by $E(t) = D - AV(t)B$. The scalar-valued norm-based error function is defined by the Frobenius norm of $E(t)$:

$$\varepsilon(t) = \varepsilon(V(t)) = \frac{1}{2} \|E(t)\|_F^2.$$

The gradient of the objective function $\varepsilon(t)$ is equal to

$$\frac{\partial \varepsilon(V(t))}{\partial V} = -A^T(D - AV(t)B)B^T = -A^T E(t)B^T.$$

According to the general GNN dynamics, the GNN model for solving $AXB = D$ is defined as

$$\frac{dV(t)}{dt} = \dot{V}(t) = \gamma A^T \mathcal{F}(D - AV(t)B)B^T. \tag{3.3}$$

The GNN model in (3.3) will be denoted by $GNN(A, B, D)$.

Proposition 1 [16]. *Assume that the real matrices $A \in \mathbb{R}^{m \times n}$, $B \in \mathbb{R}^{p \times q}$ and $D \in \mathbb{R}^{m \times q}$ satisfy*

$$AA^\dagger D B^\dagger B = D. \tag{3.4}$$

Then the state matrix $V(t) \in \mathbb{R}^{n \times m}$ of the $GNN(A, B, D)$ model (3.3) satisfies:

$$AV(t)B \to D, \quad t \to +\infty$$
$$\tilde{V}_{V(0)} = \lim_{t \to \infty} V(t) = A^\dagger D B^\dagger + V(0) - A^\dagger AV(0)BB^\dagger \tag{3.5}$$

for an arbitrary initial state matrix $V(0)$.

$GNN(A, I, I)$ model is aimed to solving $AX = I$ and uses the error matrix $E(t) = AV(t) - I$. It was proposed in [21]. Its dynamics can be expressed as

$$\dot{V}(t) = \frac{dV(t)}{dt} = -\gamma A^T \mathcal{F}(AV(t) - I), \quad V(0) = V_0. \tag{3.6}$$

According to Proposition 1, the general solution of $GNN(A, I, I)$ is

$$\tilde{V}_{V(0)} = A^\dagger + V(0) - A^\dagger AV(0).$$

The $GNN(A, I, I)$ model can be used in:

- finding the inverse of A, starting from arbitrary $V(0)$;
- approximating the left inverse of a full-column rectangular matrix A, starting from arbitrary $V(0)$;
- computing the pseudoinverse of rank-deficient matrices under the zero initial condition $V(0) = 0$.

The outer inverse $X := A^{(2)}_{\mathcal{R}(G),\mathcal{N}(G)}$ of the input matrix $A \in \mathbb{C}^{m \times n}_r$ is defined using an appropriate matrix $G \in \mathbb{C}^{n \times m}_s$ satisfying $0 < s \le r$ and $\mathrm{rank}(GA) = \mathrm{rank}(G)$. Then the matrix equation

$$\mathcal{M}_G := GAX = G \tag{3.7}$$

is satisfied. As a consequence, the $GNN(GA, I, G)$ model is defined for solving the matrix equations $\mathcal{M}_G$.

The matrix equation $\mathcal{M}_G$, defined in (3.7), initiates the error matrix $E(t)$ as

$$E(t) = GAV(t) - G, \tag{3.8}$$

where $V(t) \in \mathbb{R}^{n \times m}$ denotes the unknown matrix to be solved. Our intention is to solve one of the equations included in (3.8) with respect to the unknown matrix $V(t)$ using the dynamic-system approach in conjunction with the symbolic data processing. According to the $GNN(A, B, D)$ design (3.3), we obtain the following $GNN(GA, I, G)$ dynamical system:

$$\frac{dV(t)}{dt} = -\gamma (GA)^{\mathrm{T}} (GAV(t) - G), \quad V(0) = V_0. \tag{3.9}$$

The convergence of $GNN(A, B, D)$ is investigated in Corollary 1.

Corollary 1. *Assume that the real matrices $A \in \mathbb{R}^{m \times n}_r$, $G \in \mathbb{R}^{n \times m}_s$ satisfy $0 < s \le r$ and $\mathrm{rank}(GA) = \mathrm{rank}(G)$. Then:*

(i) *The unknown matrix $V(t)$ of the model $GNN(GA, I, G)$ is convergent when $t \to +\infty$ and has the limit value*

$$\tilde{V}_{V(0)} = (GA)^{\dagger} G + V(0) - (GA)^{\dagger} GAV(0). \tag{3.10}$$

(ii) *In particular, $V(0) = 0$ initiates*

$$\tilde{V}_0 = (GA)^{\dagger} G = A^{(2,4)}_{\mathcal{R}((GA)^*),\mathcal{N}(G)}.$$

The authors of [29] defined the dynamical system by omitting the constant term $(GA)^{\mathrm{T}}$ from (3.9):

$$\frac{dV(t)}{dt} = -\gamma (GAV(t) - G), \quad V(0) = O. \tag{3.11}$$

The dynamical evolution (3.11) will be termed as GNNATS2. The application of the dynamic evolution design (3.11) is conditioned by the properties of the spectrum of the matrix GA:

$$\sigma(GA) \subset \{z : \mathrm{Re}\,(z) \ge 0\}. \tag{3.12}$$

More precisely, the first GNN approach used in [29] fails in the case when $\text{Re}\,(\sigma(GA))$ contains negative values. Clearly, the model (3.11) is simpler than the models (3.9), but it loses global stability. An approach to resolve the requirement (3.12) and recover global stability was proposed in [29], and it is based on the replacement of G by $G_0 = G(GAG)^{\mathrm{T}}G$ in (3.11).

4   Symbolic Implementation of the GNN Dynamics

The implementation in the *Matlab* programm requires a vectorization of the system of matrix differential equations into the vector form (mass matrix) and then solving the vector of differential equations by means of one of *Matlab* solvers, such as ode45, ode15s, ode23. This requires repeated applications of one of these solvers in unpredictable time instants inside the selected time interval $[0, t_f]$.

We want to define a *Mathematica* program for solving the dynamical equation (3.11) symbolically as the opposite solution to the *Matlab* numerical implementation. The main idea is to solve the matrix differential equations in symbolic form. The solution given in symbolic form would be generated only once. The solution will be termed as "symbolic solver". It is able to define wanted values of $V(t)$ in each time instant only by means of the simple replacement of the variable t by an arbitrary time instant $t_0 \in [0, t_f]$. Also, the possibility to investigate some limiting properties of the symbolic solver by means of the *Mathematica* function Limit is available.

According to the previous discussion, we present the corresponding Algorithm 1 for computing outer inverses of the input matrix $A \in \mathbb{R}(\mathcal{S})_r^{m\times n}$ by finding exact solutions of the dynamic state equation (3.11). The algorithm assumes the choice of a matrix $G \in \mathbb{R}(\mathcal{S})_s^{n\times m}$, $0 < s \le r$. It is also assumed that the variables $s_1, \dots, s_k$ are different than the symbol t representing the time; that is $t \notin \mathcal{S}$.

Algorithm 1. Computing outer inverse of a given matrix $A \in \mathbb{R}(\mathcal{S})_r^{m\times n}$.

Require: Time-invariant matrices $A \in \mathbb{R}(\mathcal{S})_r^{m\times n}$ and $G \in \mathbb{R}(\mathcal{S})_s^{n\times m}$, $0 < s \le r$.
1: Construct the dynamic state equations contained in (3.11) in symbolic form.
    Step 1:1: Construct the symbolic matrix $V(t) = v_{ij}(t)$, $t \notin \mathcal{S}$.
    Step 1:3: Construct the matrix $-\gamma\,(GAV(t) - G)$ for GNNATS2 dynamics.
    Step 1:4: Define the symbolic matrix equation, *eqnstate*, as $\dot{V}(t) - \gamma\,(GAV(t) - G) = O$.
2: Define the initial state $V(0) = O$.
3: Solve (3.11) symbolically. Denote the output by $V(t)$.
    Step 3:1: Join vectorized lists *eqnstate* and $V(0)$ in the list *eqns*.
    Step 3:2: Solve the system of differential equations *eqns* with respect to variables v_{ij} and the time t.
4: Return the outer inverse $\overline{V} = \lim_{t\to\infty} V(t) = A^{(2)}_{\mathcal{R}(G),\mathcal{N}(G)}$.

The Algorithm 1 describes the implementation in the algebraic programming language *Mathematica*. In the following, the used *Mathematica* codes are commented; for further details on the programming language *Mathematica* we refer to [27]. Let us mention that the *Mathematica* function `DSolve` can be applied to generate a solution in Step 3:2 of Algorithm 1. More precisely, `DSolve` finds symbolic (or "pure function") solutions for the entries v_{ij} which are included in the differential equations that appear in (3.9) or (3.11). Also, the *Mathematica* function `Limit` can be applied to generate a solution in Step 5 of Algorithm 1.

In the case when the elements of $A = A(\mathcal{S})$ are rational numbers or rational expressions with respect to variables included in $\mathcal{S}$ and $t \notin \mathcal{S}$, Algorithm 1 produces the exact outer inverse $A^{(2)}_{\mathcal{R}(G),\mathcal{N}(G)}$.

The main advantages of the symbolic processing in solving dynamical system with respect to numerical processing can be emphasized as follows.

(1) The solution $V(t)$ of the dynamical system is given in symbolic form with respect to the variable representing the time. Denote this solution by
`V[t]:=SymNNInv[A,G]`.
Then the values of the generalized inverse $V[t0]$, $t0 \in [0, t_f]$, $t_f > 0$, in the time instant $t0$, can be simply and efficiently generated only by the replacement
`SymNNInv[A,G]/.t->t0`.

(2) In addition, it is possible to find limiting values of generated symbolic expressions by means of the facilities of the computer algebra systems, such as *Mathematica*. These limiting values can be generated by the expressions `SymNNInv[A,G]/.t->∞` and used in order to find exact outer inverses.

(3) As it was mentioned, RNN models in some practical applications must guarantee a response within specified time frame. But, the real CPU time spanned by the proposed dynamical systems inside the time interval $[0, t_f]$ is far from t_f. It is known that dynamical models are appropriate for hardware implementation would ensure responses in predefined time intervals. We define the software implementation `V[t]:=SymNNInv[A,G]` which is applicable for real-time applications. Instead of a hardware device we can use corresponding symbolic expression which is ready for replacements of the time variable by certain time instants.

5 Examples

Example 1. Consider the input matrix from [25]

$$
A = \begin{bmatrix}
1 & -1 & 0 & 0 & 0 & 0 \\
-1 & 1 & 0 & 0 & 0 & 0 \\
-1 & -1 & 1 & -1 & 0 & 0 \\
-1 & -1 & -1 & 1 & 0 & 0 \\
-1 & -1 & -1 & 0 & 2 & -1 \\
-1 & -1 & 0 & -1 & -1 & 2
\end{bmatrix}
$$

Using `G = A.A` and $V_G^D =$ `SymNNInv[A, G]`, the following result in the convenient matrix form can be obtained:

$$
V_G^D = \begin{bmatrix}
\frac{1}{4}\left(1-e^{-8t\gamma}\right) & \frac{1}{4}\left(-1+e^{-8t\gamma}\right) & 0 \\
\frac{1}{4}\left(-1+e^{-8t\gamma}\right) & \frac{1}{4}\left(1-e^{-8t\gamma}\right) & 0 \\
0 & 0 & \frac{1}{4}\left(1-e^{-8t\gamma}\right) \\
0 & 0 & \frac{1}{4}\left(-1+e^{-8t\gamma}\right) \\
0 & 0 & \frac{1}{12}\left(-5+2e^{-27t\gamma}-3e^{-8t\gamma}+6e^{-t\gamma}\right) \\
0 & 0 & \frac{1}{12}\left(-7-2e^{-27t\gamma}+3e^{-8t\gamma}+6e^{-t\gamma}\right)
\end{bmatrix}
$$

$$
\begin{bmatrix}
0 & 0 & 0 \\
0 & 0 & 0 \\
\frac{1}{4}\left(-1+e^{-8t\gamma}\right) & 0 & 0 \\
\frac{1}{4}\left(1-e^{-8t\gamma}\right) & 0 & 0 \\
\frac{1}{12}\left(-7-2e^{-27t\gamma}+3e^{-8t\gamma}+6e^{-t\gamma}\right) & \frac{1}{6}\left(4-e^{-27t\gamma}-3e^{-t\gamma}\right) & \frac{1}{6}\left(2+e^{-27t\gamma}-3e^{-t\gamma}\right) \\
\frac{1}{12}\left(-5+2e^{-27t\gamma}-3e^{-8t\mu}+6e^{-t\mu}\right) & \frac{1}{6}\left(2+e^{-27t\gamma}-3e^{-t\gamma}\right) & \frac{1}{6}\left(4-e^{-27t\gamma}-3e^{-t\gamma}\right)
\end{bmatrix}.
$$

The matrix V_G^D can be used as "symbolic solver".

The limit expression $\overline{V}_G^D =$ `Limit[`V_G^D `/.`$\gamma \to 10^3, t \to \infty$`]` produces the following exact Drazin inverse of A:

$$
\overline{V}_G = A^D = \begin{bmatrix}
\frac{1}{4} & -\frac{1}{4} & 0 & 0 & 0 & 0 \\
-\frac{1}{4} & \frac{1}{4} & 0 & 0 & 0 & 0 \\
0 & 0 & \frac{1}{4} & -\frac{1}{4} & 0 & 0 \\
0 & 0 & -\frac{1}{4} & \frac{1}{4} & 0 & 0 \\
0 & 0 & -\frac{5}{12} & -\frac{7}{12} & \frac{2}{3} & \frac{1}{3} \\
0 & 0 & -\frac{7}{12} & -\frac{5}{12} & \frac{1}{3} & \frac{2}{3}
\end{bmatrix}.
$$

The positive real scaling constant γ should be chosen as large as possible in order to achieve the convergence for smaller values of t. The state trajectories of the elements $x_{ij} = (V_{A^T})_{ij} \neq 0$ in the case $\gamma \to 10^9$, $t \in [0, 10^{-8}]$ are presented in Fig. 1.

Example 2. The input matrix in this example is

$$
A = \begin{bmatrix}
1 & -1 & 0 & 0 & 0 \\
-1 & 1 & 0 & 0 & 0 \\
-1 & 1 & -1 & 0 & 0 \\
-1 & -1 & -1 & 1 & 0
\end{bmatrix}.
$$

The expression $V_{A^T} =$ `SymNNInv[A,Transpose[A]]` becomes the "symbolic Simulink" for computing the Moore-Penrose inverse of A. That expression is presented in Fig. 2.

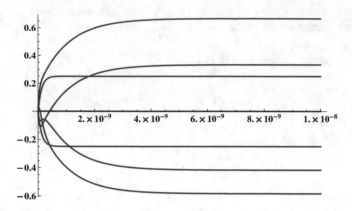

Fig. 1. The state trajectories of $x_{ij} = (V_{A^2})_{ij}$ in the case $\gamma \to 10^9$, $t \in [0, 10^{-8}]s$ in Example 1.

In[11]:= **X = SymNNInv[a, Transpose[a]] // FullSimplify**

Out[11]= $\Big\{ \Big\{ \frac{1}{1332} e^{-\left(4+\sqrt{37}\right) t \gamma}$

$\left(37 e^{\sqrt{37} t \gamma} + 8 \left(-37 + 2 \sqrt{37}\right) e^{\frac{1}{2}\left(1+\sqrt{37}\right) t \gamma} + 555 e^{\left(4+\sqrt{37}\right) t \gamma} - 8 \left(37 + 2 \sqrt{37}\right) e^{\frac{1}{2}\left(1+3\sqrt{37}\right) t \gamma}\right),$

$\frac{1}{1332} e^{-\left(4+\sqrt{37}\right) t \gamma} \left(-37 e^{\sqrt{37} t \gamma} - 8 \left(-37 + 2 \sqrt{37}\right) e^{\frac{1}{2}\left(1+\sqrt{37}\right) t \gamma} -\right.$

$\left. 555 e^{\left(4+\sqrt{37}\right) t \gamma} + 8 \left(37 + 2 \sqrt{37}\right) e^{\frac{1}{2}\left(1+3\sqrt{37}\right) t \gamma}\right),$

$\frac{1}{3} - \frac{1}{111} e^{-\frac{7 t \gamma}{2}} \left(37 \operatorname{Cosh}\left[\frac{1}{2}\sqrt{37}\ t \gamma\right] + 13 \sqrt{37}\ \operatorname{Sinh}\left[\frac{1}{2}\sqrt{37}\ t \gamma\right]\right), \frac{1}{333} e^{-\frac{1}{2}\left(15+\sqrt{37}\right) t \gamma}$

$\left(\left(37 - 2 \sqrt{37}\right) e^{4 t \gamma} + \left(37 + 2 \sqrt{37}\right) e^{\left(4+\sqrt{37}\right) t \gamma} + 37 e^{\frac{1}{2}\left(7+\sqrt{37}\right) t \gamma} - 111 e^{\frac{1}{2}\left(15+\sqrt{37}\right) t \gamma}\right)\Big\},$

$\Big\{ \frac{1}{444} \left(-37 + 37 e^{-4 t \gamma} - 8 \sqrt{37}\ e^{\frac{1}{2}\left(-7+\sqrt{37}\right) t \gamma} + 8 \sqrt{37}\ e^{-\frac{1}{2}\left(7+\sqrt{37}\right) t \gamma}\right),$

$\frac{1}{444} \left(37 - 37 e^{-4 t \gamma} + 8 \sqrt{37}\ e^{\frac{1}{2}\left(-7+\sqrt{37}\right) t \gamma} - 8 \sqrt{37}\ e^{-\frac{1}{2}\left(7+\sqrt{37}\right) t \gamma}\right),$

$\frac{1}{3} - \frac{1}{111} e^{-\frac{7 t \gamma}{2}} \left(37 \operatorname{Cosh}\left[\frac{1}{2}\sqrt{37}\ t \gamma\right] + \sqrt{37}\ \operatorname{Sinh}\left[\frac{1}{2}\sqrt{37}\ t \gamma\right]\right),$

$\frac{1}{111} \left(-37 + 37 e^{-4 t \gamma} + \sqrt{37}\ e^{\frac{1}{2}\left(-7+\sqrt{37}\right) t \gamma} - \sqrt{37}\ e^{-\frac{1}{2}\left(7+\sqrt{37}\right) t \gamma}\right)\Big\},$

$\Big\{ \frac{1}{666} e^{-\left(4+\sqrt{37}\right) t \gamma} \left(37 e^{\sqrt{37} t \gamma} - 4 \left(-37 + 8 \sqrt{37}\right) e^{\frac{1}{2}\left(1+\sqrt{37}\right) t \gamma} -\right.$

$\left. 333 e^{\left(4+\sqrt{37}\right) t \gamma} + 4 \left(37 + 8 \sqrt{37}\right) e^{\frac{1}{2}\left(1+3\sqrt{37}\right) t \gamma}\right), \frac{1}{666} e^{-\left(4+\sqrt{37}\right) t \gamma}$

$\left(-37 e^{\sqrt{37} t \gamma} + 4 \left(-37 + 8 \sqrt{37}\right) e^{\frac{1}{2}\left(1+\sqrt{37}\right) t \gamma} + 333 e^{\left(4+\sqrt{37}\right) t \gamma} - 4 \left(37 + 8 \sqrt{37}\right) e^{\frac{1}{2}\left(1+3\sqrt{37}\right) t \gamma}\right),$

$-1 + \frac{1}{37} e^{-\frac{7 t \gamma}{2}} \left(37 \operatorname{Cosh}\left[\frac{1}{2}\sqrt{37}\ t \gamma\right] + 5 \sqrt{37}\ \operatorname{Sinh}\left[\frac{1}{2}\sqrt{37}\ t \gamma\right]\right),$

$\frac{1}{333} e^{-\left(4+\sqrt{37}\right) t \gamma} \left(74 e^{\sqrt{37} t \gamma} + \left(-37 + 8 \sqrt{37}\right) e^{\frac{1}{2}\left(1+\sqrt{37}\right) t \gamma} - \left(37 + 8 \sqrt{37}\right) e^{\frac{1}{2}\left(1+3\sqrt{37}\right) t \gamma}\right)\Big\},$

$\Big\{ \frac{1}{666} e^{-\frac{1}{2}\left(15+\sqrt{37}\right) t \gamma} \left(\left(74 - 10 \sqrt{37}\right) e^{4 t \gamma} + 2 \left(37 + 5 \sqrt{37}\right) e^{\left(4+\sqrt{37}\right) t \gamma} -\right.$

$\left. 37 e^{\frac{1}{2}\left(7+\sqrt{37}\right) t \gamma} - 111 e^{\frac{1}{2}\left(15+\sqrt{37}\right) t \gamma}\right), \frac{1}{666} e^{-\left(4+\sqrt{37}\right) t \gamma}$

$\left(37 e^{\sqrt{37} t \gamma} + 2 \left(-37 + 5 \sqrt{37}\right) e^{\frac{1}{2}\left(1+\sqrt{37}\right) t \gamma} + 111 e^{\left(4+\sqrt{37}\right) t \gamma} - 2 \left(37 + 5 \sqrt{37}\right) e^{\frac{1}{2}\left(1+3\sqrt{37}\right) t \gamma}\right),$

$-\frac{1}{3} + \frac{1}{111} e^{-\frac{7 t \gamma}{2}} \left(37 \operatorname{Cosh}\left[\frac{1}{2}\sqrt{37}\ t \gamma\right] + 7 \sqrt{37}\ \operatorname{Sinh}\left[\frac{1}{2}\sqrt{37}\ t \gamma\right]\right),$

$\frac{1}{666} e^{-\left(4+\sqrt{37}\right) t \gamma} \left(-148 e^{\sqrt{37} t \gamma} + \left(-37 + 5 \sqrt{37}\right) e^{\frac{1}{2}\left(1+\sqrt{37}\right) t \gamma} +\right.$

$\left. 222 e^{\left(4+\sqrt{37}\right) t \gamma} - \left(37 + 5 \sqrt{37}\right) e^{\frac{1}{2}\left(1+3\sqrt{37}\right) t \gamma}\right)\Big\}, \{0, 0, 0, 0\}\Big\}$

Fig. 2. The matrix X=SymNNInv[A,Transpose[A]]//FullSimplify in Example 2.

The limit expression $\overline{V}_{A^{\mathrm{T}}} = \mathtt{Limit[X/.\gamma} \to 10^3, t \to \infty]$ produces the exact Moore-Penrose inverse of A, equal to

$$\overline{V}_{A^{\mathrm{T}}} = A^{\dagger} = \begin{bmatrix} \frac{5}{12} & -\frac{5}{12} & \frac{1}{3} & -\frac{1}{3} \\ -\frac{1}{12} & \frac{1}{12} & \frac{1}{3} & -\frac{1}{3} \\ -\frac{1}{2} & \frac{1}{2} & -1 & 0 \\ -\frac{1}{6} & \frac{1}{6} & -\frac{1}{3} & \frac{1}{3} \\ 0 & 0 & 0 & 0 \end{bmatrix}.$$

6 Conclusion

The present paper is a contribution to the symbolic computation of outer generalized inverses of matrices. Also, derived results are a contribution to the application of computer algebra systems in solving dynamical systems. The algorithm considered in the present paper is based on the exact solution of systems of differential equations which appear in dynamic state equations included in the GNNN modeling of generalized inverses. The implementation in the programming package *Mathematica* is described and used and tested in several examples.

The central part of our algorithm is the possibility to solve the system of ordinary differential equations. This problem is a part of the scientific research known as *Computer Algebra and Differential Equations*. Here, we used the possibility of the standard *Mathematica* function DSolve. Clearly, many different approaches are available and could be exploited in further research.

References

1. Ben-Israel, A., Greville, T.N.E.: Generalized inverses: Theory and Applications, vol. 2. Springer, New York (2003). https://doi.org/10.1007/b97366
2. Fragulis, G., Mertzios, B.G., Vardulakis, A.I.G.: Computation of the inverse of a polynomial matrix and evaluation of its Laurent expansion. Int. J. Control **53**, 431–443 (1991)
3. Greville, T.N.E.: Some applications of the pseudo-inverse of matrix. SIAM Rev. **3**, 15–22 (1960)
4. Jones, J., Karampetakis, N.P., Pugh, A.C.: The computation and application of the generalized inverse vai Maple. J. Symbolic Comput. **25**, 99–124 (1998)
5. Karampetakis, N.P.: Computation of the generalized inverse of a polynomial matrix and applications. Linear Algebra Appl. **252**, 35–60 (1997)
6. Karampetakis, N.P.: Generalized inverses of two-variable polynomial matrices and applications. Circ. Syst. Sign. Process. **16**, 439–453 (1997)
7. Karampetakis, N.P., Vologiannidis, S.: DFT calculation of the generalized and Drazin inverse of a polynomial matrix. Appl. Math. Comput. **143**, 501–521 (2003)
8. McNulty, S.K., Kennedy, W.J.: Error-free computation of a reflexive generalized inverse. Linear Algebra Appl. **67**, 157–167 (1985)
9. Luo, F.L., Bao, Z.: Neural network approach to computing matrix inversion. Appl. Math. Comput. **47**, 109–120 (1992)

10. Petković, M.D., Stanimirović, P.S., Tasić, M.B.: Effective partitioning method for computing weighted Moore-Penrose inverse. Comput. Math. Appl. **55**, 1720–1734 (2008)
11. Petković, M.D., Stanimirović, P.S.: Symbolic computation of the Moore-Penrose inverse using partitioning method. Int. J. Comput. Math. **82**, 355–367 (2005)
12. Rao, T.M., Subramanian, K., Krishnamurthy, E.V.: Residue arithmetic algorithms for exact computation of g-Inverses of matrices. SIAM J. Numer. Anal. **13**, 155–171 (1976)
13. Sendra, J.R., Sendra, J.: Symbolic computation of Drazin inverses by specializations. J. Comput. Appl. Math. **301**, 201–212 (2016)
14. Sendra, J.R., Sendra, J.: Computation of moore-penrose generalized inverses of matrices with meromorphic function entries. Appl. Math. Comput. **313**, 355–366 (2017)
15. Stanimirović, I.P., Tasić, M.B.: Computation of generalized inverses by using the LDL^* decomposition. Appl. Math. Lett. **25**, 526–531 (2012)
16. Stanimirović, P.S., Petković, M.: Gradient neural dynamics for solving matrix equations and their applications. Neurocomputing **306**, 200–212 (2018)
17. Stanimirović, P.S., Tasić, M.B.: Partitioning method for rational and polynomial matrices. Appl. Math. Comput. **155**, 137–163 (2004)
18. Stanimirović, P.S., Pappas, D., Katsikis, V.N., Stanimirović, I.P.: Symbolic computation of $A_{T,S}^{(2)}$-inverses using QDR factorization. Linear Algebra Appl. **437**, 1317–1331 (2012)
19. Tasić, M.B., Stanimirović, P.S., Petković, M.D.: Symbolic computation of weighted Moore-Penrose inverse using partitioning method. Appl. Math. Comput. **189**, 615–640 (2007)
20. Wang, G.R., Wei, Y.M., Qiao, S.Z.: Generalized Inverses: Theory and Computations. Second edition. Developments in Mathematics, vol. 53. Springer, Singapore; Science Press Beijing, Beijing (2018)
21. Wang, J.: Recurrent neural networks for solving linear matrix equations. Comput. Math. Appl. **26**, 23–34 (1993)
22. Wang, J.: A recurrent neural network for real-time matrix inversion. Appl. Math. Comput. **55**, 89–100 (1993)
23. Wang, J.: Recurrent neural networks for computing pseudoinverses of rank-deficient matrices. SIAM J. Sci. Comput. **18**, 1479–1493 (1997)
24. Wei, Y.: Recurrent neural networks for computing weighted Moore-Penrose inverse. Appl. Math. Comput. **116**, 279–287 (2000)
25. Wei, Y.: Integral representation of the generalized inverse $A_{T,S}^{(2)}$ and its applications. In: Recent Research on Pure and Applied Algebra, pp. 59–65. Nova Science Publisher, New York (2003)
26. Wei, Y., Stanimirović, P.S., Petković, M.D.: Numerical and Symbolic Computations of Generalized Inverses. World Scientific Publishing Co., Pte. Ltd., Hackensack (2018)
27. Wolfram Research Inc., Mathematica, Version 10.0, Champaign, IL (2015)
28. Yu, Y., Wang, G.: DFT calculation for the 2-inverse of a polynomial matrix with prescribed image and kernel. Appl. Math. Comput. **215**, 2741–2749 (2009)
29. Živković, I., Stanimirović, P.S., Wei, Y.: Recurrent neural network for computing outer inverses. Neural Comput. **28**(5), 970–998 (2016)

Intersecting Two Quadrics
with GeoGebra

Alexandre Trocado[1]([⊠]) [iD], Laureano Gonzalez-Vega[2] [iD],
and José Manuel Dos Santos[1] [iD]

[1] Universidade Aberta, Lisbon, Portugal
mail@alexandretrocado.com, dossantosdossantos@gmail.com
[2] Universidad de Cantabria, Santander, Spain
laureano.gonzalez@unican.es

Abstract. This paper presents the first implementation in GeoGebra
of an algorithm computing the intersection curve of two quadrics. This
approach is based on computing the projection of the intersection curve,
also known as cutcurve, determining its singularities and structure and
lifting to 3D this plane curve. The considered problem can be used to
show some of the difficulties arising when implementing in GeoGebra
a geometric algorithm based on the algebraic analysis of the equations
defining the considered objects.

Keywords: GeoGebra · Cutcurve · Intersection curve · Lifting

1 Introduction

Algorithms for computing quadrics intersection date back to the late seventies.
Computing the representation of the curve defined as the intersection of two
quadrics has been a relevant problem to solve over the last decades. Levin in
1976 and 1979 (see [6,7]) introduced a method failing when the intersection
curve is singular and even generates results that are not topologically correct.
Levin's method has been improved by Wang et al. (see [12]) making it capable
of computing geometric and structural information. Besides, Dupont et al. (see
[2]) succeeded in finding parameterizations that overcame the fact that Levin's
method generated formulas that were not suited for further symbolic processing.
On the other hand, Mourrain et al. (see [8]) studied a sweeping algorithm for
computing the arrangement of a set of quadrics in $\mathbb{R}^3$ that reduces the intersection of two quadrics to a dynamic two-dimensional problem. Dupont et al. (see
[3–5]) proposed algorithms that enable to compute in practice an exact form
of the parameterization of the intersection curve of two quadrics with rational
coefficients. These algorithms represent a substantial improvement of Levin's

Second author is partially supported by the Spanish Ministerio de Economia y Competitividad and by the European Regional Development Fund (ERDF), under the project
MTM2017-88796-P.

M. Ćirić et al. (Eds.): CAI 2019, LNCS 11545, pp. 237–248, 2019.
https://doi.org/10.1007/978-3-030-21363-3_20

method and its subsequent refinements. Another approach is based on the using of the cutcurve and resultants (see [9,11]). This method can handle all kinds of inputs including all degenerate ones where intersection curves involve cutcurves with singularities. Here we propose an implementation of this method, in GeoGebra, adapting the algorithm developed by Trocado and Gonzalez-Vega [11] to the special characteristics of GeoGebra.

GeoGebra is a software system for doing dynamic geometry and algebra in the plane. Since 2001 GeoGebra has gone from a dynamic geometry software (DGS) to a powerful computational tool in several areas of mathematics. Powerful algebraic capabilities have been introduced in GeoGebra, such as an efficient spreadsheet that can deal with many kind of objects, an algebraic and symbolic system and several graphical views that extend the possibility of multidimensional representations. The recent 3D features allow more intuitive interaction with three-dimensional objects than most existing mathematical software. However, there are still missing capabilities in GeoGebra 3D, namely the determination of the intersection curve of two quadrics.

The aim of this paper is to present a new tool that allows to compute in GeoGebra a graphical representation of the intersection curve of two quadrics and, when possible, its parameterization. The implemented algorithm uses resultants to determine the projection of the intersection curve in the plane $z = 0$ (the so called cutcurve) and the lifting of its regular and singular points is made by using only one subresultant (the index one subresultant; see [11]). The implemented algorithm presented here does not need to compute any resultant or subresultant since they are provided fully precomputed (those formulae can be found in [11]). When the Computer Algebra capabilities of GeoGebra do not allow to compute a parameterization of the cutcurve (may be involving radicals) or when such a parameterization is very complicated to deal with, a discretization of this curve is determined. The lifting is independent of how the cutcurve is presented: we get either a discretization or a parameterization (involving in some cases radicals) of the intersection curve.

2 Mathematical Tools

Quadrics are the simplest non linear surfaces used in many areas and computing their intersection is a relevant problem.

Definition 1. *Quadrics are algebraic surfaces defined by the equation ($a_{i,j} \in$ $\mathbb{R}$):*

$$a_{11}x^2 + a_{22}y^2 + a_{33}z^2 + 2a_{12}xy + 2a_{13}xz + 2a_{23}yz + 2a_{14}x + 2a_{24}y + 2a_{34}z + a_{44} = 0.$$

In order to allow GeoGebra to compute the intersection curve of two quadrics by using the algorithm in [11] we only need to use resultants and subresultants. We will compute the intersection curve of two quadrics $\mathcal{E}_1$ and $\mathcal{E}_2$ presented by their implicit equations:

$$f(x, y, z) = z^2 + p_1(x, y)z + p_0(x, y) \qquad g(x, y, z) = z^2 + q_1(x, y)z + q_0(x, y)$$

with $\deg(p_1) \leq 1$, $\deg(p_0) \leq 2$, $\deg(q_1) \leq 1$ and $\deg(q_0) \leq 2$. Since p_1 and q_1 are two polynomials of degree one:

$$p_1 = a_1 x + a_2 y + a_3 \qquad q_1 = b_1 x + b_2 y + b_3.$$

The polynomials p_0 and q_0 have degree two:

$$p_0 = a_4 x^2 + a_5 xy + a_6 y^2 + a_7 x + a_8 y + a_9 \qquad q_0 = b_4 x^2 + b_5 xy + b_6 y^2 + b_7 x + b_8 y + b_9.$$

Other cases (ie degree in z smaller than 2) can be considered too (details can be found in [11]).

As usual (see [11]), to determine the projection of the intersection curve on the plane $z = 0$, resultants will be used.

Definition 2. *Let f and g be the two polynomials in $\mathbb{R}[x, y, z]$*

$$f(x, y, z) = z^2 + p_1(x, y)z + p_0(x, y) \qquad g(x, y, z) = z^2 + q_1(x, y)z + q_0(x, y)$$

($\deg(p_1(x, y)) \leq 1$, $\deg(p_0(x, y)) \leq 2$, $\deg(q_1(x, y)) \leq 1$ and $\deg(q_0(x, y)) \leq 2$) defining the quadrics whose intersection curve is to be computed. Then the Sylvester resultant of f and g, with respect to z, is equal to:

$$\mathbf{S}_0(x, y) \overset{\text{def}}{=} \mathbf{Resultant}(f, g; z) = \begin{vmatrix} 1 & p_1(x, y) & p_0(x, y) & 0 \\ 0 & 1 & p_1(x, y) & p_0(x, y) \\ 1 & q_1(x, y) & q_0(x, y) & 0 \\ 0 & 1 & q_1(x, y) & q_0(x, y) \end{vmatrix}$$

$$= (p_0(x, y) - q_0(x, y))^2 - (p_1(x, y) - q_1(x, y)) \begin{vmatrix} p_0(x, y) & p_1(x, y) \\ q_0(x, y) & q_1(x, y) \end{vmatrix}.$$

The projection of the intersection curve is contained in the curve of $\mathbb{R}^2$ defined implicitly by $\mathbf{S}_0(x, y) = 0$.

Computing the intersection of the two quadrics defined by f and g is equivalent to solve in $\mathbb{R}$ the system of polynomial equations $f(x, y, z) = 0$, $g(x, y, z) = 0$ which is equivalent to solve

$$\mathbf{S}_0(x, y) = 0 \wedge (q_1(x, y) - p_1(x, y)) z + (q_0(x, y) - p_0(x, y)) = 0. \qquad (1)$$

These two equations correspond to the subresultants of index 0 and 1 of f and g with respect to z (see [11]).

When f or g have degree 1 in z, $\mathbf{S}_0(x, y)$ is also the resultant of f and g with respect to z and the subresultant of index 1 is one of the quadrics of degree 1. The case of both equations with degree zero reduces to the intersection of two conics and, for sake of simplicity, will not be included here.

3   Implementation of the Algorithm in GeoGebra

3.1   3D Capabilities of GeoGebra for Intersecting Two Quadrics

GeoGebra allows us to represent and determine the intersection curve of some quadrics with a plane by using the command *IntersectPath*. We can also use the command *IntersectConic* to determine the intersection of two quadrics when this can be characterised as the intersection of a plane with a quadric [1]. Figure 1 provides some examples of these cases.

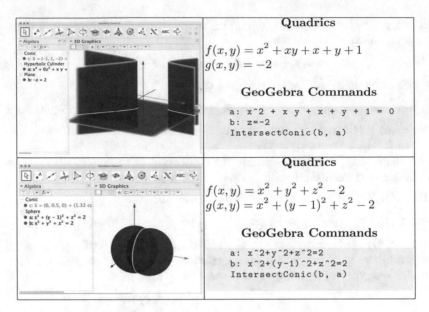

Quadrics
$f(x,y) = x^2 + xy + x + y + 1$
$g(x,y) = -2$
GeoGebra Commands
a: x^2 + x y + x + y + 1 = 0
b: z=-2
IntersectConic(b, a)

Quadrics
$f(x,y) = x^2 + y^2 + z^2 - 2$
$g(x,y) = x^2 + (y-1)^2 + z^2 - 2$
GeoGebra Commands
a: x^2+y^2+z^2=2
b: x^2+(y-1)^2+z^2=2
IntersectConic(b, a)

Fig. 1. Some cases when GeoGebra can determine the intersection of two quadrics.

3.2 The Algorithm

Let $\mathcal{E}_1$ and $\mathcal{E}_2$ be two quadrics in $\mathbb{R}^3$ defined by $f(x,y,z) = 0$ and $g(x,y,z) = 0$ respectively. For computing the intersection curve of $\mathcal{E}_1$ and $\mathcal{E}_2$ in GeoGebra we will consider three cases depending on the degree in z of f and g: both of them with degree two, one with degree two and the other one with degree one and both of them with degree one. The cases of one equation (or both) with degree 0, for sake of simplicity, will not be included here but they follow the same strategy (see [11]). For the three cases considered here, we will show how the algorithm in [11] can be implemented in GeoGebra by adapting it to the special characteristics of this software in order to get a more efficient behavior of GeoGebra when computing the intersection curve of the two considered quadrics.

The method to be implemented in GeoGebra will be as follows (see [11]): starts by computing the projection of the intersection curve of $\mathcal{E}_1$ and $\mathcal{E}_2$ (the so-called cutcurve), continues with its analysis (paying special attention to its singular points) and ends with its lifting.

Two Quadrics of Degree 2 in z

Let $\mathcal{E}_1$ and $\mathcal{E}_2$ be two quadrics defined by:

$$f(x,y,z) = z^2 + p_1(x,y)z + p_0(x,y) \qquad g(x,y,z) = z^2 + q_1(x,y)z + q_0(x,y).$$

The cutcurve of $\mathcal{E}_1$ and $\mathcal{E}_2$ is the set

$$\left\{ (x,y) \in \mathbb{R}^2 : \mathbf{S}_0(x,y) = 0, p_1(x,y)^2 - 4p_0(x,y) \geq 0, q_1(x,y)^2 - 4q_0(x,y) \geq 0 \right\}.$$

The curves defined by $p_1(x,y)^2 - 4p_0(x,y) = 0$ and $q_1(x,y)^2 - 4q_0(x,y) = 0$ are called the silhouette curves of $\mathcal{E}_1$ and $\mathcal{E}_2$, respectively.

The way of proceeding here will be the following one:

(a) Compute $\mathbf{S}_0(x,y)$.
(b) Compute the region

$$\mathcal{A}_{\mathcal{E}_1,\mathcal{E}_2} = \{(x,y) \in \mathbb{R}^2 : p_1(x,y)^2 - 4p_0(x,y) \geq 0, q_1(x,y)^2 - 4q_0(x,y) \geq 0\}$$

where the cutcurve lives: requires to compute the region defined by the two silhouette curves which are two conics.
(c) Compute the singular points of the cutcurve on the line $p_1 = q_1$.
(d) Compute the singular points of the cutcurve outside the line $p_1 = q_1$.
(e) Compute "enough" regular points of the cutcurve (either in closed form or through a discretization) and their lifting.
(f) Compute the lifting of the singular points of the cutcurve.

As seen in [11], we have:

$$\mathbf{S}_0(x,y) = \frac{1}{16}\left[(p_1 - q_1)^4 + (\Delta_{\mathcal{E}_1} - \Delta_{\mathcal{E}_2})^2 - 2(p_1 - q_1)^2 (\Delta_{\mathcal{E}_1} + \Delta_{\mathcal{E}_2}) \right] \quad (2)$$

where $\Delta_{\mathcal{E}_1} = p_1^2 - 4p_0$ and $\Delta_{\mathcal{E}_2} = q_1^2 - 4q_0$.

Fig. 2. Silhouette curves of two quadrics (in red and blue) and the cutcurve (in green)

The cutcurve and the region $\mathcal{A}_{\mathcal{E}_1,\mathcal{E}_2}$ where the cutcurve lives can be defined in GeoGebra in closed form by using the commands[1].

```
*The validation region within a list *
l1:={((a1 x+a2 y+a3)^2-4 (a4 x^2+a5 x y +a6 y^2+a7 x+a8 y+a9))>=0,((b1 x+b2 y
    +b3)^2-4 (b4 x^2+b5 x y+b6 y^2+b7 x+b8 y+b9))>=0}
*Definition of the curve defined by the resultant *
S0:=Implicitcurve(1 / 16 ((a1 x+a2 y+a3-(b1 x+b2 y+b3))^4+((a1 x+a2 y+a3)^2-4
    (a4 x^2+a5 x y+a6 y^2+a7 x+a8 y+a9)-((b1 x+b2 y+b3)^2-4 (b4 x^2+ b5 x y+
    b6 y^2+b7 x+b8 y+b9)))^2-2(a1 x+a2 y+a3-(b1 x+b2 y+b3))^2 ((a1 x+a2 y+a3)
    ^2-4 (a4 x^2+a5 x y+a6 y^2+a7 x+a8 y+a9)+(b1 x+b2 y+b3)^2-4 (b4 x^2+b5 x
    y+b6 y^2+b7 x+b8 y+b9))))
*Definition of the cutcurve*
cond(x,y):=Product(l1) *CAS*
K1:=point(S0)
K2:=If(cond(x(K1),y(K1))>0,K1)
cutcurve:=Locus(K2, K1)
```

---

1 In the GeoGebra commands presented here, we denote $x_\#$ by x#.

The lifting of the regular points of the cutcurve is determined by:

$$z = \frac{p_0(\alpha, \beta) - q_0(\alpha, \beta)}{q_1(\alpha, \beta) - p_1(\alpha, \beta)}. \tag{3}$$

All regular points of the cutcurve are outside the line $p_1(x, y) = q_1(x, y)$. The same formula is to be used for lifting the singular points of the cutcurve outside the line $p_1(x, y) = q_1(x, y)$.

Some cutcurves can not be parameterized easily in closed form even by using radicals: since the cutcurve has degree at most four, it can be parameterized by using radicals but this parameterization is quite complicated and does not take into account properly the real branches (see for example [10]). This is the reason why the lifting of the regular points of the cutcurve will be made on the discretization of the cutcurve branches (when such a parameterization is too involved). To do that in GeoGebra, we need to define a free point, A, on the cutcurve and compute its lifting by using the GeoGebra commands:

```
*Define a free point on the cutcurve*
A=Point(cutcurve)
*Lifting of point A*
P=(x(A),y(A),(a4 x(A)^2+a5 x(A) y(A)+a6 y(A)^2+a7 x(A)+a8 y(A)+a9-(b4 x(A)^2+
    b5 x(A) y(A)+b6 y(A)^2+b7 x(A)+b8 y(A)+b9))/ (b1 x(A)+b2 y(A)+b3-(a1 x(A)
    +a2 y(A)+a3)))
*Lifting of the cutcurve*
Locus(P,A)
```

In practice, if the cutcurve has singular points then Locus, close to some of these points, does not work very properly. This is the reason why the singular points will be determined separately and the lifting of the regular points close to them will be made in a different way.

Singular points of the cutcurve can be classified as follows: those lying on the intersection of the cutcurve and the line $p_1(x, y) = q_1(x, y)$ and those lying outside the line, $p_1(x, y) = q_1(x, y)$, always coming from tangential intersection points of $\mathcal{E}_1$ and $\mathcal{E}_2$. Singular points of the cutcurve on the line $p_1(x, y) = q_1(x, y)$ are determined by solving $\Delta_{\mathcal{E}_1}(x, y) - \Delta_{\mathcal{E}_2}(x, y) = 0 \wedge p_1(x, y) = q_1(x, y)$ which amounts to solve an univariate equation of degree 2. These singular points are stored in the list s. Only those singular points stored in s in $\mathcal{A}_{\mathcal{E}_1, \mathcal{E}_2}$ will be stored in the list $s2$.

```
s:=Solutions({a1 x+a2 y+a3=b1 x+b2 y+b3, (a1 x+a2 y+a3)^2-4 (a4 x^2+a5 x y+a6
    y^2+ a7 x+a8 y+a9)=(b1 x+b2 y+b3)^2-4 (b4 x^2+ b5 x y+b6 y^2+b7 x+b8 y+
    b9)},{x, y})  *CAS*
*Testing points of s that live in the validation region*
s2:=RemoveUndefined(Sequence(If(cond(Element(s, i, 1), Element(s, i, 2)) > 0,
    {Element(s, i, 1), Element(s, i, 2)}), i, 1, Length(s)))
```

The lifting of the points generated by the Locus GeoGebra function works properly on the regular points of the cutcurve and on its singular points outside the line $p_1(x, y) = q_1(x, y)$. Therefore we only need to deal with the lifting of the points in the list $s2$ but also with those regular points of the cutcurve which are close to the points in the list $s2$. For that we remove from the cutcurve the circles with center the points in the list $s2$ and radius $eps \in \mathbb{R}^+$ (defined by the user). To achieve this goal, we use the following GeoGebra command:

```
*Sequence of circles stored in a list 12*
12:=Sequence((x-Element(s2,i,1))^2+(y-Element(s2,i,2))^2>=eps^2,i,1,Length(s2
    ))
```

The points on the cutcurve to be lifted by using (3) are characterized by the conditions defining the region $\mathcal{A}_{\mathcal{E}_1,\mathcal{E}_2}$ excluding the circles around the points on the line $p_1(x,y) = q_1(x,y)$ and on the cutcurve. In order to do this we define the list $l3$ by using the conditions mentioned above, thus defining a new point, B existing only if the conditions in $l3$ are verified. Making use of the point B, we define in GeoGebra a new curve, called *adaptcutcurve*, by using the following commands:

```
*Defining a list with both conditions*
13:=Join(11,12)
*Function returning zero if and only if at least one condition is not
    verified*
d(x,y):=Product(13) *CAS*
B:=If(d(x(A),y(A))>0,A)
*Representing the cutcurve excluding the circles around the singular points
    on the line p_1(x,y)=q_1(x,y)*
adaptcutcurve:=If(Length(s2)!=0, Locus(B, A), cutcurve)
```

Let C be a point in the curve *adaptcutcurve*. This point runs along the whole cutcurve except for several small circles around the singular points on the line $p_1(x,y) = q_1(x,y)$. The lifting of the curve *adaptcutcurve* is determined by the following GeoGebra commands:

```
*Defining the lifting of the adapted cutcurve*
C:=Point(adaptcutcurve)
P:=(x(C),y(C),(a4 (x(C))^2+a5 (x(C)) (y(C)) +a6 (y(C))^2+a7 (x(C))+a8 (y(C))+
    a9-(b4 (x(C))^2+b5 (x(C)) (y(C))+b6 (y(C))^2+b7 (x(C))+b8 (y(C))+b9))/(b1
    (x(C))+b2 (y(C)) +b3-(a1 (x(C))+a2 (y(C))+a3)))
adaptlift:=Locus(P,C)
```

The singular points of the cutcurve on the line $p_1(x,y) = q_1(x,y)$, stored in $s2$, will be lifted by using:

$$z = \frac{-p_1(\alpha,\beta) \pm \sqrt{p_1(\alpha,\beta)^2 - 4p_0(\alpha,\beta)}}{2} \tag{4}$$

or

$$z = \frac{-q_1(\alpha,\beta) \pm \sqrt{q_1(\alpha,\beta)^2 - 4q_0(\alpha,\beta)}}{2} \tag{5}$$

The GeoGebra commands performing the lifting of these points are:

```
*Definition of functions*
lift1:=(-(a1 x+a2 y+a3)+sqrt((a1 x+a2 y+a3)^2-4 (a4 x^2+a5 x y+a6 y^2+a7 x+a8
    y+a9)))/2
lift2:=(-(a1 x+a2 y+a3)-sqrt((a1 x+a2 y+a3)^2-4 (a4 x^2+a5 x y+a6 y^2+a7 x+a8
    y+a9)))/2
*Definition of the lift*
sing1:=Sequence((Element(s2,i,1),Element(s2,i,2),lift1(Element(s2,i,1),
    Element(s2,i,2))),i,1,Length(s2))
sing2:=Sequence((Element(s2,i,1),Element(s2,i,2),lift2(Element(s2,i,1),
    Element(s2,i,2))),i,1,Length(s2))
```

As mentioned before (see [11]) singular points of the cutcurve not belonging to the line $p_1(x,y) = q_1(x,y)$ come from tangential intersection points of $\mathcal{E}_1$ and $\mathcal{E}_2$. The determination of these points is quite complicated but they will be very easy to lift: by continuity the lifting of the points produced by the Locus GeoGebra function around these singular points produces automatically their lifting to the intersection curve (and it is not necessary to compute them explicitly).

Regular points of the cutcurve in the silhouette curves (see Fig. 2) can be determined by solving (according to Eq. (2)):

$$2(p_0 + q_0) = p_1 q_1, \Delta_{\varepsilon_1} = 0 \quad \text{and} \quad 2(p_0 + q_0) = p_1 q_1, \Delta_{\varepsilon_2} = 0$$

which amounts, for each system, to intersect two conics. In order to determine the lifting of the regular points of the cutcurve in the silhouette curves we must use the following GeoGebra commands:

```
silh1:=Solutions({2*((a4 x^2+a5 x y+a6 y^2+ a7 x+a8 y+a9)+(b4 x^2+b5 x y+b6 y
    ^2+b7 x+b8 y+b9))=(a1 x+a2 y+a3)*(b1 x+b2 y+b3),((a1 x+a2 y+a3)^2-4 (a4 x
    ^2+a5 x y+a6 y^2+a7 x+a8 y+a9))},{x,y}) *CAS*
silh2:=Solutions({2*((a4 x^2+a5 x y+a6 y^2 +a7 x+a8 y+a9)+(b4 x^2+b5 x y+b6 y
    ^2+b7 x+ b8 y+b9))=(a1 x+a2 y+a3)*(b1 x+b2 y+ b3),((b1 x+b2 y+b3)^2-4 (b4
    x^2+b5 x y+b6 y^2+b7 x+b8 y+b9))},{x,y}) *CAS*
*Definition of the function that lifts regular points*
lift:=(a4 x^2+a5 x y+a6 y^2+a7 x+a8 y+a9-(b4 x^2 +b5 x y+b6 y^2+b7 x+b8 y+b9)
    )/(b1 x+b2 y+b3-(a1 x+a2 y+a3))
ls1:=Sequence((Element(silh1,i,1), Element(silh1,i,2), lift(Element(silh1,i
    ,1), Element(silh1,i,2))),i,1,Length(silh1))
ls2:=Sequence((Element(silh2,i,1), Element(silh2,i,2), lift(Element(silh2,i
    ,1), Element(silh2,i,2))),i,1,Length(silh2))
```

At the neighbourhood of any singular point in $p_1(x, y) = q_1(x, y)$ we will compute n points for each branch to the left and to the right of the considered singular point. Note that near the silhouette curves, GeoGebra might not represent properly the cutcurve due to precision problems: this is specially important when the cutcurve is tangent to one of the silhouette curves. This discretization for each branch to the left and to the right of every singular point is the way we use to avoid the problems brought by GeoGebra into this situation.

```
*Defining x values near to singular points*
amp=eps/n
seq:=Sequence(Sequence(Element(s2,j,1) - i amp, i, -n, n),j,1,Length(s2))
*Definition of the curve inside the validation region and near to singular
    points*
e:=If((b1 x+b2 y+b3)^2-4 (b4 x^2+b5 x y+b6 y^2+b7 x +b8 y+b9)>=0 && (a1 x+a2
    y+ a3)^2-4 (a4 x^2+a5 x y+ a6 y^2+a7 x+a8 y+a9)>=0,1/16 ((a1 x+a2 y+a3-(
    b1 x+ b2 y+b3))^4+((a1 x+a2 y+a3)^2-4 (a4 x^2+a5 x y+a6 y^2+a7 x+a8 y+a9)
    -((b1 x+b2 y+b3)^2-4 (b4 x^2+b5 x y+b6 y^2+b7 x+b8 y+b9)))^2-2(a1 x+a2 y+
    a3-(b1 x+b2 y+b3))^2 ((a1 x+a2 y+a3)^2-4 (a4 x^2+a5 x y+a6 y^2+ a7 x+a8 y
    +a9)+(b1 x + b2 y + b3)^2-4 (b4 x^2+b5 x y+b6 y^2+b7 x+b8 y+b9))))
*Compute y coordinate of every point on the cutcurve from every x value
    stored in seq *
yseq:=Sequence(Sequence(NSolve(e(Element(seq,j,i),y),y),i,1,2*n+1),j,1,Length
    (s2)) *CAS*
```

For every x value stored in *seq* and y value in *yseq* we define a list, grouping x with y, and storing them in the list *pair* defined by using the following GeoGebra commands:

```
pair:=Sequence(Sequence(Sequence(If(Element(yseq,i,j)=={},(Element(seq,i,j),
    maxim),(Element(seq,i,j),RightSide(Element(Element(yseq,i,j),k)))),k,1,If
    (Element(yseq,i,j)=={},1,Length(Element(yseq,i,j)))),j,1,2*n+1),i,1,
    Length(s2)) *CAS*
```

The lifting of this set of points, the list *fill* below, will be constructed by using the function *lift*, (3), applied to any regular point whose distance to the singular one is between *eps* and *eps/10*. On the other hand, points whose distance to the singular point is less than *eps/10* will be lift by using the functions *lift1* (4) and *lift2* (5).

```
*Definition of the lifting*
fill:=Sequence(Sequence(Sequence(If(distance((x(Element(pair, k, j, i)), y(
    Element(pair, k, j, i))),(Element(s2,k,1),Element(s2,k,2)))<eps/10,(x(
    Element(pair, k, j, i)), y(Element(pair, k, j, i)), lift1(x(Element(pair,
    k, j, i)), y(Element(pair, k, j, i)))),If(eps/10<distance((x(Element(
    pair, k, j, i)), y(Element(pair, k, j, i))),(Element(s2,k,1),Element(s2,k
    ,2)))<eps,(x(Element(pair, k, j, i)), y(Element(pair, k, j, i)), lift(x(
    Element(pair, k, j, i)), y(Element(pair, k, j, i)))))), i, 1, Length(
    Element(pair,k,j)))), j, 1, 2*n+1), k, 1, Length(s2))
```

Example 3. Let f and g be two ellipsoids defined by: $f(x,y,z) = z^2 + (-2/3x + 2/3y)z + 1/3x^2 + 2/3y^2 - 1/3$
$g(x,y,z) = z^2 + (-2/17x + 1/17y - 2/17)z + 1/4x^2 + 1/17y^2 + 2/17x - 3/17.$

Fig. 3. Two ellipsoids, the intersection curve and relevant points.

In Fig. 3, green points result from the lifting of the singular points of the cutcurve, while the red ones result from the lifting of common points between the silhouette curves and the cutcurve.

In order to create a tool, in GeoGebra, it is necessary to use the menu *Tools* and then the option *Create New Tool*. All eighteen parameters, that define the two quadrics and the parameters n and *eps* must be selected as input. For the output it is necessary to select the lists *sing1, sing2, ls1, ls2, fill* and the locus *adaptlift*.

In the particular case when $p_1(x,y) \equiv q_1(x,y)$, the cutcurve verifies:

$$\mathbf{S}_0(x,y) = \left(\Delta_{\mathcal{E}_1} - \Delta_{\mathcal{E}_2}\right)^2 /16 = (p_0 - q_0)^2.$$

The condition $\mathbf{S}_0(x,y) = 0 \wedge p_1(x,y) = q_1(x,y)$ is equivalent to $p_0(x,y) = q_0(x,y)$. Thus, all the points of the cutcurve are considered as singular and their lifting (by using (4) or (5)) is defined by:

```
part1:=If(a1 x + a2 y + a3 = b1 x + b2 y + b3,Sequence((Element(s,i,1),
    Element(s,i,2), lift1(Element(s,i,1), Element(s,i,2))),i,1,Length(s)))
```

```
part2:=If(a1 x + a2 y + a3 = b1 x + b2 y + b3,Sequence((Element(s,i,1),
    Element(s,i,2), lift2(Element(s,i,1), Element(s,i,2))),i,1,Length(s)))
```

It should be pointed out that, in this case, GeoGebra gives a parameterization of the intersection curve of the two considered quadrics.

Quadrics of Degree 2 and 1 in z

Let f and g be the polynomials in $\mathbb{R}[x,y,z]$ defined by:

$$f(x,y,z) = z^2 + p_1(x,y)z + p_0(x,y) \qquad g(x,y,z) = q_1(x,y)z + q_0(x,y)$$

In this case the cutcurve is defined by $\mathbf{S}_0(x,y) = p_0 q_1^2 - p_1 q_0 q_1 + q_0^2$ in the region defined by $p_1^2 - 4p_0 \geq 0$. The lifting of any regular point (α, β) of the cutcurve will be determined by $z = -q_0(\alpha, \beta)/q_1(\alpha, \beta)$. Points of the cutcurve determined by $q_1(x,y) = 0$ can only be lifted by using (4) and, in this case, we only have one silhouette curve. The GeoGebra commands to use start defining the following elements:

```
l1:={(a1 x+a2 y+a3)^2-4 (a4 x^2+a5 x y+a6 y^2+a7 x+a8 y+a9)>=0}
S0:=ImplicitCurve((b1 x+b2 y+b3)^2 (a4 x^2+a5 x y+a6 y^2+a7 x+a8 y+a9) - (a1
    x+a2 y+a3) (b4 x^2+b5 x y+b6 y^2+b7 x+b8 y+b9) (b1 x+b2 y+b3)+(b4 x^2 +b5
    x y+b6 y^2+b7 x+b8 y+b9)^2)
cond(x,y):=(a1 x+a2 y+a3)^2-4 (a4 x^2+a5 x y+a6 y^2+a7 x+a8 y+a9)
lift:=(-(b4 x^2+b5 x y+b6 y^2+ b7 x + b8 y + b9)) / (b1 x + b2 y + b3)
P:=(x(C),y(C),(-(b4 (x(C))^2+b5 (x(C)) (y(C))+b6 (y(C))^2+b7 (x(C))+b8 (y(C))
    + b9))/(b1 (x(C))+b2 (y(C))+b3))
s:=Solutions({(b4 x^2 + b5 x y + b6 y^2 + b7 x + b8 y + b9)=0,(b1 x + b2 y +
    b3)=0},{x, y})
silh1:=Solutions({(b1 x+b2 y+b3)^2 (a4 x^2+a5 x y+a6 y^2+a7 x+a8 y+a9) - (a1
    x + a2 y + a3) (b4 x^2+b5 x y+b6 y^2+b7 x+b8 y+b9) (b1 x+b2 y+b3)+(b4 x
    ^2+b5 x y+b6 y^2+b7 x+b8 y+b9)^2=0,(a1 x+a2 y+a3)^2 - 4 (a4 x^2+a5 x y+
    a6 y^2+a7 x+a8 y+a9)=0},{x,y})
e:=If((a1 x+a2 y+a3)^2-4 (a4 x^2+a5 x y+a6 y^2+a7 x+a8 y+a9)>=0,(b1 x+b2 y +
    b3)^2 (a4 x^2+a5 x y+a6 y^2+a7 x+a8 y+a9)-(a1 x+a2 y+a3) (b4 x^2+b5 x y+
    b6 y^2+b7 x+b8 y+ b9) (b1 x+b2 y+b3)+(b4 x^2+b5 x y+b6 y^2+b7 x+b8 y+b9)
    ^2)
```

The computation of the intersection curve in this case follows the same GeoGebra strategy shown previously.

Example 4. Let f be the hyperboloid of one sheet and g the hyperbolic paraboloid defined by:
$f(x,y,z) = z^2 + 3z - x^2 + y^2 - 3$
$g(x,y,z) = (x+y)z - 2x$
The intersection curve can be represented by GeoGebra: see Fig. 4.

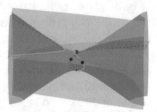

Fig. 4. Hyperboloid of one sheet and hyperbolic paraboloid intersection curve and relevant points in GeoGebra

Two Quadrics of Degree 1 in z

Let f and g be the polynomials in $\mathbb{R}[x,y,z]$ defined by:

$$f(x,y,z) = p_1(x,y)z + p_0(x,y) \qquad g(x,y,z) = q_1(x,y)z + q_0(x,y)$$

The cutcurve is defined by $\mathbf{S}_0(x,y) = p_1 q_0 - p_0 q_1$. Note that, in this case, $\mathbf{S}_0$ is a polynomial of degree three, at most. As before, in some cases GeoGebra gives us the exact parameterization of the cutcurve. However, in general, the way to proceed will be similar to the previous cases. As seen in [11], when $p_1(\alpha, \beta) \neq 0$ or $q_1(\alpha, \beta) \neq 0$, the lifting of (α, β) is given by:

$$z = -\frac{p_0(\alpha, \beta)}{p_1(\alpha, \beta)} \qquad or \qquad z = -\frac{q_0(\alpha, \beta)}{q_1(\alpha, \beta)},$$

respectively. If $p_1(\alpha, \beta) = p_0(\alpha, \beta) = 0$ then $p_0(\alpha, \beta) = q_0(\alpha, \beta) = 0$ and the line $\{(\alpha, \beta, z) : z \in \mathbb{R}\}$ is in the intersection curve.

In this case we use in GeoGebra the following commands:

```
e(x,y):=(a1 x+a2 y+a3)*(b4 x^2+b5 x y+b6 y^2+b7 x+b8 y+b9)-(a4  x^2+a5 x y+a6
      y^2+a7 x+a8 y+a9)*(b1 x+b2 y+b3)
cutcurve:=ImplicitCurve(e(x,y))

*Determine points that p1=0 && p0=0 *
s2:=Solutions({a1 x+a2 y+a3=0,(a4 x^2+a5 x y+a6 y^2+a7 x+a8 y+ a9)=0},{x, y})
      *CAS*
*Lifting points that p1!=0*
lift:=-(a4 x^2+a5 x y+a6 y^2+a7 x+a8 y+a9)/(a1 x+a2 y+a3)

*Define the region near to points that p1=0 && p0=0*
l3:=Sequence((x - Element(s2, i, 1))^2+(y - Element(s2, i, 2))^2 >=eps^2,i,1,
      Length(s2))

*Define adaptcutcurve - curve without region near to points that p1=0 & p0=0*
A=Point(cutcurve)
d(x,y):=Product(l3) *CAS*
B:=If(d(x(A),y(A))>0,A)
K1:=Point(cutcurve)
K2:=K1
adaptcutcurve:=If(Length(s2) != 0, Locus(B, A), Locus(K2,K1))
C:=Point(adaptcutcurve)
P:=(x(C),y(C),-(a4 (x(C))^2+a5 (x(C)) (y(C))+a6 (y(C))^2+a7 (x(C))+a8 (y(C))+
      a9)/(a1 (x(C))+a2 (y(C))+a3))

*Lifting of points that p1=0 && p0=0 *
lift1:=-((b4 x^2 + b5 x y + b6 y^2 + b7 x + b8 y + b9)/(b1 x + b2 y + b3))
sing1:=Sequence((Element(s2,i,1),Element(s2,i,2),lift1(Element(s2,i,1),
      Element(s2,i,2))),i,1,Length(s2))

*Determine points that p1=0 && p0=0 && q1=0 && q0=0 *
s3:=Solutions({a1 x + a2 y + a3 = 0,(a4 x^2+a5 x y+a6 y^2+a7 x+a8 y+a9)=0,(b1
      x + b2 y + b3)=0,b4 x^2+b5 x y+b6 y^2+b7 x+b8 y+b9=0},{x, y})

*Define vertical lines*
line:=Sequence(Line((Element(s3,i,1), Element(s3,i,2)),zAxis),i,1,Length(s3))
```

Example 5. Let f be the hyperbolic paraboloid and g the hyperboloid of one sheet defined by:
$$f(x, y, z) = xz + x^2 + 2y - 1$$
$$g(x, y, z) = yz + x^2 + y^2 - 2x$$
The parameterization of the intersection curve can be computed by GeoGebra by using the presented approach (Fig. 5).

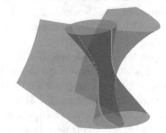

Fig. 5. Intersection between the hyperboloid of one sheet and the hyperbolic paraboloid.

4   Conclusion

For computing in GeoGebra the intersection curve of two quadrics, we have adapted to GeoGebra the algorithm in [11]. Applying this implementation to fifty examples randomly generated, it produced the intersection curve in a few seconds for each case. Despite some limitations of the Locus function in GeoGebra the implementation worked efficiently in all cases. Sometimes the performance of the software was a little slow namely when there were many singular points. Another topic to be considered deals with the values of the parameters n and *eps* which need to be carefully chosen by the user. In conclusion, our goal was attained, producing an implementation in GeoGebra of the algorithm in [11].

References

1. Dos Santos, J.M.: Intersection of two surfaces in GeoGebra. Revista do Instituto GeoGebra de São Paulo **6**(2), 04–09 (2017)
2. Dupont, L., Lazard, S., Lazard, D., Petitjean, S.: Near-optimal parameterization of the intersection of quadrics. In: Proceedings of the Annual Symposium on Computational Geometry (2003). https://doi.org/10.1145/777829.777830
3. Dupont, L., Lazard, D., Lazard, S., Petitjean, S.: Near-optimal parameterization of the intersection of quadrics: II. A classification of pencils. J. Symbolic Comput. (2008). https://doi.org/10.1016/j.jsc.2007.10.012
4. Dupont, L., Lazard, D., Lazard, S., Petitjean, S.: Near-optimal parameterization of the intersection of quadrics: III. Parameterizing singular intersections. J. Symbolic Comput. (2008). https://doi.org/10.1016/j.jsc.2007.10.007
5. Dupont, L., Lazard, D., Lazard, S., Petitjean, S.: Near-optimal parameterization of the intersection of quadrics: I. The generic algorithm. J. Symbolic Comput. (2008). https://doi.org/10.1016/j.jsc.2007.10.006
6. Levin, J.: A parametric algorithm for drawing pictures of solid objects composed of quadric surfaces. Commun. ACM **10**, 555–563 (1976). https://doi.org/10.1145/360349.360355
7. Levin, J.Z.: Mathematical models for determining the intersections of quadric surfaces. Comput. Graph. Image Process. **11**(1), 73–87 (1979)
8. Mourrain, B., Tecourt, J.P., Teillaud, M.: On the computation of an arrangement of quadrics in 3D. Comput. Geom. Theory Appl. (2005). https://doi.org/10.1016/j.comgeo.2004.05.003
9. Schomer, E., Wolpert, N.: An exact and efficient approach for computing a cell in an arrangement of quadrics. Comput. Geom. Theory Appl. (2006). https://doi.org/10.1016/j.comgeo.2004.02.007
10. Sendra, J.R., Sevilla, D.: Radical parametrizations of algebraic curves by adjoint curves. J. Symbolic Comput. **46**(9), 1030–1038 (2011)
11. Trocado, A., Gonzalez-Vega, L.: On the intersection of two quadrics (2018, Submitted). http://arxiv.org/abs/1903.06983
12. Wang, W., Goldman, R., Tu, C.: Enhancing Levin's method for computing quadric-surface intersections. Comput. Aided Geom. Des. (2003). https://doi.org/10.1016/S0167-8396(03)00081-5

Randomized Nyström Features for Fast Regression: An Error Analysis

Aleksandar Trokicić[(⊠)] and Branimir Todorović

Faculty of Sciences and Mathematics, University of Niš, Niš, Serbia
aleksandar.trokicic@pmf.edu.rs

Abstract. We consider the problem of fast approximate kernel regression. Since kernels can map input features into the infinite dimensional space, kernel trick is used to make the algorithms tractable. However on large data set time complexity of $O(n^2)$ is prohibitive. Therefore, various approximation methods are employed, such as randomization. A Nyström method (based on a random selection of columns) is usually employed. Main advantage of this algorithm is its time complexity which is reduced to $O(nm^2 + m^3)$. Space complexity is also reduced to $O(nm)$ because it does not require the computation of the entire matrix. An arbitrary number $m \ll n$ represents both the size of a random subset of an input set and the dimension of random feature vectors. A Nyström method can be extended with the randomized SVD so that l (where $l > m$) randomly selected columns of a kernel matrix without replacement are used for a construction of m dimensional random feature vectors while keeping time complexity linear in n. Approximated matrix computed in this way is a better approximation than the matrix computed via the Nyström method. We will prove here that the expected error of the approximated kernel predictor derived via this method is approximately the same in expectation as the error of the error of kernel predictor. Furthermore, we will empirically show that using the l randomly selected columns of a kernel matrix for a construction of m-dimensional random feature vectors produces smaller error on a regression problem, than using m randomly selected columns.

Keywords: Kernel methods · Nyström method ·
Randomized algorithms · Random features · Regression

1 Introduction

Kernel methods have been applied to different real world problems such as computer vision, text mining, computational biology etc. When kernels are applied to a linear method they produce a nonlinear prediction. An intrinsic part of every kernel method is the kernel matrix, a positive semi definite matrix of size $n \times n$, when n is the size of the input set. However they usually require time complexity that is cubic in the number of data points which is too expensive for large data sets. Time complexity required just for computation of a kernel matrix

© Springer Nature Switzerland AG 2019
M. Ćirić et al. (Eds.): CAI 2019, LNCS 11545, pp. 249–257, 2019.
https://doi.org/10.1007/978-3-030-21363-3_21

is $O(n^2)$. Also space complexity is $O(n^2)$ which is intractable for a number of applications. Well known solution to this problem involve an approximation of a kernel matrix.

Randomization methods which approximate kernel matrix using a random subset of an input set, represent a popular solution to both time and space complexity of kernel methods. Of these methods Nyström method achieves good results both in practice and theoretically. In Nyström method [10] we randomly select m (where $m \ll n$) columns of a kernel matrix without replacement and approximate the entire matrix based on this columns. Main advantage of this algorithm is its time complexity which is reduced to $O(nm^2 + m^3)$. Space complexity is also reduced to $O(nm)$ because it does not require the computation of the entire matrix.

Recently [8] proposed a fast method of computing random features which in turn gives rise to the fast learning algorithms. Specifically an input space is mapped using some random function and its elements are called random features. This approach appears through different formulations in [4,7,9,11]. For example, a consequence of a Nyström method applied on a kernel matrix is an m-dimensional random feature vector computed for each input vector, and this random vectors are called random Nyström features [11]. An arbitrary number $m \ll n$ represents both the size of a random subset of an input set and the dimension of random feature vectors.

Logically, idea is formed that using l (where $l > m$) randomly selected columns of a kernel matrix without replacement for a construction of m-dimensional random feature vectors will produce better results then using only m columns, all the while keeping time complexity linear in n. This is in contrast to the Nyström method which uses m selected columns to derive m-dimensional features. Li et al. [5] used this idea, and combined the Nyström method with a randomized SVD [3]. Using only Nyström method will require performing SVD on a $l \times l$ symmetric submatrix of a kernel matrix and it will produce l-dimensional random features. However if a randomized SVD is applied as in [5] algorithm will produce m-dimensional random features. This also allows the algorithm to keep the time complexity linear in n. In [5] it is empirically shown that using l (where $l > m$) randomly selected columns of a kernel matrix is better than using only m columns. In this paper we will perform theoretical analysis of this approximation method as applied on a least squares regression problem. Additionally, we will show theoretically that this algorithm with sub quadratic complexity exhibits the same predictive performance as the kernel regression. We will show that we can choose l so that the expected error of approximate kernel regression is approximately the same as the kernel regression error. Furthermore, we demonstrate on real world data sets that m-dimensional random features derived from l randomly selected input points produce without replacement better results than random Nyström features.

This paper is structured as follows. In the Sect. 2, we review the method of random Nyström features with and without a randomized SVD and its application to a linear regression. Proof that the estimator learned on random features defined in the Sect. 2 is close to the kernel estimator learned on the original input set is presented in the Sect. 3. Finally, empirical results are described in the final section.

2   Fast Kernel Regression Based on Random Nyström Views

Let assume that the input data set is the following set:

$$T = \{(x_i \in \mathbf{R}^d, y_i \in \mathbf{R})\}_{i=\overline{1,n}} \tag{1}$$

In this section we will describe the algorithm for fast approximate kernel regression. The algorithm consists of two main steps. In the first step each feature vector is mapped into a m-dimensional random feature vector, called random view, via the Nyström method. In the second step linear regression is applied. Bach [1] showed that the prediction error achieved using an approximate kernel matrix (derived from an ordinary kernel matrix) is within an ϵ distance from the prediction error achieved using an entire kernel matrix. Li et al. [5] proposed the use of l sample points for the computation of the approximate kernel matrix and empirically showed that the computed matrix is better approximation. Here we will show how to derive random features from their method.

2.1   Random Nyström Features

Random features are low dimensional vectors derived from an input set using some random mapping. Every input vector is mapped into its corresponding random feature vector. Several types of random features are present in the literature such as Random Fourier features [8] or random Nyström features [11]. Their main advantage is the use of kernels with time complexity linear in the number of points.

Assume that we have data set $\{(x_i \in \mathbf{R}^d)_{i=1}^n\}$ and a kernel (positive semi definite function) $k : \mathbf{R}^d \times \mathbf{R}^d \to \mathbf{R}$. Gram matrix $K_{ij} = k(x_i, x_j) = \langle \Phi(x_i), \Phi(x_j) \rangle$ represents a $n \times n$ positive semi definite matrix. Function $\Phi(x)$ maps data from $\mathbf{R}^d$ into the high dimensional feature space. In random feature method, every input vector x_i is mapped into the m-dimensional random feature vector r_i so that $r_i^T r_j$ approximates $k(x_i, x_j)$. Nyström method is a random matrix approximation method and when applied on a Gram matrix K its consequence are random features. Specifically for a given $m \ll n$ Nyström algorithm samples m data vectors $\{\hat{x}_i\}_{i=\overline{1,m}}$ from the input set. Approximation matrix $\tilde{K}$ is computed in the following way:

$$K \approx \tilde{K} := k(x_{1:n}, \hat{x}_{1:m})k(\hat{x}_{1:m}, \hat{x}_{1:m})^+ k(x_{1:n}, \hat{x}_{1:m})^T \tag{2}$$

where $k(\hat{x}_{1:m}, \hat{x}_{1:m})^+ = \hat{V}\hat{D}^{-1}\hat{V}^T$ is a pseudo inverse of $k(\hat{x}_{1:m}, \hat{x}_{1:m})$, where columns of $\hat{V}$ are eigenvectors and diagonal elements of matrix $\hat{D}$ are eigenvalues of the matrix $k(\hat{x}_{1:m}, \hat{x}_{1:m})$. We define a random feature vector r_i in the following way

$$r_i = \hat{D}^{-1/2}\hat{V}^T k(x_{1:n}, \hat{x}_{1:m})^T. \tag{3}$$

According to 2 $\tilde{K}_{ij} = r_i^T r_j$ and therefore $k(x_i, x_j) = r_i^T r_j$. Therefore, random vector r_i approximates $\Phi(x_i)$. Furthermore, McWilliams et al. [7] called the vector r_i random Nyström feature vector or random Nyström view. Mapping $z(x_i) = r_i$ is called a random Nyström mapping.

Training linear regression on this data set is the same as training it on the kernel regression model where instead of the kernel matrix its Nyström approximation is used.

Algorithm 1. XNV algorithm.

 procedure SC(T, k, m)
  Input: data set $T = \{(x_i \in \mathbf{R}^d, y_i \in \mathbf{R})\}_{i=\overline{1,n}}$;
  Input: kernel k ;
  Input: number of sampled feature vectors for Nyström method m ;
  Sample m data vectors $S = \{\hat{x}_i\}_{i=\overline{1,m}}$;
  Compute random features (random Nyström features) r_i $i = \overline{1,n}$;
  Perform linear regression on (r_i, y_i) $i = \overline{1,n}$
  Output: Linear regressor.
 end procedure

2.2 Randomized Eigenvalue Decomposition

Assume that we have a real symmetric matrix $W \in \mathbf{R}^{l \times l}$. In our paper we will apply this algorithm on matrix derived during Nyström method from sampled rows and columns ($W = k(\hat{x}_{1:l}, \hat{x}_{1:l})$ where $\hat{x}_{1:l}$ are l sampled data points). Our goal is to perform an eigenvalue decomposition of W. Time complexity of eigenvalue decomposition is $O(l^3)$. Using the algorithm from [3] time complexity can be reduced to $O(l^2 m + m^3)$ using a randomized algorithm for a fixed integer number M. Randomized algorithm for eigenvalue decomposition consists of several steps:

- We generate Gaussian random matrix $\Omega \in \mathbf{R}^{l \times m}$
- Construct a matrix $Y = W\Omega \in \mathbf{R}^{l \times m}$
- We perform a QR decomposition on a matrix Y. Matrix Q gives us the following approximation $A \approx QQ^T A$ from which follows $A \approx QQ^T AQQ^T$.
- Generate a matrix $B = Q^T AQ$ and perform eigenvalue decomposition on a matrix $B = V\Lambda V^T$.
- Matrix $U = QV$ gives approximate eigenvalue decomposition of a matrix $A = U\Lambda U^T$

In the next section we will show how to apply this algorithms into the Nyström method. Instead of sampling m data vectors from input data set we sample $l > m$ data vectors and perform combination of a Nyström method and a randomized eigenvalue decomposition to derive m-dimensional random feature vectors.

2.3 Random Feature Vectors Using a Combination of a Nyström Method and a Randomized Eigenvalue Decomposition

Assume that we have data set $\{(x_i \in \mathbf{R}^d)_{i=1}^n\}$ and a kernel (positive semi definite function) $k : \mathbf{R}^d \times \mathbf{R}^d \to \mathbf{R}$. For a given $l \ll n$ Nyström algorithm samples l data vectors $\{\hat{x}_i\}_{i=\overline{1,m}}$ from the input set. Recall that from a Nyström method, the following matrix is an approximation of a Gram matrix ($K_{ij} = k(x_i, x_j)$):

$$\tilde{K} := k(x_{1:n}, \hat{x}_{1:l}) k(\hat{x}_{1:l}, \hat{x}_{1:l})^+ k(x_{1:n}, \hat{x}_{1:l})^T.$$

From it we derive l-dimensional random Nyström features

$$r_i = \hat{D}^{-1/2} \hat{V}^T k(x_{1:n}, \hat{x}_{1:l})^T.$$

Random Nyström features, derived using l sampled columns, map input vectors into the l-dimensional space. However our goal is to map input vectors into the m-dimensional space while steel using l-sampled data points, where $l > m$. Our assumption is that using larger number of data points for approximation will produce better results than using m points.

In order to produce m-dimensional random vectors we propose to use the randomized eigenvalue decomposition of kernel submatrix $k(\hat{x}_{1:l}, \hat{x}_{1:l})$ and apply it to a Nyström method. Gram matrix K is now approximated as follows:

$$K \approx L := k(x_{1:n}, \hat{x}_{1:l}) k(\hat{x}_{1:l}, \hat{x}_{1:l})^* k(x_{1:n}, \hat{x}_{1:l})^T$$

where $k(\hat{x}_{1:l}, \hat{x}_{1:l})^* = \hat{V} \hat{D}^{-1} \hat{V}^T$ is an approximated pseudo inverse of $k(\hat{x}_{1:l}, \hat{x}_{1:l})$, where columns of $\hat{V} \in \mathbb{R}^{l \times m}$ are approximate eigenvectors and diagonal elements of matrix $\hat{D} \in \mathbb{R}^{m \times m}$ are approximate eigenvalues of the matrix $k(\hat{x}_{1:l}, \hat{x}_{1:l})$. Approximated eigenvalues and eigenvectors are computed using randomized SVD (Sect. 2.2). Therefore random features are computed as follows:

$$r_i = \hat{D}^{-1/2} \hat{V}^T k(x_{1:n}, \hat{y}_{1:l})^T$$

Finally we apply linear regression algorithm. Putting it all together we get an Algorithm 2.

3 Analysis of an Approximate Error

In this section, we will show that the approximated predictor (predictor learned on random features derived from a combination of a Nystom method and a randomized SVD) is close enough to the unapproximated one (the best overall predictor). Bach [1] showed the same result for the ordinary Nyström method. We will present here an outline of a proof, in which we use the following results: matrix concentraton inequalities from [2,6], ideas about kernel regression analysis from [1], and analysis of matrix approximation error from [5].

In the theorem we will use the following notation: The diagonal vector of a square matrix A is denoted by the $diag(A)$. Moreover, $||A|| = \{\sqrt{\lambda} \mid \lambda$ is an eigenvalue of $A^T A\}$ denotes a spectral norm of a matrix A, $||x||_\infty = \max(|x_1|, \ldots, |x_d|)$ denotes a max norm of a d-dimensional vector $x = (x_1, \ldots, x_d)$.

Algorithm 2. RNV algorithm.

---

procedure $SC(T, k, l, m)$
  Input: data set $T = \{(x_i \in \mathbf{R}^d, y_i \in \mathbf{R})\}_{i=\overline{1,n}}$;
  Input: kernel k ;
  Input: number of sampled feature vectors for Nyström method l ;
  Sample l data vectors $S = \{\hat{y}_i\}_{i=\overline{1,l}}$ from the input set T;
  Compute random features (random Nyström features via the randomized SVD)
r_i $i = \overline{1, n}$;
  Perform linear regress on (r_i, y_i) $i = \overline{1, n}$
  Output: Linear regressor.
end procedure

Theorem 1. *Let $\lambda > 0$ and let $z \in \mathbb{R}^n$ and $K \in \mathbb{R}^{n \times n}$ be a vector of output observations and a kernel matrix derived from input data points respectively. Assume $d = n \|diag(K(K+n\lambda I)^{-1})\|_\infty$ and $R_1^2 = \|diag(K)\|_\infty$ and $R_2^2 = \|K\|$. Define the estimate $z_K = (K+n\lambda I)^{-1}Kz$. Assume I is a uniform random subset of $p > 10$ indices in $\{1, 2, \ldots, n\}$ and consider L as approximate kernel matrix based on Nyström method and randomized eigenvalue decomposition, with the approximate estimate $z_L = (L + n\lambda I)^{-1}Lz$. For every $\delta \in (0, 1)$ there is p_0 (dependent on δ) such that for $p \geq p_0$ the following is true*

$$\frac{1}{n}E[\|z_L - z\|^2] \leq (1 + 6\delta)\frac{1}{n}\|z - z_K\|^2$$

Proof. Because K is a kernel matrix there exist a matrix $\Phi \in \mathbb{R}^{n \times n}$ such that $K = \Phi\Phi^T$. Approximate kernel matrix L based on a combination of a Nyström method and randomized eigenvalue decomposition can be written in the following way $L = K(:, I)Q(Q^T K(I, I)Q)^{-1}Q^T K(I, :)$ where Q is derived from a QR decomposition on a matrix $K(I, I)\Omega$ (where Ω is a Gaussian random matrix of $p \times l$ dimension) as in RSVD Sect. 2.2. Matrix $K(:, I)$ can be written in the following way $K(:, I) = \Phi\Phi_I^T$ where $\Phi_I = \Phi(I, :)$. Let

$$L_\gamma = \Phi\Phi_I^T Q(Q^T \Phi_I \Phi_I^T Q + p\gamma I)^{-1}Q^T \Phi_I \Phi^T$$

be a regularized kernel matrix approximation. We can write $L_\gamma = \Phi N_\gamma \Phi^T$ where $N_\gamma = \Phi_I^T Q(Q^T \Phi_I \Phi_I^T Q + p\gamma I)^{-1}Q^T \Phi_I$.

Using Sherman—Morrison—Woodbury identity approximate in sample error can be computed in the following way:

$$\frac{1}{n}\|z - z_{L_\gamma}\| = n\lambda^2 z^T(\Phi N_\gamma \Phi^T + n\lambda I)^{-2}z \tag{4}$$

Both function $\gamma \to N_\gamma$ and in sample prediction error are matrix non decreasing functions. Therefore in order to find an upper bound for on error $\|z - z_L\|^2$ it is enough to find an upper bound for $\|z - z_{L_\gamma}\|$ for any $\gamma > 0$ because $L = L_0$. Furthermore in order to find an upper bound for $\|z - z_{L_\gamma}\|$ for any $\gamma > 0$ it is enough to find a matrix lower bound for N_γ.

We define $t_{I,Q} = \lambda_{max}(\frac{1}{n}\Psi^T\Psi - \frac{1}{p}\Psi_I^T QQ^T\Psi_I)$.Using the matrix manipulation we get the following semi definite inequality:

$$I - N_\gamma \preceq \frac{\gamma}{1 - t_{I,Q}}(\frac{1}{n}\Phi^T\Phi + \gamma I)^{-1} \tag{5}$$

Applying this to the in sample approximate error we get

$$\frac{1}{n}||z - z_{L_\gamma}||^2 = \lambda^2 z^T(L_\gamma + n\lambda I)^{-2}z$$

$$\leq n\lambda^2(1 - \frac{\gamma}{\lambda(1 - t_{I,Q})})^{-2}z^T(K + n\lambda I)^{-2}z$$

$$= (1 - \frac{\gamma}{\lambda(1 - t_{I,Q})})^{-2}\frac{1}{n}||z - z_K||^2$$

$$\leq (1 + 6\frac{\gamma}{\lambda(1 - t_{I,Q})})\frac{1}{n}||z - z_K||^2 \tag{6}$$

This implies:

$$\frac{1}{n}E[||z - z_{L_\gamma}||^2] \leq (1 + 6\frac{\gamma}{\lambda(1 - E[t_{I,Q}])})\frac{1}{n}||z - z_K||^2$$

Now we need to find an upper bound for $E[t_{I,Q}]$.

$$E[t_{I,Q}] = E[||\frac{1}{n}\Psi^T\Psi - \frac{1}{p}\Psi_I^T QQ^T\Psi_I||]$$

$$= E[||\frac{1}{n}\Psi^T\Psi - \frac{1}{p}\Psi_I^T\Psi_I + \frac{1}{p}\Psi_I^T\Psi_I - \frac{1}{p}\Psi_I^T QQ^T\Psi_I||]$$

$$\leq E[||\frac{1}{n}\Psi^T\Psi - \frac{1}{p}\Psi_I^T\Psi_I||] + \frac{1}{p}E[||\Psi_I^T\Psi_I - \Psi_I^T QQ^T\Psi_I||] \tag{7}$$

Using the results from [1, 2, 5, 6] we prove that there is p_0 such that or $p \geq p_0$ the following inequality holds $E[t_{I,Q}] \leq 1 - \frac{\gamma}{\delta\lambda}$ from which the desired result follows.

4   Empirical Results

In this section we evaluate the performance of our algorithm **RNV** (Algorithm 2) that uses the combination of a Nyström method and a randomized SVD against the algorithm **XNV** (Algorithm 1) that uses the ordinary Nyström method. We compare algorithms on real world data sets[1,2,3], see Table 1.

---

[1] http://www.dcc.fc.up.pt/ltorgo/Regression/DataSets.html.
[2] http://www.gaussianprocess.org/gpml/data/.
[3] https://archive.ics.uci.edu/ml/datasets.html.

Table 1. Data sets

Data set	Instances	Attributes
Ailerons	7154	40
Cal housing	10320	8
Elevators	8152	18
Sarcos5	44484	21
Sarcos7	44484	21

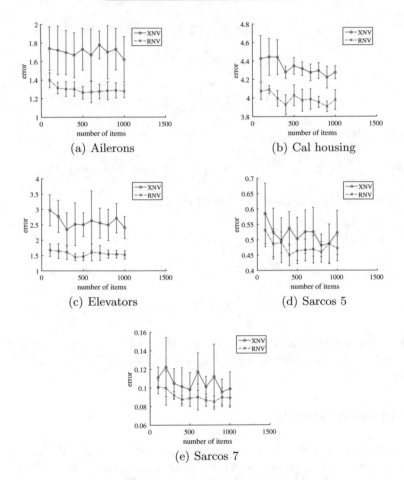

(a) Ailerons

(b) Cal housing

(c) Elevators

(d) Sarcos 5

(e) Sarcos 7

Fig. 1. Performance comparison of **RNV** and **XNV** algorithms on real world data sets.

We used Gaussian kernel whose hyper parameters are learned using 5-fold cross validation. For the value of m we put 10 and for l we put 50. Random features are derived from the entire data set, and the regression is applied on a part of the data set. We show mean predictive error and its standard deviation and report results in a Fig. 1. Across all data sets **RNV** outperforms **XNV**. Therefore we note that better results are achieved when using the larger number of randomly selected columns for random view construction while at the same time the time complexity remains linear in the size of the data set.

References

1. Bach, F.: Sharp analysis of low-rank kernel matrix approximations. In: 2013 Proceedings of Machine Learning Research, 26th Annual Conference on Learning Theory (COLT), vol. 30, pp. 185–209, Princeton (2013)
2. Gross, D., Nesme, V.: Note on sampling without replacing from a finite collection of matrices. arXiv preprint arXiv:1001.2738 (2010)
3. Halko, N., Martinsson, P.-G., Tropp, J.A.: Finding structure with randomness: probabilistic algorithms for constructing approximate matrix decompositions. SIAM Rev. **53**(2), 217–288 (2011)
4. Le, Q., Sarlós, T., Smola, A.: Fastfood-approximating kernel expansions in loglinear time. In: 2013 Proceedings of the 30th International Conference on Machine Learning, JMLR: W&CP, vol. 28, Atlanta, Georgia, USA (2013)
5. Li, M., Bi, W., Kwok, J.T., Lu, B.-L.: Large-scale Nyström kernel matrix approximation using randomized SVD. IEEE Trans. Neural Netw. Learn. Syst. **26**(1), 152–164 (2015)
6. Mackey, L., Jordan, M.I., Chen, R.Y., Farrell, B., Tropp, J.A., et al.: Matrix concentration inequalities via the method of exchangeable pairs. Ann. Probab. **42**(3), 906–945 (2014)
7. McWilliams, B., Balduzzi, D., Buhmann, J.M.: Correlated random features for fast semi-supervised learning. In: Advances in Neural Information Processing Systems (NIPS 2013), vol. 26, pp. 440–448 (2013)
8. Rahimi, A., Recht, B.: Random features for large-scale kernel machines. In: Advances in Neural Information Processing Systems (NIPS 2007), vol. 20, pp. 1177–1184 (2008)
9. Rahimi, A., Recht, B.: Weighted sums of random kitchen sinks: replacing minimization with randomization in learning. In: Advances in Neural Information Processing Systems (NIPS 2008), vol. 21, pp. 1313–1320 (2009)
10. Williams, C., Seeger, M.: Using the Nyström method to speed up kernel machines. In: Advances in Neural Information Processing Systems (NIPS 2000), vol. 13, pp. 682–688 (2001)
11. Yang, T., Li, Y.-F., Mahdavi, M., Jin, R., Zhou, Z.-H.: Nyström method vs random fourier features: a theoretical andempirical comparison. In: Advances in Neural Information Processing Systems, vol. 25 (NIPS 2012), pp. 476–484 (2012)

Author Index